Water Infrastructure Planning, Management and Design Under Climate Uncertainty: Methods and Approaches in Support of the UN High-Level Experts and Leaders Panel on Water and Disasters (HELP)

In Focus – Special Book Series

Water Infrastructure Planning, Management and Design Under Climate Uncertainty: Methods and Approaches in Support of the UN High-Level Experts and Leaders Panel on Water and Disasters (HELP)

Editor

Eugene Z. Stakhiv

WORLD WATER COUNCIL

IWA PUBLISHING

With the support of:

ICIWaRM

International Center for Integrated Water Resources Management

unesco

Category 2 Centre under the auspices of UNESCO

Published by **IWA Publishing**
Unit 104–105, Export Building
1 Clove Crescent
London E14 2BA, UK
Telephone: +44 (0)20 7654 5500
Fax: +44 (0)20 7654 5555
Email: publications@iwap.co.uk
Web: www.iwapublishing.com

First published 2022

British Library Cataloguing in Publication Data
A CIP catalogue record for this book is available from the British Library

ISBN: 9781789063097 (paperback)

Contents

Water Infrastructure Planning, MDCU doi: 10.2166/wp.2021.109

Preface

Water Policy has published two Special Issues in support of the UN 'High-level Experts and Leaders Panel on Water and Disasters' (HELP), one on Droughts and one on floods. This Special Issue, Water infrastructure planning, management and design under climate uncertainty, is the third. It is the product of a worldwide collaboration among UN water resources programs, prominent national water resources operating agencies and selected academic specialists. This Special Issue deals directly with some of the most difficult problems water engineers, managers and decision makers face when designing and implementing water resources infrastructure under new patterns of climate variability. Projecting the probabilities and risks associated with such events has traditionally been central to water resources planning and modeling. Such projections have been keys to social, environmental, economic, and financial investment decisions making. However, our contemporary world has called into question some of the traditional methods and approaches to formulating such projections. This collection provides practical grounding in selected advances already in place and practiced by leading institutions; advances that could be readily adapted by the water resources engineering and design professionals to better deal with risk and uncertainty in a non-stationary climate world. We believe the articles in this Special Issue open important venues which can help to enhance the water and climate dialogues worldwide.

Jerome Delli Priscoli
Editor in Chief of *Water Policy*

Water Infrastructure Planning, MDCU doi: 10.2166/wp.2021.268

Special Issue for UN HELP: 'Water infrastructure planning, management and design under climate uncertainty'

Eugene Z. Stakhiv[a,b,*] and Kenzo Hiroki[c,d]

[a] UNESCO International Center for Integrated Water Resources Management (ICIWaRM), Alexandria VA, USA
[b] Johns Hopkins University, Baltimore, MD, USA
[c] National Graduate Institute for Policy Studies (GRIPS), Tokyo, Japan
[d] Coordinator, UN High Level Experts and Leaders Panel on Water and Disasters (HELP), Tokyo, Japan
*Corresponding author. E-mail: estakhi1@jhu.ed; eugene.z.stakhiv@usace.army.mil

INTRODUCTION AND OVERVIEW OF PAPERS IN THIS SPECIAL ISSUE OF *WATER POLICY*

Alfred North Whitehead, a co-author of 'Principia Mathematica' with Lord Bertrand Russell, once stated an obvious pragmatic principle for dealing with change and uncertainty: 'the art of progress is to preserve order amidst change – and change amidst order' (Whitehead, 1929). Climate change, and its numerous associated uncertainties, impinges on virtually every aspect of water resources planning, management, design and operations.

At the same time, many other profound societal changes are simultaneously superimposed on the water resources management community such as changing demographics and demands, technological innovation, along with evolving environmental, economic and social well-being goals. Under these multiple sources of change, professional water management practitioners strive to maintain an orderly pace of innovation and integration of modern analytical techniques amidst rapid societal changes and political pressures emerging from the climate mitigation community. On the other hand, public and politicians demand rapid changes in decision processes and outcomes, within what they perceive are fairly slow and relatively inflexible institutional decision frameworks of agencies and ministries.

Water resources and coastal protection infrastructure systems represent the basic components of structural and non-structural adaptation to future uncertain impacts of climate change. It is evident from the most recent IPCC report (2021), and the outcomes of COP 26 (UNFCCC Conference of Parties, 2021) that many nations are having difficulties fulfilling their greenhouse gas reduction commitments made at the Paris Climate Convention in 2015. Temperatures will continue to rise, as greenhouse gas reduction mitigation efforts falter. Adaptation, then, stands as a practical response to the international community's failure to coalesce around their climate mitigation strategy and goals.

It is estimated by the Inter-American Development Bank (IDB, 2017) that between 2015 and 2030, $90 trillion in new infrastructure investments will be needed. This amount is equivalent to the world's total existing stock. Much of the world's new infrastructure will be built in developing nations. As these nations work to address widespread poverty and build thriving economies, they will be facing daunting challenges such as responding to natural hazards and rapid urbanization. These circumstances present opportunities to design sustainable and robust infrastructure that also serves fundamental climate adaptation objectives, forming the basis of a 'no regrets', cost-effective response to climate change.

The issues examined within this volume are essentially policy-driven issues that require engineering solutions. The articles offer practical responses to the fundamental question: How does society, and its numerous institutions and decision-making authorities, go about making communities more 'resilient' and 'robust' in the face of large uncertainties associated with climate change, along with other major determinants of growth and development such as population increases, globalization and economy, and regional population shifts? Upgrading resilience and robustness relies, inherently, on a series of innovative climate adaptation measures that can accommodate the uncertainties associated with these major external forces that affect every community throughout the globe.

The major determinants of growth and development are largely external in nature such as climate, population shifts and economic forces. Most decisions affecting community resilience and robustness are endogenous forces such as local land use regulations, zoning decisions, building codes and other agreed-to goals, incentives and constraints that shape the political response of communities to a variety of factors related to preferred growth and development pathways, as well as to natural phenomena and socio-economic trends.

Many agencies, ministries and multilateral lending institutions are all struggling with the similar problem of how to effectively deal with numerous uncertainties without spending excessive sums on peripheral studies and needlessly protracting the time for decisions. Usually, many critical infrastructure investment decisions such as rebuilding after each natural disaster need to be made presently. However, the quantitative tools required for such decisions are not yet ready for 'prime time' (Kundzewicz & Stakhiv, 2010). More importantly, the information generated by such complex technical analyses is often not suitable or readily understandable by the ultimate decision-makers and politicians who authorize and fund large infrastructure projects and systems. In short, institutions and agencies have struggled to find sensible ways to 'work around' the many conceptual and technical problems posed by the lack of knowledge. And, paradoxically, also contending with a myriad of untested methods and decision frameworks that have sprung up to deal with climate uncertainties.

Institutions and their administrative procedures are the governors and gyroscopes of change. Institutions possess grave responsibilities to the societies and communities that they serve. Hence, 'public engineering' is built not only on the foundations of science, data, empirical evidence and models, but also on a millennia-long evolutionary experience with natural disasters, engineering failures and addressing societal needs in a reliable manner (Petroski, 2006). Experience based on failures is the most fundamental and reality-based peer review that guides the engineering profession.

Planning procedures, engineering design standards and codes represent the core of institutional procedures that reflect a continuous evolutionary trend of learning from disasters and their accompanying infrastructure failures (Petroski, 2006). Natural disasters and extreme hydroclimatic conditions form the basis of engineering responses by the water resources management sectors, and that are also required for a pragmatic climate adaptation strategy. They comprise the core of contemporary 'best management practices' that are constantly evolving, keeping pace with societal adaptation mechanisms to accelerating global changes linked to trade, regional conflicts, energy shortages, poverty, pandemics and economic instabilities.

Infrastructures are key components that underly the stability and sustainability of communities. They are means to enhance reliability and robustness, to cope with unexpected natural hazards, and to build resilience to bounce back in the aftermath of a disaster. Without water treatment facilities, reliable power grids, adequate transportation networks, communications network or adequate flood control, how do communities grapple with and endure the effects of a natural disaster? How do societies go about retrofitting existing infrastructure to deal with future unknown climate vagaries, combined with uncertain economic and demographic changes? Who determines when and how to undertake innovative analyses that will be required for planning for future hazards? What should be the new standards for infrastructure design to increase future capacity and reliability, as well as robustness and resilience, the ability to withstand unknown forces and the ability to rebound more quickly after a disaster? What does 'climate-resilient' infrastructure look like, and how does it differ from contemporary infrastructure design that is inherently aimed at withstanding a wide range of external forces (Hill *et al.*, 2019)?

These are the questions that the papers in this volume address. Multi-national lending institutions, such as the World Bank and the Asian Development Bank, who work mostly in developing nations, together with leading national water infrastructure engineering agencies, such as the U.S. Army Corps of Engineers, the Bureau of Reclamation, Japan's Ministry of Land, Infrastructure and Transportation, and Holland's Rijkswaterstaat, have been addressing these issues as part of their updated approaches to climate adaptation. They stand in the forefront of finding practical solutions to a broad range of perplexing analytical problems. Together, they have developed several comparable variants of a tiered approach to planning, design and decision-making for practically considering the superposition of climate uncertainties over the ever-present suite of demographic, economic and socio-cultural uncertainties that normally form the basis of long-term planning in the nations with which they work.

ORIGINS OF THE PROJECT

The UN Secretary General's Advisory Board (UNSGAB) and High Level Exert Panel on Water and Disasters (HELP) was mobilized to provide expert advice to the many nations, particularly developing nations, confronted with the increased frequency and intensity of climate-related disasters. The HELP 'Action Plan', developed in 2009, had 6 major themes and 40 actions. UNSGAB Action 29 reads as follows:

> *'National and international hydrological institutes must take the initiative to identify underlying analytical and data requirements to meet climate changes that are likely to be highly uncertain and so as to support structural and non-structural measures for disaster risk deduction'.*
>
> *'UNESCO-ICIWaRM and U.S. Army Corps of Engineers Institute for Water Resources are given responsibility for implementing Action 29, in conjunction with ICHARM and other world-renowned Institutes'.*

The overall goal of this collaborative effort between UNSGAB HELP and UNESCO-based institutions was to examine and recommend new practical risk-based planning methods, analytical approaches and criteria associated with engineering design standards and guidelines for water resources infrastructure planning and design. Specifically, the goal was to focus on those procedures currently in use, that could be readily adapted, upgraded or extended by various agencies and ministries, such as the Bureau of Reclamation, Rijkswaterstaat, Corps of Engineers or the Bangladesh Water Development Board, to better accommodate realistic considerations of future climate uncertainties and other rapid societal ecological, and economic changes.

This project simultaneously fulfilled an important component of the eighth phase of the International Hydrological Program (IHP-VIII) conducted by UNESCO. This IHP (2014–2021) program has the overall theme of: '*Water security: Responses to local, regional, and global challenges*' http://www.iciwarm.org/en/docs/Appendix_EXECUTIVE_SUMMARY_OF_IHP_VIII.pdf.

As part of the IHP-VIII initiative, UNESCO-ICIWaRM engaged several other UNESCO Centers to carry out this effort, including ICHARM (International Center for Water Hazard and Risk Management in Japan), and ICWRGC (International Centre for Water Resources and Global Change, Cologne, Germany) and WMO. Three technical workshops were held at ICIWaRM and ICWRGC as part of this effort, designed to lay out the key issues and methods available for a risk-based approach to analysis. Nearly 50 participants discussed the major themes that were of concern to the engineering profession and practitioners. Papers were solicited from many participants and they reflect the major themes that were discussed at these workshops.

OVERVIEW OF CONCEPTS

The overall aim of this collection of papers is to provide additional practical intellectual grounding to selected advances that are already in place and practiced by leading institutions, and others that could readily be adapted by the water resources engineering profession to better deal with risk and uncertainty in a non-stationary climate world. This volume does not present original research, as such, since the focus is on synthesizing promising best management practices and existing ideas that have been developed over the past decade, since UNSGAB's directives, and relevant published research that has not yet been incorporated into the body of institutional 'best management practices'. These practices exist in legally binding forms of engineering regulations, manuals and design standards of the water resources engineering institutions, agencies, ministries and professional societies. The papers represent an explication of, and search for appropriate engineering solutions to fundamental policy questions surrounding climate change and adaptation.

A wide range of existing hydrologic planning and design frameworks and methods were considered, ranging from deterministic to probabilistic-based design approaches and criteria for water supply, flood control infrastructure, hydropower, hurricane protection, etc. There are three overlapping themes, reflecting the title of the special volume. How the planning and design practices of the three primary functions of water resources, such as planning, design and management, can be modified to reinforce climate adaptation solutions. The three clusters of papers in this Special Issue reflect those themes.

The first group of papers examines several comparable project planning and evaluation frameworks that treat climate uncertainties and non-stationarity within existing traditional project justification frameworks used by the World Bank, the Asian Development Bank, the Corps of Engineers and other comparable national agencies. The planning frameworks are important because they reflect a common set of policies related to climate adaptation, which are transformed into principles and guidelines for dealing with climate uncertainties and how practically and cost-effectively to analyze the reliability, resilience and robustness of proposed systems and projects.

The second group of papers reviews existing engineering design standards and risk-based procedures for updating those standards by various agencies in the EU, the US and Japan. This set of papers provides case study examples of variations of risk-based approaches to upgrade design standards. One of the central confounding issues is how to quantitatively define the risks of an essentially unknowable non-stationary climate. The principal way that the engineering profession has dealt with this dilemma is to focus on quantifying hydroclimatic extremes, and assigning probabilities to empirically derived estimates. Designs are then based on both extremes and added safety factors to accommodate known and unknown uncertainties. Hurricane Katrina reconstruction is offered as a prime case study example of innovative risk-based design methods used to upgrade the New Orleans hurricane protection system.

The third cluster of papers examines additional analytical innovations that should be considered or need to be developed in the near future to improve planning and design. They cover different decision criteria, such as *satisficing*, which would be required for evaluating projects under climate uncertainty, instead of relying on traditional economics and the benefit–cost paradigm. Also presented is the need for drought frequency analytical standards to complement existing flood frequency analysis standards. Another paper discusses adaptation pathways needed for capacity expansion of existing infrastructure. In the vitally important management function of operating existing water storage systems, the viability of methods is discussed for incorporating decadal climate variability (DCV) information as an aid to seasonal forecasts.

Most papers in this Special Issue are focused on specific methodologies and quantitative techniques for making better estimates of risk and uncertainty for key underlying hydrologic phenomena and decision criteria. However, these analytical methods are necessary inputs to broader evaluation and decision frameworks that societies have structured to assist in making sensible, normative, unbiased, and uniformly applied and socially acceptable decisions. Such frameworks are underlain by traditional and well-accepted economic and engineering decision criteria, many of which are embedded within law and agency regulations.

INSTITUTIONAL EVALUATION AND DECISION FRAMEWORKS

The first series of papers spotlight the encompassing project evaluation and decision frameworks of major infrastructure planning and financing institutions, such as the World Bank and the Asian Development Bank, and of national approaches, such as those of Japan and the US. These frameworks evolved from a series of policy directives which attempted to integrate climate change and its uncertainties within each institution's existing planning and design guidelines. The important distinction is that the multilateral international lending institutions deal with the developing world. It is there that innovative, cost-effective and climate-resilient adaptation technologies will have the greatest beneficial economic impacts. New methods and technological solutions can more easily 'leap-frog' outdated traditional approaches in these settings. In contrast, retrofitting existing infrastructure in densely populated urban areas is very costly, and will be a challenge for the developed nations of the EU, Japan or the US.

The papers are:

1. 'Towards Pragmatism in Climate Risk Analysis and Adaption', by R.L. Wilby, X. Lu, P. Watkiss and C.A. Rodgers
2. 'Water-Related Infrastructure Investments in a Changing Environment: A Perspective from the World Bank', by D.J. Rodriguez, H.A. Paltan, L.E. Garcia, P. Ray and S.S.G. Freeman
3. 'Climate Risk-Informed Decision Analysis [CRIDA]: "Top-down" vs "Bottom-up" Decision Making for Planning Water Resources Infrastructure', by J. Manous and E. Stakhiv
4. 'Evolution of Japan's Flood Control Planning and Policy in Response to Climate Change and Social Changes', by T. Koike

Except for Japan, the papers present fairly similar evaluation frameworks that have practically sorted through a bewildering array of various climate modelling approaches, analytical techniques for non-stationary hydrologic analysis, and planning evaluation approaches that were devised to elucidate climate uncertainties. In turn, this array of methods has created additional challenges with associated uncertainties (Slater *et al*., 2021), that include the following:

- Downscaling a multitude of climate scenario precipitation and temperature outputs and transforming into hydrological outputs (runoff and flood and drought frequencies)
- Dealing with non-stationary climate signals with standard frequency-based hydrologic analysis that are currently the basis for project economic justification and design
- Adapting existing planning evaluation approaches that are not readily suited for dealing with large magnitude climate uncertainties
- Designing and economically justifying robustness and resilience for water projects under conventional economic project evaluation and justification procedures and criteria.

Each of the papers in this group covers each institution's transition from a 'top-down' climate scenario-driven approach to a more traditional 'bottom-up' vulnerability assessment decision approach. Each institution has developed a hierarchical tiered approach to deciding how much climate-related analysis is required, and under what circumstances. The paper by Wilby *et al.* focuses on Asian Development Bank's (ADB) practical approaches that are focused on 'delivering project outcomes *despite* climate change'. ADB developed a climate risk management framework that screens projects for climate risks across the entire project development cycle.

The paper by Rodriguez *et al.* reflects the World Bank's approach to support climate-informed project investment decision making. The core of the World Bank approach is a hierarchical four-phase 'Decision-Tree Framework (DTF)' that helps to screen projects for vulnerabilities and analyze plans and determine whether and which risk management options may be suitable. It is a similar process to the one described by the ADB, using a 'bottom-up' approach and a series of 'stress tests' that determine how much detail is warranted for 'top-down' climate-related analysis. The Rodriguez *et al.* paper provides several case studies as examples of the application of their 'DTF'.

The paper by Manous and Stakhiv presents another variant of the ADB and WB approach, termed as Climate Risk-Informed Decision Analysis (CRIDA). It is comparable to both, but follows the World Bank DTF model more closely. However, the paper delves more deeply into specific evaluation and decision characteristics that are associated with the traditional U.S. water resources planning paradigm used by federal agencies, and provides more qualitative guidance on the circumstances under which 'bottom-up' or 'top-down' climate analysis may be required for the various traditional functions of water management, such as project planning, design, operations and river basin portfolio analysis. CRIDA also promotes a variant of a pragmatic climate adaptation approach that expands on the notion of incremental capacity expansion, through a process called 'adaptation pathways'.

Koike's paper overviews a more traditional evolution of Japan's policies and planning approaches in response to climate change risks and social changes. Japan's approach was to base their new flood risk guidelines on traditional empirical data and recalculated extreme floods, conditioned on the potential impacts associated with climate model representative concentration pathway scenarios of RCP 2.6 (low forecast) and RCP 8.5 (high forecast). Japan, which is subjected to frequent extreme floods that are accentuated by its steep topography, undertook a thorough analysis of its watersheds, updating hydrologic information and revised its design criteria for 15 regions.

RISK-BASED ENGINEERING DESIGN STANDARDS

The second group of papers highlights the important role and evolution of risk-based engineering design standards in dealing with the extremes of contemporary climate, and how they could be updated to encompass climate non-stationarity. Several of the authors have argued that these updated standards can readily serve as the basis for dealing with many of the uncertain trends associated with climate non-stationarity. The practices of the EU, Japan and the US are highlighted, with case studies of engineering design progressively adapting to the growing extremes of climate non-stationarity, with significant adjustments made after each natural disaster.

5. 'Climate Change Adjustments in Engineering Design Standards. European perspective', by Z.W. Kundzewicz and Paul Licznar
6. 'The Centrality of Engineering Codes and Risk-Based Design Standards in Climate Adaptation Strategies', by E.Z. Stakhiv
7. 'Anticipated Maximum Scale Precipitation for Calculating the Worst-Case Floods', by K. Takeuchi and S. Tanaka
8. 'Screening for Non-Stationary Analysis', by P.H. Kirshen
9. 'Hurricane Katrina and New Orleans: A Forensic Assessment – Part 1', by L.E. Link
10. 'Designing the Hurricane Storm Damage Reduction System and Resulting Long-Term Engineering Guidance and Practice Changes – Part 2', by L.E. Link

Kundzewicz and Licznar's paper provides an overview of progress made by the European Union to grapple with developing a common approach to climate uncertainties and adaptation. A review of progress shows that though EU member states have devised qualitative national flood risk management objectives and have delineated adaptation measures that address those objectives, they are still far from implementing a consistent and uniform system among member states. The authors provide an array of readily attainable advances in climate-driven risk-based adjustments of urban and transport drainage design standards and 'design floods' for major flood protection infrastructure. They conclude that advances in hydrological sciences and climate science must ultimately be 'translated into acceptable and replicable engineering design standards' and commonly used methods for dealing with climate non-stationarity.

Stakhiv's paper emphasizes the same message as Kundzewicz and Licznar: that since engineering codes, design standards and analytical criteria for hydraulic structures are the final determinative specifications for designing and constructing a water resources project, they are the authoritative and legally accepted standards for project design and construction. Stakhiv stresses that these standards are updated regularly by most developed nations, especially after major disasters. Because they are based on considering extreme climate conditions, they can serve as a reliable platform for the next phase of updated risk-based standards to deal with future climate uncertainties. This is important, because the standards of such developed nations as the US, Japan and the Netherlands, which experience numerous hydrologic hazards and hydraulic failures, serve as the basis for international standards. Stakhiv concludes that contemporary risk analysis methods and risk-based standards, codes and methods comprise an important part of a progressive autonomous adaptation to climate change. They represent an essential component of 'no-regrets' climate adaptation.

The paper by Takeuchi and Tanaka details how Japan's new flood risk standards were developed in support of Japan's 2015 Flood Risk Management Act. This Act was also covered in Koike's paper. As a policy statement, the Japanese government decided that saving lives was the top priority, so the hydrologists and engineers developed a 'worst-case scenario', based mostly on empirically calculated precipitation maxima, anticipated maximum scale precipitation, for 15 climatologically uniform regions. Using this new data, Japan revised all their flood hazard maps. The hydro-climatological outputs of climate scenarios were referenced to better understand where the empirically derived maximum precipitation estimates lay within the envelope of possible future climate outcomes.

Kirshen's paper addresses the important question of how and when to formally consider climate non-stationarity as part of project planning. This key analytical issue was raised in many of the papers, with different perspectives. Kirshen provides a set of practical common-sense suggestions on identifying the circumstances for either side-stepping the need for employing specific methods to quantitatively represent non-stationarity, or situations where it is essential to explicitly consider non-stationarity.

Link's two papers on the post-flood engineering analysis associated with Hurricane Katrina describes a major inflection point in the evolution of risk-based engineering standards for flood protection. Hurricane Katrina was a four-fold extreme event: storm surges, wind damage, severe flooding and considerable loss of life. Part 1 describes the damage to the existing system, through the various stages of flooding and storm surges. Part 2 covers the extensive and innovative risk and reliability analysis that was undertaken as part of the design for a substantially upgraded storm surge and flood protection system. It resulted in a major overhaul of many engineering regulations and design manuals of the Corps of Engineers, which the author enumerates.

FURTHER ADVANCES AND CHALLENGES

The third group of papers addresses additional evaluation perspectives and prospective analytical needs for continuing research, development and applications that will further assist planners and design engineers in dealing with climate change uncertainties.

11. 'Infrastructure Capacity Planning for Reducing Risks of Future Hydrologic Extremes', by J.J. Boland and D.P. Loucks
12. 'Flood Risk Management in Changing Time', by G.B. Baecher and G.E. Galloway
13. 'On the Need for Guidelines for Streamflow Drought Frequency Analysis', by R.M. Vogel and C. N. Kroll
14. 'Incorporating Decadal Climate Variability Information in Operation and Design of water Infrastructure' by J.R. Olsen, V.M. Mehta and H. Hill

Boland and Loucks tackle the important concept of infrastructure capacity expansion as a practical way of coping with climate uncertainties associated with the extremes of floods and droughts and keeping pace with climate signals as they become evident. Capacity expansion is inherently a central feature of adaptive management, and ways of determining the best time sequence for implementing the increments of capacity expansion. The authors review several traditional approaches to capacity expansion, along with more innovative approaches to adaptive capacity planning.

In a similar vein, Baecher and Galloway consider the traditional regulatory and policy approach to flood risk management, and find it deficient in dealing with climate uncertainties. They argue that because of the uncertainties involved, it is unreasonable to assign probabilities to future events, making it very difficult to employ traditional economic optimization methods. Instead, they advocate using a *satisficing* approach. These are satisfactory solutions that do not undergo economic optimization tests, but address a series of risk reduction factors that increase robustness and resilience in a cost-effective manner.

Vogel and Kroll tackle the need for streamflow drought frequency guidelines to complement existing flood frequency guidelines and standards. Existing flood frequency guidelines provide uniformity for U.S. federal agencies in design, flood insurance programs and management of existing flood control systems. The authors argue that drought frequency guidelines are needed to manage comparable multi-billion dollar crop insurance programs of the U.S. Department of Agriculture and many other existing drought compensation programs. Coordinating the U.S. Federal Government and State government drought management programs for pollution control, ecological purposes and municipal and industrial water supply require a consistent analytical approach to drought analysis.

Olsen, Mehta and Hill cover the important topic of seasonal forecasts that rely on Decadal Climate Variability (DCV) projections for improving reservoir management and operations, especially a system of reservoirs. While substantial advances have been made in drought forecasting, because of the persistence of droughts, the authors conclude that much more research is needed to overcome Type I (false positive) and Type II (false negative) errors associated with such forecasts. If DCV indices indicate that the climate has a higher probability of dry conditions, drought contingency plans could be triggered earlier. Understanding DCV phenomena could also improve long-range water resources planning. The authors provide several case studies about the relative effectiveness of employing DCV indices, and the research gaps that still exist.

DATA AVAILABILITY STATEMENT

All relevant data are included in the paper or its Supplementary Information.

REFERENCES

Hill, A., Mason, D., Potter, J., Hellmuth, M., Ayyub, B. & Baker, J. (2019). *Ready for Tomorrow: Seven Strategies for Climate-Resilient Infrastructure.* Hoover Institution, Washington DC.

Inter-American Development Bank (IDB) (2017). *Inter-American Development Bank Sustainability Report 2017.* p. 68. http://dx.doi.org/10.18235/0001034. Available at: https://publications.iadb.org/publications/english/document/Inter-American-Development-Bank-Sustainability-Report-2017.pdf.

IPCC (2021). Climate change 2021: the physical science basis. In *Contribution of Working Group I to the Sixth Assessment Report of the Intergovernmental Panel on Climate Change.* Masson-Delmotte, V., Zhai, P., Pirani, A., Connors, S. L., Péan, C., Berger, S., Caud, N., Chen, Y., Goldfarb, L., Gomis, M. I., Huang, M., Leitzell, K., Lonnoy, E., Matthews, J. B. R., Maycock, T. K., Waterfield, T., Yelekçi, O., Yu, R. & Zhou, B., (eds). Cambridge University Press, Cambridge, UK, In Press.

Kundzewicz, Z. & Stakhiv, E. (2010). Are climate models 'ready for prime time' in water resources applications, or is more research needed? *Hydrological Sciences Journal* 55(7), 1085–1089. doi:10.1080/02626667.2010.51311.

Petroski, H. (2006). *Success Through Failure: The Paradox of Design.* Princeton University Press, Princeton, NJ, p. 256.

Slater, L., Anderson, B., Buechel, M., Dadson, S., Han, S., Harrigan, S., Kelder, T., Kowal, K., Lees, T., Matthews, T., Murphy, C. & Wilby, R. (2021). Nonstationary weather and weather extremes: a review of methods for their detection, attribution and management. *Hydrology and Earth System Sciences 25*, 3897–3935. https://doi.org/10.5194/hess-25-3897-2021.

UNFCCC Conference of Parties (2021). 26th UN Climate Change Conference of the Parties (COP26), Glasgow, 31 October–13 November 2021.

Whitehead, A. N. (1929). *Process and Reality: An Essay in Cosmology*. Available at: https://en.wikipedia.org/wiki/Process_and_Reality

First received 23 November 2021; accepted in revised form 23 November 2021. Available online 1 December 2021

WORLD WATER COUNCIL IWA PUBLISHING

Water Infrastructure Planning, MDCU doi: 10.2166/wp.2021.232

Towards pragmatism in climate risk analysis and adaptation

Robert Leonard Wilby [a,*], Xianfu Lu [b], Paul Watkiss [c] and Charles Andrew Rodgers [d]

[a] Geography and Environment, Loughborough University, Loughborough LE11 3TU, UK
[b] Independent Adaptation Specialist, Somerton OX25 6NB, UK
[c] Paul Watkiss Associates, Oxford OX2 7SN, UK
[d] Asian Development Bank, Metro Manila 1550, Philippines
*Corresponding author. E-mail: r.l.wilby@lboro.ac.uk

RLW, 0000-0002-4662-9344; XL, 0000-0002-7150-3882; CAR, 0000-0001-9264-6471

ABSTRACT

The Asia-Pacific region is extremely vulnerable to climate variability and change. This reflects high exposure to hydroclimatic hazards such as tropical cyclones, floods, droughts, and heatwaves. Rapidly growing cities and low-lying coastal zones/estuaries also face threats from sea level rise and storm surges. However, climate model projections remain very uncertain about most of these risks, so water infrastructure and operations need to consider a range of plausible futures. Against this background, the Asian Development Bank (ADB) has been developing frameworks, tools, and capacities in climate risk and adaptation assessment and management. Project teams are often operating in data-scarce situations and under significant time constraints, so the emphasis has been on creating pragmatic guidance and training resources. This paper charts the transition of climate risk management (CRM) within the ADB from a predominantly scenario-led to decision-led approach to adaptation. Examples are given of light-touch procedures for screening climate risks, strengthening the transparency and rigour of scenario analysis, raising awareness of a broad range of adaptation options, streamlining identification of CRM options, and embedding allowances for climate change in detailed engineering designs. Such practical innovations would benefit communities of practice beyond the Asia-Pacific region.

Key words: Adaptation, Asia, Climate allowance, Climate change, Risk management

HIGHLIGHTS

- Water infrastructure and operating rules must be adapted such that they continue to deliver intended benefits despite climate change.
- The Asian Development Bank has been developing frameworks, tools, and capacity for climate risk assessment and adaptation investments.
- Pragmatic procedures for screening climate risk are particularly helpful in information- and resource-constrained situations.

INTRODUCTION

The Asia-Pacific region is extremely vulnerable to the impacts of climate variability and change – threats that could undermine progress towards sustainable and inclusive economic growth of the region and beyond (Auffhammer, 2019). Over recent decades, some areas have experienced a significant rise in direct physical losses caused by extreme weather events such as tropical cyclones, floods, droughts, and heat waves. Annual average economic losses from floods, droughts, and tropical cyclones in the Asia-Pacific region are ~US$575 billion per year, and the number of people impacted has more than doubled in 40 years (UNESCAP, 2019). Economic costs from climate change in the region, as a percentage of GDP, are expected to be twice the global average by the 2050s (OECD, 2015a). This partly stems from high exposure to hydroclimatic hazards (Jongman *et al.*, 2012). Many cities and megacities – Shanghai, Mumbai, Ho Chi Minh City, and Jakarta among them – are in low-lying coastal areas exposed to rising sea levels, tropical cyclones, and extremely humid heat (Hallegatte *et al.*, 2013; Matthews *et al.*, 2017). In addition, rapid urbanization in many parts of Asia is exacerbating this vulnerability by placing more people, their assets, and infrastructure potentially in harm's way. Even small changes in climate hazards could translate into discernible changes in impacts within these 'hot spots' (Schleussner *et al.*, 2018).

Asia's vulnerability to climate change impacts also matters internationally because it is a key region for multinational businesses and industries to manufacture in, source from, and sell to. Therefore, with the global economy increasingly dependent on the free flow of goods along international supply chains, impacts of climate change in the Asia-Pacific region have attracted global attention. For example, the 2011 floods in Thailand highlighted the fragility of international supply chains for the electronics and automotive industries under extreme weather (Haraguchi & Lall, 2015).

The gravity of the adverse impacts of climate change and disaster risks in Asia-Pacific and the need for concerted action to address them are fully recognized at all levels. For example, the Asian Development Bank (ADB) Strategy 2030 (ADB, 2018a) identifies 'Tackling climate change, building climate and disaster resilience, and enhancing environmental sustainability' as one among several operational priorities. All countries in the region have committed to taking action to reduce greenhouse gas emissions and to adapt to climate change, as outlined by their Nationally Determined Contributions (NDCs), in line with the 2015 Paris Agreement of the United Nations Framework Convention on Climate Change. This is within the broad context of ADB's mission to achieve prosperity, inclusiveness, and resilience while sustaining efforts to eradicate extreme poverty in the region.

Delivering economic growth, eradicating remaining poverty, and responding to climate change in Asia and the Pacific will require significant investment in infrastructure – an estimated $26 trillion, or $1.7 trillion per year, over the period 2016–2030 (ADB, 2017a). This is concentrated in power, transport, telecommunications, water, and sanitation. There is an urgent need to ensure that this investment takes account of rising and future climate risks, because of the long lifetime of many investments, and because infrastructure decisions can lock in development patterns for decades. At the same time, there will be an increasing demand for investment in adaptation projects. However, there are formidable challenges which have previously hindered progress in managing climate change risks in the region.

First, there are often limited policy requirements or technical guidance in countries, and across or within sectors, for factoring climate risk management (CRM) into investment planning and infrastructure design. Second, climate change involves deep uncertainty and, until relatively recently, there has been a lack of technical guidance and capacity on how to address this in CRM (Hallegatte *et al.*, 2012). This is critical when it comes to the planning and design of water resource infrastructure, given the long lifetime of such investments, but also the profound uncertainties related to the regional and local characteristics of the hydrological cycle in a changing climate (Clark *et al.*, 2016). Third, the appraisal of different climate risks and adaptation options, as part of investment project analysis, is hampered by the challenges of economic appraisal of options (OECD, 2015b). Finally, financial resources are not always available to implement CRM interventions: the additional investments required to make water resource infrastructure climate-resilient may be contingent on the success of efforts to secure concessional or grant finances from 'third-party' entities, such as the Green Climate Fund (Tanner *et al.*, 2019).

Meanwhile, there has been a marked shift in the conceptual approach to CRM and adaptation planning within the water sector (Brown & Wilby, 2012; Ray & Brown, 2015). Formerly, the opening question might have been 'What are the expected climate risks and associated impacts at the project site, and what can be done to counter these threats?' Invariably, this approach focused effort on quantifying regional climate changes, and less so on evaluating adaptation options. A decade ago, this science-first approach was the preferred method of researchers and practitioners alike (IEG, 2012).

More recent practice is to invert this thought process by asking 'What are the project objectives and measures needed to ensure intended development outcomes are achieved, regardless of climate change?' In this new framework, climate model-based information is used to stress test and evaluate the efficacy of different adaptation options and project designs (e.g., Ray & Brown, 2015; Yates *et al.*, 2015; Poff *et al.*, 2016; IHA, 2019). The approach has been variously described as 'bottom-up', 'decision-centric', or 'vulnerability-led' but is essentially a decision-making procedure that embraces the inherent uncertainty of climate projections (Wilby & Murphy, 2018). Similar decision-making under uncertainty approaches is centred on principles of flexibility, robustness, and hedging strategies (e.g., real options analysis and portfolio analysis) or dynamic adaptation decision pathways and adaptive management (Watkiss *et al.*, 2015). Other considerations are important too – such as the evaluation principles used for project design/justification (Stakhiv, 2011).

Given the above context, this paper traces the evolution of pragmatic approaches to climate risk analysis and management by the ADB. This progress reflects a deeper sense of the development context and climate risk profile of the Asia-Pacific region, as well as the gradual assimilation of a new conceptual framework for adaptation planning. We begin by charting advances in CRM procedures within the ADB, including efforts to strengthen the enabling environment. We then explain how the ADB has moved to a more decision-centric approach that is focused on delivering project outcomes *despite* climate change. This is being achieved through the development of guiding principles and tools that are intended to embed climate resilience within the designs of new investment projects, i.e., to ensure that the ADB investment portfolio is 'climate smart'. However, implementing such activities may require policy and institutional reforms, and an example is provided by emerging techniques for adjusting national standards used in detailed engineering designs. We conclude by identifying opportunities for transferring insights about CRM from the Asia-Pacific region to wider communities of practice.

EVOLVING TO CRM IN THE ADB

The ADB *Strategy 2030* states that 75% of its committed operations will be supporting climate change mitigation and adaptation by 2030 (ADB, 2018a). To ensure this, climate finance from ADB's own resources will reach $80 billion cumulatively from 2019 to 2030. This commitment builds on an existing major investment base. In the year before the pandemic began, the ADB invested $1,536 million in adaptation from its own resources (Figure 1). The ADB also plays a key strategic role in the region in understanding and addressing the potential impacts of disaster risk and climate change in developing member countries (DMCs).

In 2014, it became mandatory for all ADB projects to be screened for climate risks, and a CRM framework was developed to facilitate these efforts at the corporate level (ADB, 2014). As illustrated by Figure 2, activities related to CRM are carried out across the project development cycle (with distinct concepts, preparation, and implementation phases). The CRM framework consists of the following steps that match this cycle:

1. Context-sensitive climate risk screening during the concept phase to identify projects that may be at medium or high risk.
2. For those projects identified as being at medium or high risk, a climate risk and adaptation (CRA) assessment is undertaken during the preparation phase of projects.

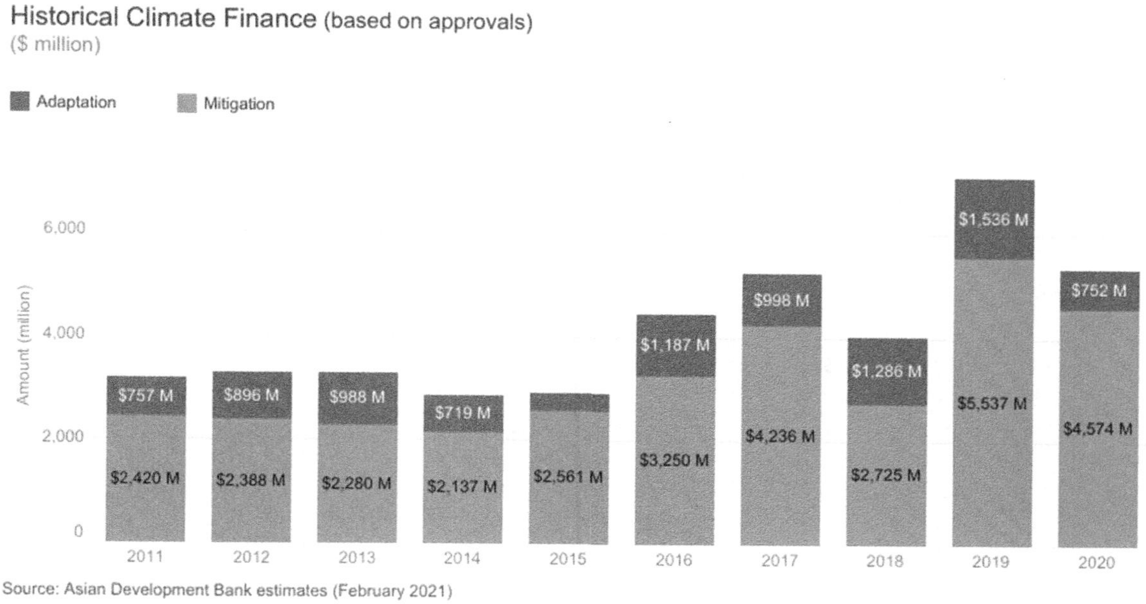

Fig. 1 | ADB finance for climate adaptation and mitigation projects 2011–2020. *Source*: https://data.adb.org/dashboard/climate-change-financing-adb.

3. The application and monitoring of selected climate risk measures are undertaken during the implementation phase.

Project teams have also been estimating and recording the level of adaptation finance, in line with the MDB methodology for tracking adaptation finance[1]. Furthermore, disaster risk concerns, including near-term disaster risks and risk emanating from geophysical hazards, have been incorporated into the CRM framework. Details on project climate risk assessment and associated risk management measures are captured in key project document(s), predominantly prior to project approval. These are made publicly available through the ADB portal[2].

For ADB projects, the Report and Recommendation of the President to the Board of Directors (RRP) includes specific placeholders for (1) reporting the assessed level of climate change impacts on the project and the amount of adaptation finance associated with the project (in 'Project at a Glance'); (2) describing the nature and severity of climate risks to the project and the proposed strategy to address these risks (usually in the 'Due Diligence' section); and (3) presenting a climate change assessment (as a 'Linked Document' to the RRP) for projects classified as at medium or high climate risk.

To support the implementation of the CRM framework, technical and financial support is provided, to the extent possible and mostly in a coordinated fashion. Technical and financial provisions for project-level CRM activities at the ADB are largely coordinated through the Climate Change team within the Sustainable Development and Climate Change Department. Financial support is made available internally (e.g., through the Climate Change Fund and various trust funds) to support the preparation of CRA assessments and related tasks. A large, searchable body of technical and sector-specific guidance notes has now been assembled to support ADB climate

[1] https://www.ebrd.com/documents/climate-finance/-2018-joint-report-on-multilateral-development-banks-climate-finance.pdf.

[2] http://www.adb.org/projects

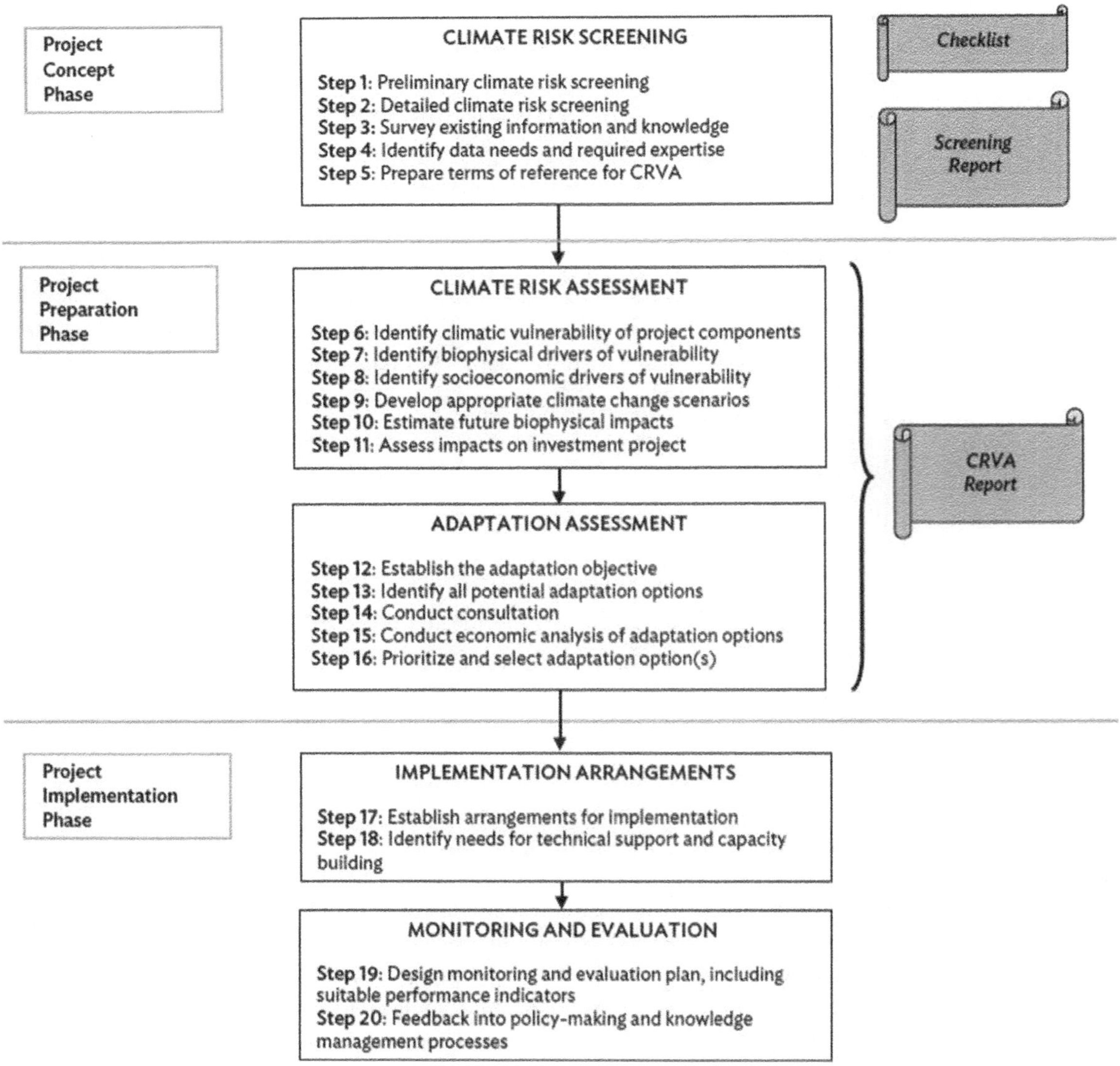

Fig. 2 | The ADB (2015) CRM framework for investment projects. CRVA, climate risk and vulnerability assessment.

risk assessments and adaptation planning[3]. For example, *Guidelines for Climate Proofing Investment in the Water Sector* (ADB, 2016) helps manage climate risks in water supply and sanitation projects. Similarly, knowledge products, such as *Information Sources to Support ADB Climate Risk Assessments and Management* (ADB, 2018b), have been developed to raise awareness of public domain data among project team leads and their CRA consultants (see Strengthening enabling environments for risk analysis, below). In addition, there are ongoing interactions and technical co-operations between the Climate Change team and project teams in operations departments within the ADB. Figure 3 illustrates the various engagements of the different teams across the

[3] https://www.adb.org/themes/climate-change-disaster-risk-management/publications-documents.

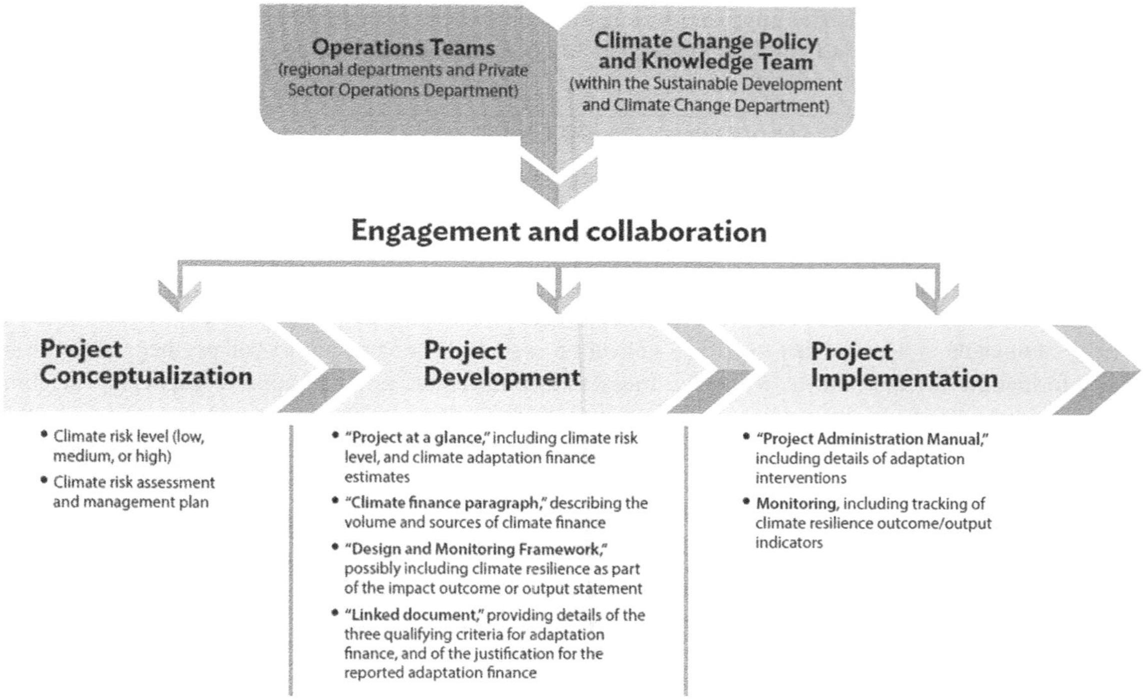

Fig. 3 | Collaboration and coordination for climate-resilient project development at ADB. *Source*: Lu (2019).

organization. This largely centralized CRM support mechanism enhances resource efficiency as well as technical consistency.

To take the stock of ADB processes and practices in implementing the CRM framework, a comprehensive review was carried out to assess how effective CRAs had been in helping teams to design and implement more climate-resilient projects (ADB, 2017b). This included a detailed technical evaluation of a large sample of CRAs for ADB investments. The primary criterion by which these CRAs were evaluated was *fitness for purpose.* In other words, to what extent had the CRAs assisted operations teams in designing and implementing more climate-resilient projects? The review found many positive aspects, even though the conclusions from the selected CRAs were often challenging to implement by project teams. Several specific insights were gained through this exercise. These included, among others: (1) ADB's approach was judged to be relatively pragmatic and proportionate; however, (2) there were opportunities to improve procedures and to strengthen the overall quality of project design to deliver climate-resilient projects; (3) the uptake of economic analysis of adaptation in CRA analysis could be strengthened; (4) bespoke guidance might be beneficial for adapting investments in the coastal zone and built environment given their high exposure and vulnerability to climate hazards; and (5) longitudinal analysis of completed projects could trace the value added and lessons learned by CRAs within the eventual design and delivery of resilience investments.

Following the recommendations of the above review, various initiatives were launched to strengthen CRM practices within the ADB, through streamlining procedures and building internal capacities through guidance and training. Work was carried out to re-orientate CRAs such that they focus more on project objectives and outcomes and consider risk management interventions, as opposed to top-down climate model-driven assessments. This was pursued through the development of *Principles of Climate Risk Management for Climate Proofing*

Projects (Watkiss *et al.*, 2020). The approach was light-touch (where feasible) and pilot-tested with live ADB projects before rolling out to staff via training sessions in 2018 and 2019.

STRENGTHENING ENABLING ENVIRONMENTS FOR RISK ANALYSIS

Project teams face many challenges when undertaking CRAs, including the climate risk assessment, adaptation options identification, and economic analysis ADB (2017c). Obtaining baseline climate data and contextual information at the project scale can be a significant obstacle to adaptation decision-making (Bhave *et al.*, 2016). Solutions may involve the use of unconventional data types, including technical reports, newspapers, indigenous knowledge, and even social media (e.g., Thompson *et al.*, 2021). Sometimes it is necessary to fill data gaps with brief field campaigns or to establish new data collection systems. The capacity to collate then interpret such diverse information streams and to recognize inherent quality issues or uncertainties (especially in climate model output) may also be in short supply. Likewise, tools for quantifying climate risks and evaluating adaptation option sets may be difficult to source or non-existent. Here, we recount some of the steps being taken by the ADB to address these practicalities.

Considerable attention has been paid to the procurement of climate risk information for adaptation and development planning more generally (e.g., Wilby *et al.*, 2009) as well as for specific sectors such as water (e.g., Fowler *et al.*, 2007). 'Country profiles' have been made available by organizations such as the World Bank and United Nations Development Programme for some time (see McSweeney *et al.*, 2010). Indeed, during the time of writing a new set of country profiles was released by the ADB and the World Bank[4]. These draw on climate scenarios from the CMIP5 ensemble associated with the IPCC Fifth Assessment Report, as well as indicators of climate risk and vulnerability. High-level profiles are helpful for situating projects within a national context and signposting useful resources but are typically insufficient for local-scale CRAs.

Portals like the KNMI Climate Explorer significantly improve access to baseline meteorological data and climate model projections for some variables (e.g., monthly mean temperatures and precipitation totals), extremes (e.g., annual maximum/minimum temperatures and 1-day rainfall totals), and impact indices (e.g., cooling and heating degree days). However, much of this information is also at low spatial and temporal resolutions – certainly too coarse to resolve past and projected conditions for a Pacific island or the specific locality of a project. Most Pacific islands are too small to be resolved as land surfaces (so are misrepresented climatologically as ocean spaces in climate models). More generally, there is still a disconnect between the outputs of the climate modelling community and the information needs of users, not least in terms of the formats and accessibility of climate model products. The Australian Bureau of Meteorology and CSIRO (2014) report on *Climate Variability, Extremes and Change in the Western Tropical Pacific* is a good example of a document that bridges this gap, so has received widespread uptake across the region.

Moreover, climate models struggle to represent important features of the local climatology (such as El Niño signatures in regional climate patterns of rainfall and temperature) (Yun *et al.*, 2016; Chen *et al.*, 2017). Key design variables such as storm surge height (for flood projects), potential evapotranspiration, and drought indices (for water resource and irrigation schemes) depend on post-processing and subsequent modelling with climate model outputs. Historic data that allow such assessments are scarce, difficult, or costly to obtain, or of questionable quality (especially for extreme values such as heavy rainfall). Compounding these concerns, for future climate projections, there is deep uncertainty from the combination of alternative representative concentration pathways (RCPs), and a large number of alternative climate models and

[4] https://www.adb.org/publications/series/climate-risk-country-profiles and https://climateknowledgeportal.worldbank.org/country-profiles.

simulations are available for each RCP. There is also uncertainty about future socio-economic change, which is usually at least as important as climate in determining future climate risks or levels of vulnerability. Hence, there may be a need to manage expectations about what climate experts can realistically supply to project teams. At the same time, there is scope to mainstream the use of climate risk information within the project life cycle (see below).

Efforts were made by the ADB to consolidate open information sources into a single document to save time and resources for project teams. As mentioned before, ADB (2018b) provides material to support CRA experts undertaking early stages of project development in the Asia-Pacific region. Approximately 70 sources of public domain data were compiled and evaluated. These resources were grouped into four categories covering: (1) inventories of national emissions, climate risks, vulnerability, and impacts; (2) historic weather, climate, and environmental change; (3) regional climate change projections; and (4) climate change impacts and adaptation. The ADB compendium also called for more investment in capacities for geospatial analysis, data testing and post-processing, climate downscaling, and impact assessment. Moreover, through closer cooperation between scientific and practitioner communities, access to publicly-funded, research-grade climate information could be much improved and made available to CRA teams.

Gridded satellite and reanalysis products are becoming more widely available (e.g., Beck *et al.*, 2017). These information sources can be used in data-sparse situations or combined with available local measurements to infill and extend records (e.g., Wilby & Yu, 2013). However, remotely sensed data have recognized limitations – not least poor performance at estimating heavy rainfall (e.g., Alijanian *et al.*, 2017). Furthermore, the skill of different precipitation products varies with information source, post-processing algorithm, region, season, and terrain (Sun *et al.*, 2018). Where available, locally observed, point meteorological data can be used to bias correct gridded data but these procedures are not without uncertainties either (Teutschbein & Seibert, 2012). Hence, a key tenet when collecting and analysing baseline data for CRAs is complete transparency about all information sources and working assumptions. Giving the impression of undue precision should be avoided by rounding values and presenting confidence intervals as standard practice.

Specialist technical assistance in data gathering, post-processing, and downscaling is routinely procured by ADB project teams. This can help to fill skills gaps but presupposes that the in-house capacity exists to critically evaluate the quality of work undertaken by consultants (or indeed to develop the terms of reference to commission appropriate information). Simple steps can be taken to raise standards. For example, to encourage greater consistency and rigour in technical analysis of climate model information, the ADB could require that climate service providers complete a quick checklist (Table 1). This would drive greater transparency of implicit and explicit assumptions about climate information sources, so project teams are aware of any issues that could materially affect project design at later stages. This goes alongside the expected norms of carefully managing and quality assuring data at each stage of the information flow (see Wilby *et al.*, 2017).

In-house training and capacity building are being pursued to better equip ADB project teams when developing terms of reference for consultants and quality assuring their deliverables. Pools of specialist advisers are used to provide independent peer reviews of CRA documents and ensure quality. Staff training has been provided on: shaping more explicit climate risk objectives, adaptation decisions, and eligibility criteria; understanding the project and what really matters (in terms of climate risks); smarter use of climate risk information in CRAs; hands-on experience of climate information portals; adaptation typologies; and economic analysis of adaptation costs and benefits. These elements have been integrated within a framework that is more focused on project outcomes than climate risk analyses. Looking beyond this, as the ADB contracts

Table 1 | Climate projection checklist for service providers.

Contextual information	**Check**
State the purpose(s) of the chosen climate projections at the start of the report	
Justify the choice(s) of emissions pathways and climate model sensitivity	
Justify the choice(s) of GCM(s) or climate model ensemble	
State and justify the choice of the baseline period	
Explain the choice of short-listed variables	
Historic data quality assurance and analysis	
State any recognized strengths and weaknesses of the data sources	
Provide complete details for historic data (e.g., site locations and years of record)	
Document any known changes in site, instrument, or observer practices	
Perform quality assurance, homogeneity, and stationarity checks on all data	
Sanity check extreme values against independent sources (e.g., media reports)	
State and justify any extreme value distributions (including for GCM output)	
Provide confidence intervals for all extreme value estimates	
Regional climate downscaling and derived indices	
State any recognized strengths and weaknesses of the chosen GCMs/RCMs	
Avoid mixing different scale GCM and RCM projections in the same ensemble	
Justify the choice of downscaling methods in terms of fitness for purpose(s)	
Provide full details and supporting references for all downscaling methods	
Provide full equations and supporting references for all derived variables	
Use GCM outputs in ways that are internally consistent (avoid pick-and-mix)	
Acknowledge when GCM/RCM ocean grid cells are used for terrestrial sites	
Itemize all components of regional sea level rise, justify omitted components	
Acknowledge uncertainty in derived indices (e.g., modelled evapotranspiration)	
State the percentile of extreme value estimates	
Compare historic data and GCM/RCM output using periods of overlap	
Presentation	
Proof-read documents for typographic errors and cite all literature sources	
State climate model details and emissions scenarios in figure and table legends	
Apply most appropriate quantiles (not just the mean) for the intended use	
Use 'could' rather than 'will' when referring to uncertain climate model projections	
Avoid undue precision when reporting climate change projections	

much of its technical work with consultants, there are opportunities to strengthen the climate service community more generally, whether by training, some process of certification[5], and/or by setting professional standards (e.g., Adams *et al.*, 2015).

[5] Such as the Global Center on Adaptation and World Bank 'Knowledge Module on PPPs for Climate-Resilient Infrastructure' https://gca.org/programs-infrastructure-and-nbs-knowledge-module-ppp/.

STREAMLINING AND REFOCUSING PROJECT-LEVEL CLIMATE RISK ANALYSIS

As discussed above, the ADB (2017b) review of CRM processes and practices revealed several recurrent issues within a sizeable sample of CRAs.

First, early CRAs tended to focus on model projections of the future climate for a limited number of variables (annual precipitation and temperature), i.e., climate hazard assessment, with significantly less time and resources devoted to project vulnerability, and even less on identifying adaptation strategies and options. This was due in part to the practice of engaging climate specialists, as opposed to sector specialists with CRA training. It also reflects the timing and nature of CRA studies within the project cycle, which are typically conducted in parallel with, but often independently from, project feasibility studies before the final design.

Second, these CRAs typically applied an impact assessment ('predict-then-act') approach. In such cases, the climate change analysis attempts to identify a set of climate change projections that 'best' characterizes the project location, which can lead to over-interpretation of projections from climate models, and time periods that rarely match the project design lifetime (such as the use of scenarios for mid- or late-21st century even in projects with a 30-year lifetime). As noted in the Introduction, this 'top-down' perspective is being superseded by approaches that pay greater attention to project objectives and outcomes with decision-making under uncertainty from outset.

Third, the nature of the ADB project cycle, and allocation of tasks across different teams, hinders the ability of CRA consultants to implement detailed recommendations concerning climate-sensitive design parameters. In many cases, the type and location of a project, and even some of the design elements, is decided upon by the time the CRA is underway. This means there is generally limited scope to re-design or alter a project significantly (to increase climate resilience) once the concept phase is complete. At the other end of the process, as the CRA consultants' work is completed prior to project Board approval, they are no longer available to consult with implementing agencies and design firms once project activities begin. To address these issues, there is now greater emphasis on improving the CRM process upstream (before the concept phase) and downstream (in the implementation phase).

Fourth, it is clearly desirable for CRAs to provide a 'menu' of risk management options to project teams, with each option evaluated (in terms of, e.g., economic costs and benefits, effectiveness, and technical feasibility). In practice, CRAs have tended to offer a narrow range of risk management options (in many cases only one), and these are usually focused on technical design options. This is being addressed through the development of searchable inventories of sector-specific adaptation options, with supporting literature and case studies (Georgi *et al.*, 2021). Other material showcases the integration of adaptation (and mitigation) solutions at city scales (e.g., ADB, 2018c). Even a one-page typology of adaptation options can be a powerful training resource and a reminder to project teams of the wider possibilities (Table 2). Importantly, this broadens the analysis of adaptation from hard engineering options to a suite of technical and non-technical options, including consideration of no- and low-regret activities, operational risk mitigation, green solutions, and other alternatives such as risk spreading, not just civil works and structures.

As mentioned before, the ADB is re-orientating CRM procedures to focus on achieving intended project outcomes *regardless* of climate change. This is part of wider efforts to simplify and streamline the existing CRM framework (Figure 2). The new workflow (Figure 4) is founded on several guiding principles that are elaborated in detail by Watkiss *et al.* (2020). In summary, they are discussed below.

Principle 1: move upstream

This entails activities before the development of project concepts, including (1) screening sector strategies and roadmaps to identify potential climate risks in relevant thematic and country operations to enable adequate time and resource allocation for climate-resilient project preparations; (2) identifying key climate risks in

Table 2 | Typology of adaptation actions: modified from Biagini *et al.* (2014).

Category	Sub-category	Description	Examples
Human and social capital	*Capacity building (C)*	Developing human resources, institutions, and communities, equipping them with the capability to adapt to climate change	Training/workshops for knowledge/skills development, public outreach and education, dissemination of information to decision-makers/stakeholders; identification of best practices, training materials
	Practice and behaviour (B)	Revision or expansion of practices and on the ground behaviour that are directly related to building resilience	Soil/land management techniques; climate-resilient crops or livestock practices, post-harvest storage, rainwater collection, expanding integrated pest management
Institutions and governance	*Management and planning (M)*	Incorporating understanding of climate science, impacts, vulnerability, and risk into government and institutional planning and management	Developing an adaptation plan, livelihood diversification, drought planning, coastal planning, ecosystem-based planning, changing natural resource management
	Policy (P)	Creation of new policies or revisions to policies or regulations to allow flexibility to adapt to changing climate	Mainstreaming adaptation into development policies, land-use specific policies, improvement of water resource governance, revised design parameters, ensuring compliance with existing regulations
Information and communication tools or technology	*Information (I)*	Systems for communicating climate information to help build resilience towards climate impacts (other than communication for early warning systems)	Decision support tools, communication tools, data acquisition efforts, digital databases, remote communication technologies
	Warning or observing systems (W)	Implementation of new or enhanced tools and technologies for communicating extreme weather and climate risks, and for monitoring changes in the climate system	Developing, testing, and deploying monitoring systems; upgrading weather or hydromet services
Infrastructure	*'Grey' infrastructure (HI)*	Any new or improved hard physical infrastructure aimed at providing direct or indirect protection from climate hazards	Climate-resilient buildings, reservoirs for water storage, irrigation systems, canal infrastructure, sea walls
	'Green' infrastructure (SI)	Any new or improved soft, natural infrastructure aimed at providing direct or indirect protection from climate hazards	Revegetation, afforestation, woodland management, increased landscape cover, wetland/mangrove restoration
Financing (F)		New financing or insurance strategies to prepare for future climate disturbances	Insurance schemes, microfinance, contingency funds for disasters
Technology (T)		Develop or expand climate-resilient technologies	Technologies that improve water use or water access, solar energy capacity, biogas, water purification, solar salt production

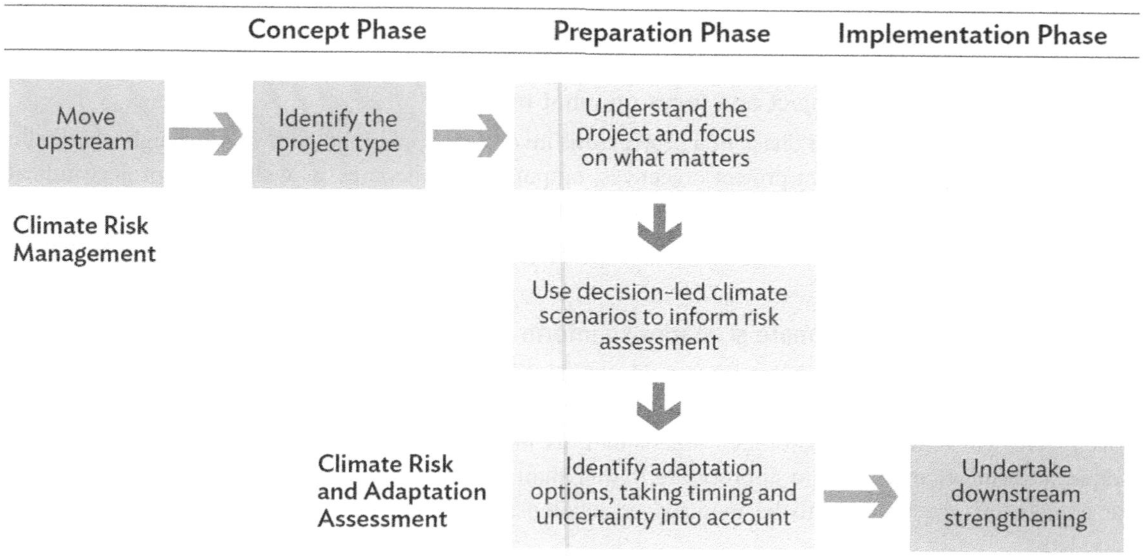

Fig. 4 | Guiding principles of CRM for climate-proofing projects. *Source*: Watkiss *et al.* (2020).

DMCs or sectors and hence opportunities to invest in building climate resilience in development plans and through targeted investments; and (3) enhancing the provision of key technical resources to support the identification of potential climate risks and opportunities for climate resilience investment (e.g., ADB's Urban Climate Change Resilience Trust Fund). Country Partnership Strategies (CPS) developed by the ADB with DMCs provide key entry points for mainstreaming climate adaptation at this more strategic and integrated level.

Principle 2: identify the project type

As projects are formulated and move towards the preparation phase, there is a need to differentiate investments and develop CRAs that match the materiality (proportionality) of climate change risks and adaptation needs. An initial checklist of five criteria has been developed to make these distinctions. The criteria are (i) amount of climate finance; (ii) project lifetime; (iii) risk of lock-in; (iv) warranted level of precaution; and (v) potential economic impact. The resulting project categories are the following:

Type 1a climate-proofing projects: where climate is a relatively minor risk to project performance, so the objective is to make the project more resilient with marginal adjustments for which a lighter-touch CRA approach would be sufficient. An example might be a standard roads project where drainage is a primary concern.

Type 1b climate-proofing projects: where climate change is a significant threat to project performance, so a greater level of adaptation and a more detailed CRA analysis are warranted. An example might be a large hydroelectric power and reservoir project.

Type 2-targeted adaptation investment projects: where addressing climate risks is the primary objective. An example might be a new coastal protection project to counter floods caused by rising sea levels. These types of investment may become a larger part of the ADB investment portfolio in the future with moves to scale up support to DMCs on climate action. Such projects require a more integrated approach in which CRA analysis are part of the core project design and outcomes. In many cases, this will involve more detailed climate and adaptation analysis, but it may also include lighter-touch situations (e.g., technical assistance projects or investments focusing on non-structural resilience solutions such as early warning systems and data analytics).

Principle 3: understand the project and focus on what matters

Moving into the main preparation phase of a project, this involves mapping the pathways through which climate variability and change could affect project objectives, outputs, and outcomes, as well as key project indicators. This can be used to narrow the list of risks for consideration, based on how material they are to the project performance and key indicators (including the economic and financial performance) and to identify adaptation objectives at an early stage.

Principle 4: use decision-led climate scenarios to inform risk assessment

Once climate impact pathways and key risks have been identified, the analysis of climate information and climate model projections can begin. This focuses on the appraisal of relevant climate impact and risk metrics. The approach here will differ. For light-touch assessment, as in Type 1a projects (Principle 2), sensitivity testing and rapid assessment methods are appropriate, rather than detailed analysis. However, for Type 1b projects, where climate change is likely to be material, a detailed assessment will be warranted but should still be decision-led.

Principle 5: identify adaptation options taking account of timing and uncertainty

It is recognized that the adaptation of long-lived infrastructure is challenging because of uncertainty. However, it is possible to identify promising measures by matching the type and timing of adaptation options to climate risks identified and placing greater emphasis on the economic rationale for adaptation (the costs and benefits). For Type 1a projects, this is likely to focus on low and no-regret regret options. A key focus is to introduce a wider consideration of adaptation options that extend beyond technical design changes (while noting these remain an important option). For projects that have more significant climate risks (Type 1b projects), low- and no-regret options are still important, but a greater focus on decision-making under uncertainty approaches, including iterative adaptive management, is recommended.

Principle 6: strengthen downstream (implementation, monitoring, and evaluation)

To ensure that benefits from the CRM are delivered, there is a need to enhance the programming of adaptation as projects move into the implementation phase (of the project cycle). This may require some changes in existing practice, especially to address additional challenges with flexible and iterative programming due to long-term implementation and governance arrangements. There is also a need to move towards implementation, monitoring, evaluation, and learning frameworks. This will require that resources be allocated for post-project climate evaluation and that institutional processes are in place to benefit from lessons learned. Occasionally, there may be covenants in loan agreements to ensure that climate-resilient features remain in contracts and bidding documents for civil works.

To illustrate these principles, consider a hypothetical irrigation project in southeast Asia. The overall objective is to improve rural incomes by increasing farm output from improved irrigation efficiency, leading to higher agricultural productivity, developed within an integrated water resources management approach that considers ecosystem services. Such a project might involve upgrading existing canals, installing new piped distribution systems and more efficient supply of water, introducing an improved management and operation (with water charging), and supporting farmers with extension services to enhance productivity and/or grow higher-value crops, as well as watershed management (e.g., slope stabilization and protected forests). The expected operating lifetime of the new infrastructure is 20–30 years, so the focus should be on risks from *climate variability* and emerging climate trends (rather than on long-term climate change). However, the project should not be in a water-scarce region and/or area where future climate or socio-economic development could exceed sustainable water use (i.e., to avoid lock-in and potential maladaptation). We may assume that the region faces possible climate risks from changes in average, minimum, and maximum temperatures (and evapotranspiration), average and seasonal precipitation, dry and wet season onset, and intensity and duration; changes in the frequency and intensity of tropical cyclones, the frequency and intensity of heavy rainfall, flash flooding, and erosion; and changes in the frequency and intensity of hot days and heatwaves. These would have implications for both water availability and irrigation demand, as well as for agricultural production (and crops and varieties).

Against this backdrop, there are prospects for mainstreaming adaptation thinking at an earlier stage (Principle 1) such as more strategic consideration of national and regional climate change and integrated land-water management in the ADB CPS and the Country Operations Business Plan. Since climate risks/opportunities are highly material to the project, an in-depth CRA is warranted (Principle 2) – this is a Type 1b project requiring more detailed analysis and closer collaboration with the core design team. The key risks affecting project outputs are identified (Principle 3) as changes in: (a) climate variability and abstraction levels, and thus the ability to meet the performance criteria for year-round irrigation supply – potentially affecting infrastructure design; (b) increasing temperature and evapotranspiration, as well as changing climate variability, affecting irrigation demand (total, seasonal, and peak) and design criteria (tolerance levels for inter-annual variability); (c) agroclimatic shifts and particularly possible changes in dry season and heat extremes, that might affect the choice of crops, timing of planting and harvest dates, crop pests, and diseases – potentially affecting the management and operation of the scheme and the farmer extension service components; (d) drought and fire risks to headwater forests, soil/slope stability, and downstream river flows – potential affecting the watershed management options; and (e) flooding and physical damage to irrigation systems, fields (and crops), and access roads from increased intensity of heavy precipitation (Table 3).

Next, the climate information needed to evaluate these risks is identified (Principle 4). This might begin with a detailed appraisal of historic temperature, evapotranspiration, and wet and dry seasonal rainfall totals, along with their inter-annual variability, compared to the water requirements of different crops. Indicators of extreme events, such as annual maximum daily rainfall amounts, dry-spell duration, and extreme heat could be analysed where data exist. Some basic water balance modelling and flood risks estimation might be performed too. The operating lifetime of some planned infrastructure could reach the 2050s, so changes in the same extreme values and water balance components could also be assessed using sensitivity testing or marker scenarios or from the climate model ensemble (such as CMIP5 or CMIP6).

A portfolio of adaptation options could then be developed and mapped against the identified climate risks and vulnerabilities (Principle 5) (Table 3). The options should include a mix of adaptation types. Importantly, these must target key risks, noting these affect project engineering as well as the services that are also included in the project (irrigation operation and farm extension services). The immediate priority is to identify no- and low-regret measures for early adoption (such as improved weather forecasts). Other measures, notably for possible changes to civil works and structures, should enable decision-making under uncertainty. For example, engineered options might include an appropriate allowance for climate change to harden critical assets, but alternatives that introduce the flexibility of design for later upgrade, or options that build robustness such as enhanced demand-side efficiency, should be considered too (see section Incorporating climate change within engineering design standards). Alongside this, it is important to strengthen monitoring within the project to enable adaptive management of emerging risks. For example, this might include a greater focus on measuring and correlating irrigation supply and demand with climate information but could also extend to enhanced pest and disease surveillance. Finally, roles and responsibilities should be assigned among project partners for

Table 3 | Intended project outcomes, climate risks, and adaptation options for a hypothetical irrigation scheme in southeast Asia.

Project outcomes	Climate risks	Adaptation options
Improved efficiency and reliability of irrigation supply	Change in precipitation and water (availability), increasing crop evapotranspiration (demand), change in variability, affecting system dimensions to meet peak supply (1 in 5 years)	• Increased irrigation capacity • Increased efficiency (e.g., drip irrigation) • Increased water storage • Supplementary water sources (conjunctive) • Improved demand-side management
Increased market-oriented agricultural production	Agroclimatic shift and change in the dry season, risk of weather extremes, impacts on the choice of crops, timing of planting and harvest dates, crop pests, and diseases, in turn affecting crop productivity and quality	• Weather and seasonal climate forecasts • Climate-smart extension services • Crops or varieties with shorter growing cycles or more resilient choices • Pest monitoring and management • 'Tall rice'
Enhanced watershed ecological services	Change in drought and fire risk affects headwater forests, soil/slope instability, downstream river flows	• Land-use zoning or buffer zones to deter encroachment into forest areas • Replant deforested areas with fast-growing tree species • Install live check dams
More productive rural infrastructure	Physical damage to irrigation system and access roads by flash floods, erosion, and sedimentation	• Early warning systems • Water management plans • Increase water storage (ponds and cisterns) • Strengthen emergency spillways, sluice gates, site drainage • Increase design standard with climate uplift for key access roads

long-term monitoring of project performance against expected outcomes and climate pressures (Principle 6). Taken together, these encourage adaptive management of emergent climate threats through a cycle of review that could be built into the management contracts for the project.

INCORPORATING CLIMATE CHANGE WITHIN ENGINEERING DESIGN STANDARDS

Although there is widespread expectation that extreme rainfall could become more intense (IPCC, 2018), and that the pace of global mean sea level rise will accelerate over coming decades (IPCC, in press), there is considerable uncertainty about the local values of these important design variables – even for the present climate. Nonetheless, engineers have to produce detailed designs for infrastructure that could have operating lifetimes of decades to centuries. Hence, the ADB, along with a few other agencies (e.g., CSIRO and Bureau of Meteorology, 2015; Environment Agency, 2016; United States Army Corps of Engineers, 2019), has been developing step-by-step procedures for incorporating adjustments for climate change within the detailed engineering design stage of a project. (Here, attention is on adjustments to *individual* project designs, rather than on facilitating an adaptive pathway comprised of *multiple* adaptation investments over time (e.g., Ranger *et al.*, 2013)).

The ADB approach seeks to strengthen the enabling environment (policy and regulatory) for adaptation upstream, as well as for individual projects. This may be achieved through national standards for detailed engineering design – noting that in the absence of such changes, these can act as a barrier to project-level adaptation (because projects comply with current, non-climate adjusted national standards). The ADB toolkit includes a technical report, user manual, and spreadsheets to guide users through a sequential process of updating design standards, with worked examples and data to aid understanding. In the ADB, these procedures were piloted in Viet Nam where national standards for roads and bridges require a design flood level of 1 in 100 years for expressways, and 1 in 25 years for categories 3, 4, or 5 rural roads. In Viet Nam, the 1-day annual maximum rainfall total (Rx1day) is a critical variable in provincial-level road design formulae for hydrologic calculations of design water discharge, flood level, and flow velocity. Parts of the Viet Nam coastline are also highly susceptible to storm surge, flooding, and erosion (Duc *et al.*, 2017). Therefore, a parallel set of procedures was devised for estimating local sea level for specified design periods. The two overarching philosophies are transparency and accountability – all sources of evidence and each calculation is made explicit. This means that as scientific understanding and climate modelling evolve, key data entries can be periodically reviewed and revised accordingly. For example, the IPCC (in press) *Special Report on the Ocean and Cryosphere in a Changing Climate* provides higher upper-bound estimates of global mean sea level rise than the previous point of reference which was the Fifth Assessment Report. These and other lines of evidence beyond the scope of IPCC reports are kept under review to ensure that guidance issued to project teams on design standards is sufficiently precautionary.

The ADB (2020) procedure for estimating design values for future extreme rainfall involves seven stages. Step 1: project objectives are specified as clearly as possible. These define the site(s), required design variable(s), return period, and operating life of the project. Step 2: a check is made for contextual risks. For instance, the design may be for a climate-resilient culvert, bridge, or road, but a coastal road might be more at risk from storm surge than flash flooding from the land. Step 3: design values are estimated from the annual maximum series of historic rainfall using established extreme value analysis. Quality assurance checks are applied to data and the choice of extreme value distribution (with confidence intervals) is made clear, along with reasons why. Step 4: equivalent climate model projections are obtained for the design variable (such as Rx1day) and operating lifetime of the asset. Again, decisions about the baseline period, choice of emissions scenario, climate model ensemble, and spatial domain of the scenarios are all made explicit. Step 5: the same extreme value distribution (as in step 3) is fit to the annual maximum series from the climate models for the baseline and future period. The design value is estimated at the required return period for these two periods. Step 6: the percent change in the design value is calculated by comparing the future and baseline estimates. To avoid undue precision and include some contingency, this change factor is rounded up to the nearest 5% (as in Table 4). Step 7: the change factor is applied to the historic extreme value estimate (from step 3) to give the new design value for the future, at the specified return period and confidence level.

This procedure captures non-stationarity in extreme rainfall (Cheng & AghaKouchak, 2014), but there are several points where the engineer makes an expert judgement. The most important decisions likely concern the choice of extreme value distribution (if any) guided by considerations of distribution fit, the short-listed climate model ensemble, and confidence interval. It is well known that extreme rainfall estimates are sensitive to the assumed distribution, geographic location, and length of record used (Papalexiou & Koutsoyiannis, 2013; Hu

Table 4 | Climate change adjustment factors for the 1-day annual maximum rainfall total (Rx1day) in Viet Nam.

	Return period (years)				
Future period	**2**	**5**	**10**	**20**	**25**
2016–2035	15	20	25	25	25
2036–2055	35	25	30	30	35
2056–2075	50	45	45	45	45
2076–2095	80	75	75	70	70

All values are expressed as percentage changes from 1986 to 2005 based on climate model experiments (CMIP5) under RCP8.5. Return-period estimates assume a Gumbel distribution and the 97.5th percentile of the credible ensemble, rounded up to the nearest 5%. *Source*: ADB (2020).

et al., 2020). There is little consensus about how best to judge the accuracy of a future scenario, but the ability to simulate key features and processes of the present climate system is regarded as a necessary but insufficient test (e.g., McSweeney *et al.*, 2015; Vogel *et al.*, 2018). The confidence interval should reflect tolerance for risk and economic considerations, which will vary by project. Since the results are sensitive to all these factors, the default position is to be totally explicit about what was done and why. At the very least, this leaves the calculations open to scrutiny and there is a clear audit trail leading to the design value.

A similar approach is proposed for the estimation of local mean sea level and high-water levels (ADB, 2020). As before, the first step is to be clear about the project objectives (S1). This specifies the design variable, time horizon, and location(s) for the flood and/or coastal erosion assessment. Once more, contextual risks should be identified (S2). For example, local rates of erosion may be influenced by changes in sediment supply from nearby rivers, as well as by sea level rise and storm surges. The *lower bound* (most optimistic) projection of future sea level rise is probably defined by an extrapolation of the historic trend from the nearest tide gauge (S3). A *central estimate* of sea level is derived by summing published values of global (e.g., thermal expansion and ice melt contributions), regional (e.g., gravitational fingerprint and vertical land movement), and local (e.g., tidal-range, wave, and surge) sea level components (S4) from sources such as IPCC (in press). A *high estimate* of sea level is obtained from the same components as in step S4, but now upper-bound uncertainty values for each constituent are blended into the calculation (S5). Finally, an *extreme estimate* of sea level is produced from the highest plausible values for each component/uncertainty bound in steps S4 and S5 to give a credible maximum water level under a high-end sea level change scenario with tide, storm, and wave effects (S6).

A key advantage of the components-based method of sea level estimation is that individual elements are accounted for and visible (Nicholls *et al.*, 2014). Moreover, the approach has been applied previously in some high-profile projects, not least in the long-term coastal protection plan of the Netherlands (Katsman *et al.*, 2011) and for planned nuclear power stations in the UK (Wilby *et al.*, 2011). Even so, the ADB (2020) procedures may encounter lack of information for some local components such as wave and storm surge heights – in which case, nearest neighbour or published values may be adopted. Choices must also be made about the numerical methods used to combine uncertainties in high and extreme sea level estimates. Simplifying assumptions may require that: maximum surge and wave heights remain unchanged (unless model evidence suggests otherwise); extreme surge, wave, and tide behave independently of each other; vertical land movements linked to changes in coastal sediment budgets can be neglected; and coastal morphology and beach profiles will be the same as present.

Finally, there are differing views about what constitutes a 'credible' upper-bound global mean sea level rise scenario. One expert elicitation attached a 5% probability of global mean sea level rise exceeding 2 m by 2100, and 7.5 m by 2200 in +5 °C world (Bamber *et al.*, 2019). Such judgements are clearly pertinent to risk assessments of long-lived assets such as power stations, ports, major flood defence assets, or other critical infrastructures in the coastal zone. It is also important to acknowledge that sea level rise is likely to continue well beyond 2100. Such high-end, multi-decadal scenarios are not needed for short-lived investments and/or where the risk of lock-in is low but, regardless of the project lifetime, all working assumptions should still be laid bare as a matter of principle.

CONCLUSIONS

The Asia-Pacific region has achieved major economic growth over recent decades. However, the region frequently experiences extreme weather events and is one of the most vulnerable globally to future climate change. The ADB has recognized these risks and become a regional leader on the CRM of its investment operations, by undertaking climate impact assessments from as early as 2005, then introducing mandatory risk screening of projects from 2014. Over time, the acquired experience has yielded valuable insights and led to an evolution of progressively more pragmatic CRM and adaptation within the organization.

Recent review work by the ADB, as part of a cycle of evaluation, led to a reappraisal of the current CRM processes, and a set of guiding principles for climate risk assessment have been developed, as summarized in this paper. These make the case for updating the CRM process and CRAs in the ADB. The tenets also provide a foundation for developing strategic and project-level guidance, as well as for enhancing the frameworks and processes needed to deliver the step change in climate finance envisaged by *Strategy 2030*. Alongside, there are new operational guidance and tools to embed climate change allowances in detailed project designs. These advances represent a welcome shift towards development outcome-orientated CRM. The latest work is reviewing the future possibilities offered by regional re-analyses (Mahmood *et al.*, 2018), ways of simulating unprecedented extreme events (Jain *et al.*, 2020), or the use of storylines to address highly uncertain climate risks (Shepherd *et al.*, 2018).

Many of the practical innovations made by the ADB would benefit communities of practice beyond the Asia-Pacific region. For example, there is a catalogue of resources aimed at CRA consultants, including project-level guidance, compendia of information sources, checklists for climate model scenarios and adaptation options, plus training materials. There are also manuals and tools for calculating hydrological extremes and sea levels for detailed engineering design – instruments that could fill a glaring gap internationally. Procedures are emerging too that will assist portfolio – rather than project-level adaptation strategies and thereby offer more integrated solutions. Above all, there are the conceptual advances that seek to rationalize the whole process of CRM and to focus resources on what really matters.

ACKNOWLEDGEMENTS

This work benefited from the authors' associations with the Asian Development Bank and from the helpful remarks made by Arghya Sinha Roy.

DISCLAIMER

The views expressed in this paper are those of the authors and do not necessarily reflect the views and policies of the ADB or its Board of Governors or the governments they represent.

DATA AVAILABILITY STATEMENT

All relevant data are included in the paper or its Supplementary Information.

REFERENCES

Adams, P., Hewitson, B., Vaughan, C., Wilby, R. L., Zebiak, S. & Eitland, E. & WMO Secretariat (2015). Call for an ethical framework for climate services. *WMO Bulletin 64*(2), 51–54.

ADB (2014). *Climate Risk Management in ADB Projects.* Asian Development Bank, Manila, Philippines.

ADB (2015). *Economic Analysis of Climate-Proofing Investment Projects.* Asian Development Bank, Manila, Philippines.

ADB (2016). *Guidelines for Climate Proofing Investment in the Water Sector: Water Supply and Sanitation.* Asian Development Bank, Manila, Philippines.

ADB (2017a). *Meeting Asia's Infrastructure Needs.* Asian Development Bank, Manila, Philippines.

ADB (2017b). *Strengthening ADB Business Processes, Project Level Guidance and Operational Frameworks for Climate Resilient Investments.* Asian Development Bank, Manila, Philippines.

ADB (2017c). *Guidelines for the Economic Analysis of Projects.* Asian Development Bank, Manila, Philippines.

ADB (2018a). *Strategy 2030: Achieving a Prosperous, Inclusive, Resilient and Sustainable Asia and the Pacific.* Asian Development Bank, Manila, Philippines.

ADB (2018b). *Information Sources to Support ADB Climate Risk Assessment and Management.* Technical Note. Asian Development Bank, Manila, Philippines.

ADB (2018c). *50 Climate Solutions from the People's Republic of China: Best Practices from Cities Taking Action on Climate Change.* Asian Development Bank, Manila, Philippines.

ADB (2020). *Climate Change Adjustments for Detailed Engineering Design: Step-by-Step Manual Using Worked Examples from Viet Nam.* Asian Development Bank, Manila. In press.

Alijanian, M., Rakhshandehroo, G. R., Mishra, A. K. & Dehghani, M. (2017). Evaluation of satellite rainfall climatology using CMORPH, PERSIANN-CDR, PERSIANN, TRMM, MSWEP over Iran. *International Journal of Climatology 37*, 4896–4914.

Auffhammer, M. (2019). *The (Economic) Impacts of Climate Change: Some Implications for Asian Economies. ADBI Working Paper 1051.* Asian Development Bank Institute, Tokyo.

Australian Bureau of Meteorology and CSIRO (2014). *Climate Variability, Extremes and Change in the Western Tropical Pacific: New Science and Updated Country Reports.* Pacific-Australia Climate Change Science and Adaptation Planning Program Technical Report. Australian Bureau of Meteorology and Commonwealth Scientific and Industrial Research Organisation, Melbourne, Australia.

Bamber, J. L., Oppenheimer, M., Kopp, R. E., Aspinall, W. P. & Cooke, R. M. (2019). Ice sheet contributions to future sea-level rise from structured expert judgment. *Proceedings of the National Academy of Sciences 116*, 11195.

Beck, H. E., Van Dijk, A. I., Levizzani, V., Schellekens, J., Gonzalez Miralles, D., Martens, B. & De Roo, A. (2017). MSWEP: 3-hourly 0.25 global gridded precipitation (1979–2015) by merging gauge, satellite, and reanalysis data. *Hydrology and Earth System Sciences 21*, 589–615.

Bhave, A. G., Conway, D., Dessai, S. & Stainforth, D. A. (2016). Barriers and opportunities for robust decision making approaches to support climate change adaptation in the developing world. *Climate Risk Management 14*, 1–10.

Biagini, B., Bierbaum, R., Stults, M., Dobardzic, S. & McNeeley, S. M. (2014). A typology of adaptation actions: a global look at climate adaptation actions financed through the Global Environment Facility. *Global Environmental Change 25*, 97–108.

Brown, C. & Wilby, R. L. (2012). An alternate approach to assessing climate risks. *Eos 92*, 401–403.

Chen, C., Cane, M. A., Wittenberg, A. T. & Chen, D. (2017). ENSO in the CMIP5 simulations: life cycles, diversity, and responses to climate change. *Journal of Climate 30*, 775–801.

Cheng, L. & AghaKouchak, A. (2014). Nonstationary precipitation intensity-duration-frequency curves for infrastructure design in a changing climate. *Scientific Reports 4*, 7093.

Clark, M. P., Wilby, R. L., Gutmann, E. D., Vano, J. A., Gangopadhyay, S., Wood, A. W., Fowler, H. J., Prudhomme, C., Arnold, J. R. & Brekke, L. D. (2016). Characterizing uncertainty of the hydrologic impacts of climate change. *Current Climate Change Reports 2*, 55–64.

CSIRO and Bureau of Meteorology (2015). *Climate Change in Australia Information for Australia's Natural Resource Management Regions: Technical Report.* CSIRO and Bureau of Meteorology, Australia. Available at: https://www.climatechangeinaustralia.gov.au/en/publications-library/technical-report/ (accessed 6 December 2019).

Duc, D. M., Yasuhara, K., Hieu, N. M. & Lan, N. C. (2017). Climate change impacts on a large-scale erosion coast of Hai Hau district, Vietnam and the adaptation. *Journal of Coastal Conservation 21*, 47–62.

Environment Agency (2016). *Adapting to Climate Change: Advice for Flood and Coastal Erosion Risk Management Authorities.* Environment Agency, Bristol. Available at: https://www.gov.uk/government/publications/adapting-to-climate-change-for-risk-management-authorities (accessed 6 December 2019).

Fowler, H. J., Blenkinsop, S. & Tebaldi, C. (2007). Linking climate change modelling to impacts studies: recent advances in downscaling techniques for hydrological modelling. *International Journal of Climatology 27*, 1547–1578.

Georgi, B., Carpenter, C., Droogers, P., Jonassen, R. G., Kenny, G., Kiem, A., Story, R., von Hippel, D. F., Wade, S. & Wilby, R. L. (2021). *Good Practice Guidance on Climate-Resilient Infrastructure Design.* Asian Development Bank, Manila, Philippines.

Hallegatte, S., Shah, A., Brown, C., Lempert, R. & Gill, S. (2012). *Investment Decision Making under Deep Uncertainty – Application to Climate Change.* World Bank Policy Research Working Paper No. 6193. World Bank, Washington, DC.

Hallegatte, S., Green, C., Nicholls, R. J. & Corfee-Morlot, J. (2013). Future flood losses in major coastal cities. *Nature Climate Change 3*, 802–806.

Haraguchi, M. & Lall, U. (2015). Flood risks and impacts: a case study of Thailand's floods in 2011 and research questions for supply chain decision making. *International Journal of Disaster Risk Reduction 14*, 256–272.

Hu, L., Nikolopoulos, E. I., Marra, F. & Anagnostou, E. N. (2020). Sensitivity of flood frequency analysis to data record, statistical model, and parameter estimation methods: an evaluation over the contiguous United States. *Journal of Flood Risk Management 13*, e12580.

Independent Evaluation Group (IEG) of the World Bank (2012). *Adapting to Climate Change: Assessing the World Bank Group Experience, Phase III of the World Bank Group and Climate Change.* World Bank, Washington, DC.

International Hydropower Association (IHA) (2019). *Hydropower Sector Climate Resilience Guide.* International Hydropower Association, London.

IPCC (2018). Summary for policymakers. In *Global Warming of 1.5 °C.* Masson-Delmotte, V., Zhai, P., Pörtner, H. O., Roberts, D., Skea, J., Shukla, P. R., Pirani, A., Moufouma-Okia, W., Péan, C., Pidcock, R., Connors, S., Matthews, J. B. R., Chen, Y., Zhou, X., Gomis, M. I., Lonnoy, E., Maycock, T., Tignor, M. & Waterfield, T. (eds). World Meteorological Organization, Geneva, Switzerland, p. 32.

IPCC (in press). Technical summary. In *IPCC Special Report on the Ocean and Cryosphere in a Changing Climate.* Pörtner, H.-O., Roberts, D. C., Masson-Delmotte, V., Zhai, P., Tignor, M., Poloczanska, E., Mintenbeck, K., Alegría, A., Nicolai, M., Okem, A., Petzold, J., Rama, B. & Weyer, N. M. (eds). IPCC, Geneva.

Jain, S., Scaife, A. A., Dunstone, N., Smith, D. & Mishra, S. K. (2020). Current chance of unprecedented monsoon rainfall over India using dynamical ensemble simulations. *Environmental Research Letters 15*, 094095.

Jongman, B., Ward, P. J. & Aerts, J. C. (2012). Global exposure to river and coastal flooding: long term trends and changes. *Global Environmental Change 22*, 823–835.

Katsman, C. A., Sterl, A., Beersma, J. J., van den Brink, H. W., Church, J. A., Hazeleger, W., Kopp, R. E., Kroon, D., Kwadijk, J., Lammersen, R., Lowe, J., Oppenheimer, M., Plag, H. -P., Ridley, J., von Storch, H., Vaughan, D. G., Vellinga, P., Vermeersen, L. L. A., van de Wal, S. W. & Weisse, R. (2011). Exploring high-end scenarios for local sea level rise to develop flood protection strategies for a low-lying delta – the Netherlands as an example. *Climatic Change 109*, 617–645.

Lu, X. (2019). *Building Resilient Infrastructure for the Future.* Background Paper for the G20 Climate Sustainability Working Group. Asian Development Bank, Manila.

Mahmood, S., Davie, J., Jermey, P., Renshaw, R., George, J. P., Rajagopal, E. N. & Rani, S. I. (2018). Indian monsoon data assimilation and analysis regional reanalysis: configuration and performance. *Atmospheric Science Letters 19*, e808.

Matthews, T., Wilby, R. L. & Murphy, C. (2017). Communicating the deadly consequences of global warming for human heat stress. *Proceedings of the National Academy of Sciences 114*, 3861–3866.

McSweeney, C., New, M., Lizcano, G. & Lu, X. (2010). The UNDP climate change country profiles: improving the accessibility of observed and projected climate information for studies of climate change in developing countries. *Bulletin of the American Meteorological Society 91*, 157–166.

McSweeney, C. F., Jones, R. G., Lee, R. W. & Rowell, D. P. (2015). Selecting CMIP5 GCMs for downscaling over multiple regions. *Climate Dynamics 44*, 3237–3260.

Nicholls, R. J., Hanson, S. E., Lowe, J. A., Warrick, R. A., Lu, X. & Long, A. J. (2014). Sea level scenarios for evaluating coastal impacts. *WIRES Climate Change 5*, 129–150.

OECD (2015a). *The Economic Consequences of Climate Change.* OECD Publishing, Paris.

OECD (2015b). *Climate Change Risks and Adaptation: Linking Policy and Economics.* OECD Publishing, Paris.

Papalexiou, S. M. & Koutsoyiannis, D. (2013). Battle of extreme value distributions: a global survey on extreme daily rainfall. *Water Resources Research 49*, 187–201.

Poff, N. L., Brown, C. M., Grantham, T. E., Matthews, J. H., Palmer, M. A., Spence, C. M., Wilby, R. L., Haasnoot, M., Mendoza, G. F., Dominique, K. C. & Baeza, A. (2016). Sustainable water management under future uncertainty with eco-engineering decision scaling. *Nature Climate Change 6*, 25–34.

Ranger, N., Reeder, T. & Lowe, J. (2013). Addressing 'deep' uncertainty over long-term climate in major infrastructure projects: four innovations of the Thames Estuary 2100 Project. *EURO Journal on Decision Processes 1*, 233–262.

Ray, P. A. & Brown, C. M. (2015). *Confronting Climate Uncertainty in Water Resources Planning and Project Design: The Decision Tree Framework*. World Bank, Washington, DC.

Schleussner, C. F., Deryng, D., D'haen, S., Hare, W., Lissner, T., Ly, M., Nauels, A., Noblet, M., Pfleiderer, P., Pringle, P. & Rokitzki, M. (2018). 1.5°C hotspots: climate hazards, vulnerabilities, and impacts. *Annual Review of Environment and Resources 43*, 135–163.

Shepherd, T. G., Boyd, E., Calel, R. A., Chapman, S. C., Dessai, S., Dima-West, I. M., Fowler, H. J., James, R., Maraun, D., Martius, O. & Senior, C. A. (2018). Storylines: an alternative approach to representing uncertainty in physical aspects of climate change. *Climatic Change 151*, 555–571.

Stakhiv, E. Z. (2011). Pragmatic approaches for water management under climate change uncertainty. *JAWRA Journal of the American Water Resources Association 47*, 1183–1196.

Sun, Q., Miao, C., Duan, Q., Ashouri, H., Sorooshian, S. & Hsu, K. L. (2018). A review of global precipitation data sets: data sources, estimation, and intercomparisons. *Reviews of Geophysics 56*, 79–107.

Tanner, T., Bisht, H., Quevedo, A., Malik, M., Nadiruzzaman, M. & Biswas, S. (2019). *Enabling Access to the Green Climate Fund: Sharing Country Lessons from South Asia*. Action on Climate Today, Oxford Policy Management, Oxford.

Teutschbein, C. & Seibert, J. (2012). Bias correction of regional climate model simulations for hydrological climate-change impact studies: review and evaluation of different methods. *Journal of Hydrology 456*, 12–29.

Thompson, J. J., Wilby, R. L., Matthews, T. K. R. & Murphy, C. (2021). The utility of Google Trends as a tool for evaluating flooding in data-scarce places. *Area*. doi:10.1111/area.12719.

UNESCAP (2019). *The Disaster Riskscape Across Asia-Pacific*. Asia-Pacific Disaster Report 2019, United Nations Economic and Social Commission for Asia and the Pacific, Bangkok.

United States Army Corps of Engineers (2019). *Incorporating Sea Level Change in Civil Works*. ER 1110-2-8162. Sea Level Change Calculator. United States Army Corps of Engineers, Washington DC.

Vogel, M. M., Zscheischler, J. & Seneviratne, S. I. (2018). Varying soil moisture–atmosphere feedbacks explain divergent temperature extremes and precipitation projections in central Europe. *Earth System Dynamics 9*, 1107–1125.

Watkiss, P., Hunt, A., Blyth, W. & Dyszynski, J. (2015). The use of new economic decision support tools for adaptation assessment: a review of methods and applications, towards guidance on applicability. *Climatic Change 132*, 401–4016.

Watkiss, P., Wilby, R. L. & Rodgers, C. A. (2020). *Principles of Climate Risk Management for Climate Proofing Projects*. Asian Development Bank, Manila, Philippines.

Wilby, R. L. & Murphy, C. (2018). Decision making by water managers despite climate uncertainty. In: *Oxford Handbook of Planning for Climate Change Hazards*. Pfeffer, W. T., Smith, J. B. & Ebi, K. L. (eds). Oxford University Press, Oxford.

Wilby, R. L. & Yu, D. (2013). Rainfall and temperature estimation for a data sparse region. *Hydrology and Earth System Sciences 17*, 3937–3955.

Wilby, R. L., Troni, J., Biot, Y., Tedd, L., Hewitson, B. C., Smith, D. G. & Sutton, R. T. (2009). A review of climate risk information for adaptation and development planning. *International Journal of Climatology 29*, 1193–1215.

Wilby, R. L., Nicholls, R. J., Warren, R., Wheater, H. S., Clarke, D. & Dawson, R. J. (2011). Keeping nuclear and other coastal sites safe from climate change. *Proceedings of the Institution of Civil Engineers: Civil Engineering 164*, 129–136.

Wilby, R. L., Clifford, N. J., De Luca, P., Harrigan, S. O., Hillier, J. K., Hodgkins, R., Johnson, M. F., Matthews, T. K. R., Murphy, C., Noone, S. J., Parry, S., Prudhomme, C., Rice, S. P., Slater, L. J., Smith, K. A. & Wood, P. J. (2017). The 'dirty dozen' of freshwater science: detecting then reconciling hydrological data biases and errors. *WIREs Water 4*, e1209. doi:10.1002/wat2.1209.

Yates, D., Miller, K. A., Wilby, R. L. & Kaatz, L. (2015). Decision-centric adaptation appraisal for water management across Colorado's Continental Divide. *Climate Risk Management 10*, 35–50.

Yun, K. S., Yeh, S. W. & Ha, K. J. (2016). Inter-El Niño variability in CMIP5 models: model deficiencies and future changes. *Journal of Geophysical Research: Atmospheres 121*, 3894–3906.

First received 18 October 2021; accepted in revised form 18 October 2021. Available online 28 October 2021

Water Infrastructure Planning, MDCU doi: 10.2166/wp.2021.265

Water-related infrastructure investments in a changing environment: a perspective from the World Bank

Diego J. Rodríguez[a,*], Homero A. Paltán[a], Luis E. García[b], Patrick Ray[c] and Sarah St. George Freeman[d]

[a] World Bank Group, Global Water Practice, Washington DC, USA
[b] Independent Consultant, Washington DC, USA
[c] Department of Chemical and Environmental Engineering, University of Cincinnati, Cincinnati, USA
[d] Civil and Environmental Engineering Department, University of Massachusetts Amherst, Amherst, USA
*Corresponding author. E-mail: drodriguez1@worldbank.org

ABSTRACT

At present, there is a global deficit in infrastructure and the World Bank Group (WBG) is one of the major sources of financing to reduce this gap worldwide. The WBG has policies and protocols for approving investments taking into consideration financial and economic indicators while ensuring social and environmental safeguards. In recent years, these safeguards have been updated to include the effects of climate change and robustness and resilience to support climate-informed project investment decision-making. A series of tools for screening projects for climate vulnerabilities and identification of risk management options have been developed to help project teams comply with these requirements. One of these tools is the hierarchical four-phased Decision Tree Framework (DTF) that, beyond screening, helps to analyze plans and project vulnerabilities, climate-related or otherwise, using a decision scaling approach, and explore risk management options, if necessary. The four phases of the DTF are (i) project screening, (ii) initial analysis, (iii) stress test, and (iv) climate risk management. This paper reviews applications of the DTF from the climate change screening phase to non-climate uncertainty screening and decision-making for project investments and prioritization. A peek into work in progress for incorporating resilience in the decision-making process, both for projects and through projects, is also provided, as well as next steps, looking forward.

Key words: Climate change, Climate-informed investments, Decision-making, Development, Infrastructure, Uncertainty

HIGHLIGHTS

- Since 2010, the World Bank Group has developed several tools that facilitate climate risk assessment and evaluation to inform decision-making.
- The Decision Tree Framework (DTF) integrates them within a bottom-up approach which accounts for various uncertainties that challenge the water sector while narrowing the gap between science and its practical applications.
- The DTF permits the consideration of resilience aspects beyond the physical dimensions of water infrastructure including the integration of human, economic, and institutional-governance systems.
- The DTF pilot studies suggest its practicality, flexibility, and scientific coherence for its application in the wide diversity of water projects across various geographies.
- Apart from its progress, the DTF still needs to adequately refine the incorporation of green infrastructure, water quality risks, energy efficiency and renewable, sea-level rise, and others.

1. INTRODUCTION

Managing water is a key factor in reducing poverty and promoting shared prosperity, playing an important role in achieving the sustainable development goals (SDGs) (Merrey & Awulachew, 2015; United Nations, 2019). Our ability to manage water is at the same time facing a critical point in human–water relations that merits urgent attention and action (Wilby, 2019). The contributions and limitations of both gray and green water infrastructure for this purpose have been highlighted around the world over the years, so has the social and economic benefits of closing the investment gap for building new, and maintaining existing, water infrastructure. This is needed to confront the challenges posed by climate variability and projected changes, growing populations, intensification of water use activities, increasing pollution, and rising social and economic costs caused by floods and droughts (Rodríguez *et al.*, 2012; Cervigni, *et al.*, 2015; Value of Water Campaign, 2017; UN Water/WBG, 2018).

Never has this been more evident than in present times. Important uncertainties in the changing conditions of the natural and climatic system, as well as human, social, economic, and institutional factors, are likely to alter the principal characteristics for which a project was originally designed, jeopardizing its success. The hydrological response of a catchment to accelerated changes in the historic climatic conditions, alongside land-use changes and population growth, casts a shadow of doubt about the reliability of long-cherished traditional hydrological methods based on erstwhile stationarity working assumptions. In what cases are these methods still applicable? When are new approaches and methods needed? What are the desired characteristics of new methodologies? Recognizing and assessing these uncertainties is only the first step. They must be incorporated into the institutional decision-making investment process framed by financial, economic, social, and environmental conditions of existing policies.

Aware of these circumstances, since the 2010s the World Bank Group (WBG) in collaboration with its partners has been working to quantify these uncertainties and incorporate them in the decision-making process for water-related investments. With its current portfolio of annual water investments on the order of US$41.5 billion, the WBG is presently an important multilateral source of financing and investment for water projects globally. It benefits, therefore, from an excellent opportunity to incorporate the latest developments in investment decision-making under uncertainty (climate and other), applied to water-related infrastructure and planning, and transmit them to its client countries for application at the local level.

Section 1.1 of this paper overviews the WBG's framework for water-related project financing and how this framework has been gradually incorporating climate change, robustness, and resilience considerations. Section 2 summarizes some of the tools available for implementing major climate change-related commitments. Section 4 describes in more detail the origin, evolution, and potential of one of these tools for decision-making under uncertainty - the Decision Tree Framework (DTF) (Ray & Brown, 2015). Section 4 reviews piloting applications of the DTF as it has evolved from the climate change screening phase to non-climate uncertainty screening and decision-making for project investments and prioritization in planning, incorporating robustness and resilience in the decision-making process, work in progress, and next steps. Section 6 includes potential DTF roles not yet developed.

1.1. World Bank Investment Project Financing

The WBG's Investment Project Financing Policy calls for the assessment of a proposed project based on various country and project-specific considerations, project objectives and technical, economic, financial management and procurement, environmental and social considerations, and related risks. The environmental and social considerations (known as Safeguard Policies for Investment Project Finance operations) included environmental assessments and action plans as well as safeguards related to natural habitats, water resources management, pest management, indigenous peoples, physical and cultural resources, involuntary resettlements, forests, and safety of dams, as well as other considerations for each specific project, related to location in international waterways and in disputed areas, as applicable.

Echoing the conclusion of the IPCC in its 2007 Fourth Assessment Report, that warming of the climate system is unequivocal, and that the effects of climate change are already visible in terms of higher average air and ocean temperatures, widespread melting of snow and ice, rising sea levels, floods becoming more common in some regions, while in others droughts becoming more intense, the WBG made climate change the central theme of its 2010 World Development Report '*Development and Climate Change*' (World Bank, 2010). Among key conclusions were that climate change threatened all, but particularly developing countries, and that these countries were more exposed and less resilient to climate hazards. Nevertheless, it also concluded that reducing climate risk was affordable. However, further climate change is unavoidable and decision-making processes needed to evolve to tackle a riskier and more complex environment. This brought to the forefront the need for alternative frameworks for decision-making to address uncertainties that are difficult to evaluate and to minimize vulnerabilities not just to a single but to a range of possible risks. The 2010 World Development Report put forward a policy prescription, stating that this approach should be complemented by decision support techniques that looked for 'robust' (encompassing a wider range of unanticipated variability) rather than just economically optimal strategies, where costs, benefits, and the trade-offs inherent in climate policies were assessed under all scenarios.

In 2016 the WBG's Climate Change Action Plan 2016–2020 was published (World Bank, 2016), supported by five strategic shifts for the Bank's climate work: (i) support for the implementation of plans, (ii) convergence of the WBG's climate and development agendas into strategies and operations, (iii) shaping national investment policies and programs to maximize the impact, (iv) rebalance the WBG climate portfolio increasing the focus on adaptation and resilience, (v) and facilitating transformation from business as usual to achieving global climate commitments. The activities under the Climate Change Action Plan were organized around four priorities: (i) support transformational policies and institutions, (ii) leverage resources, (iii) scale-up climate action (in renewable energy and energy efficiency, sustainable mobility, sustainable and resilient cities, climate-smart land use, water and food security, green competitiveness, and other actions such as early warning systems, climate and health, and others), and (iv) align internal processes and work with others.

By 2017, the Bank priority areas included climate change and infrastructure alongside gender, social and political fragility, and public–private partnerships. The Environment and Social Framework or ESF (World Bank, 2017a) was launched, offering broader and more systematic coverage of environmental and social risks, including climate change. The ESF will progressively replace the current Safeguard Policies for Investment Project Finance operations, with the two operating in parallel for about 7 years (World Bank, 2018a). The Adaptation and Resilience Action Plan (World Bank, 2018b) prioritizes adaptation and resilience, elevating them to an equal footing with climate mitigation actions. Besides boosting adaptation financing and mainstreaming the incorporation of climate risk in planning, design, and implementation, the Plan includes developing a new resilience rating system for development and infrastructure projects on the ground and creating incentives for governments, project developers, and investors to select higher-resilience options.

Over the past few years, the WBG has also made important climate change-related corporate commitments, establishing targets to integrate climate change into operations and strategies. Guidance notes have been produced and periodically revised and adjusted as necessary to help operational teams move toward a climate-smart portfolio and maximize the project's climate adaptation and mitigation co-benefits (the portion of financing in a project that contributes to climate adaptation and/or mitigation).

Currently, the WBG aims to implement four major commitments related to climate change: (i) climate and disaster risk screening; (ii) greenhouse gas (GHG) accounting; (iii) shadow price of carbon application (to GHG emissions and reductions); and (iv) climate adaptation and mitigation co-benefits tracking. Guidelines for designing climate-informed projects around the four major climate change commitments applicable to the WBG's Water Global Practice would include three interconnected steps:

1. Gain insight into climate change vulnerability context and GHG emissions associated with the project, identifying current and expected climate and disaster risks, emission implications, and opportunities to improve the project.
2. Informed by the results of step 1, design project components to be climate-prepared, particularly for adaptation and resilience.
3. Ensure that the mitigation and adaptation co-benefits are articulated according to the Joint Multilateral Development Bank agreed methodologies for tracking climate change adaptation and mitigation finance. Adaptation co-benefits are assigned to a project only if it incorporates the project's climate change vulnerability context, the intent to address the identified vulnerabilities is explicitly stated, and a clear and explicit link between the identified vulnerabilities and specific actions is established.

To address each of these steps, the guidelines provide a wide list of tools and resources that the project team can use. For the purpose of this paper, only the following will be discussed in the next section: the Climate Change Knowledge Portal (CCKP), the WBG Climate and Disaster Risk Screening Tools (CDRST), and the DTF.

2. TOOLS AVAILABLE FOR IMPLEMENTING MAJOR CLIMATE CHANGE-RELATED COMMITMENTS

The first two tools, described below, provide a rich guide to initially identify whether a project is vulnerable to climate variability and other hazards, and the level at which a project may be impacted. Yet they do not provide by themselves a detailed description of risks, nor do they provide enough bases for risk management or to design strategies to increase the robustness or resilience of a project. This can be done with the third tool – the DTF – described in following sections.

2.1. Climate Change Knowledge Portal

This web-based platform (http://sdwebx.worldbank.org/climateportal/) runs on a Google Maps interface so the user can query any location, country, or region on the globe. Historical climate conditions are derived from the global network of weather stations which have been quality-corrected as well as complemented with gridded global historical products. Future conditions are calculated from global circulation model (GCM) outputs from several models utilized in the IPCC's 5th Assessment Report. Examples of statistics derived from the main climate variables shown in this platform include projected changes in maximum daily temperature, monthly maximum temperature, largest 5-day cumulative rainfall, annual maximum daily rainfall, and others. Similarly, the tool provides projected impacts in various indicators describing change in the energy, agriculture, health, drought, and water variables. The change in these variables can be visualized and downloaded for various future climatic scenarios and decades in the 21st century. An example of an output of this tool is shown in Figure 1.

The standardized spatial resolution is 2° (about 200 km). The tool has also aggregated these results at the basin level for various major rivers of the globe. Although information provided by the CCKP cannot be taken as precise predictions of future climate change, water-planners may be able to use them to get an initial idea of the range of future climate, as part of a first-order examination of the projected change of the historic climate conditions which could be anticipated for a specific water project.

2.2. Climate and Disaster Risk Screening Tools

These specific tools are intended to support Bank Project Teams to understand if and why further studies and consultations may be required during the design of climate-vulnerable projects but may be of use also for

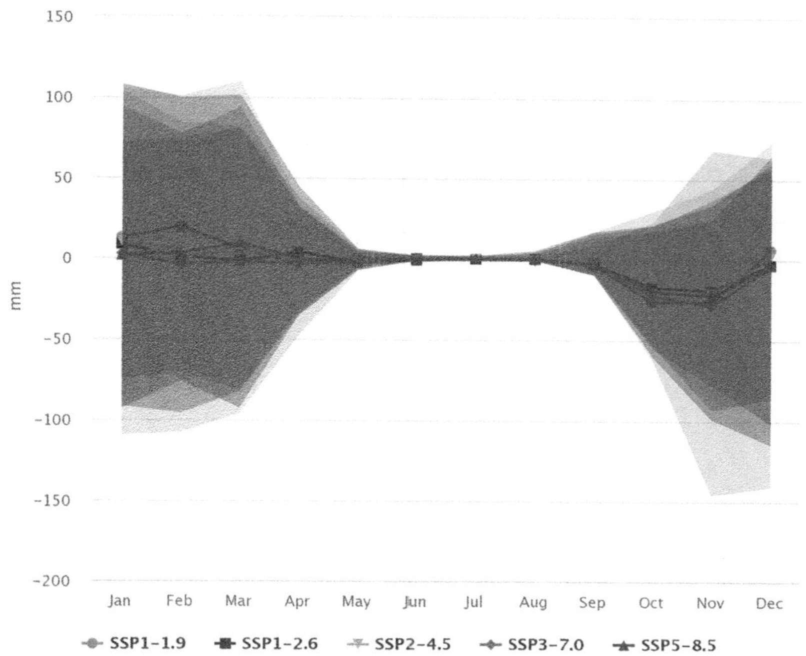

Fig. 1 | Projected precipitation anomaly for a catchment in Southern Africa, 2060–2079.

other decision-makers. This set of screening tools is best applied at the early stages of project evaluation, as they consider climate and disaster risks at the high level (in both the short and long term, and at the country and project scales). The aim of these screening tools is, thus, to ensure that decision-makers properly incorporate anticipated climate variability and particularly extreme events that are the basis of potential natural disasters, in the development and design of policies, programs, and projects. At a country-level scope, the user can apply this tool to examine the way in which climate, and other natural hazards, may affect national strategies and policies or institutional reforms during their design phase. At the project level, the tool provides specific instruments to systematically explore risks and support investments in the health, transportation, energy, coastal flood protection, agriculture, and the water sector.

The CDRS for water projects can be accessed via its online WBG platform (https://climatescreeningtools.worldbank.org). First, it allows the user to detail the basic characteristic of a water project, namely title, type, location, and others. From here, the user can evaluate the project's perceived or qualitatively estimated exposure to the range of climate and geophysical conditions and hazards which are given by the location of the project. The climatic conditions and hazards, which can be identified, include extreme temperatures, extreme precipitation and flooding, drought, sea level, storm surge, and extreme winds. This climatic information is given by the CCKP platform. The user is then able to categorize the estimated level of exposure of a project. Next, this set of tools permits the exploration of potential impacts by combining exposure to multiple hazards and the vulnerability of the project, which depends on the user understanding of the project. In this stage, the users can rate the expected level of impact to one or various sources of present or future hazards. Then, the tool permits one to assess and rate the proposed project's capacity to adapt to the range of impacts identified. Lastly, these data are used to develop a summary report which identifies the overall estimated risk of a project (see Figure 2).

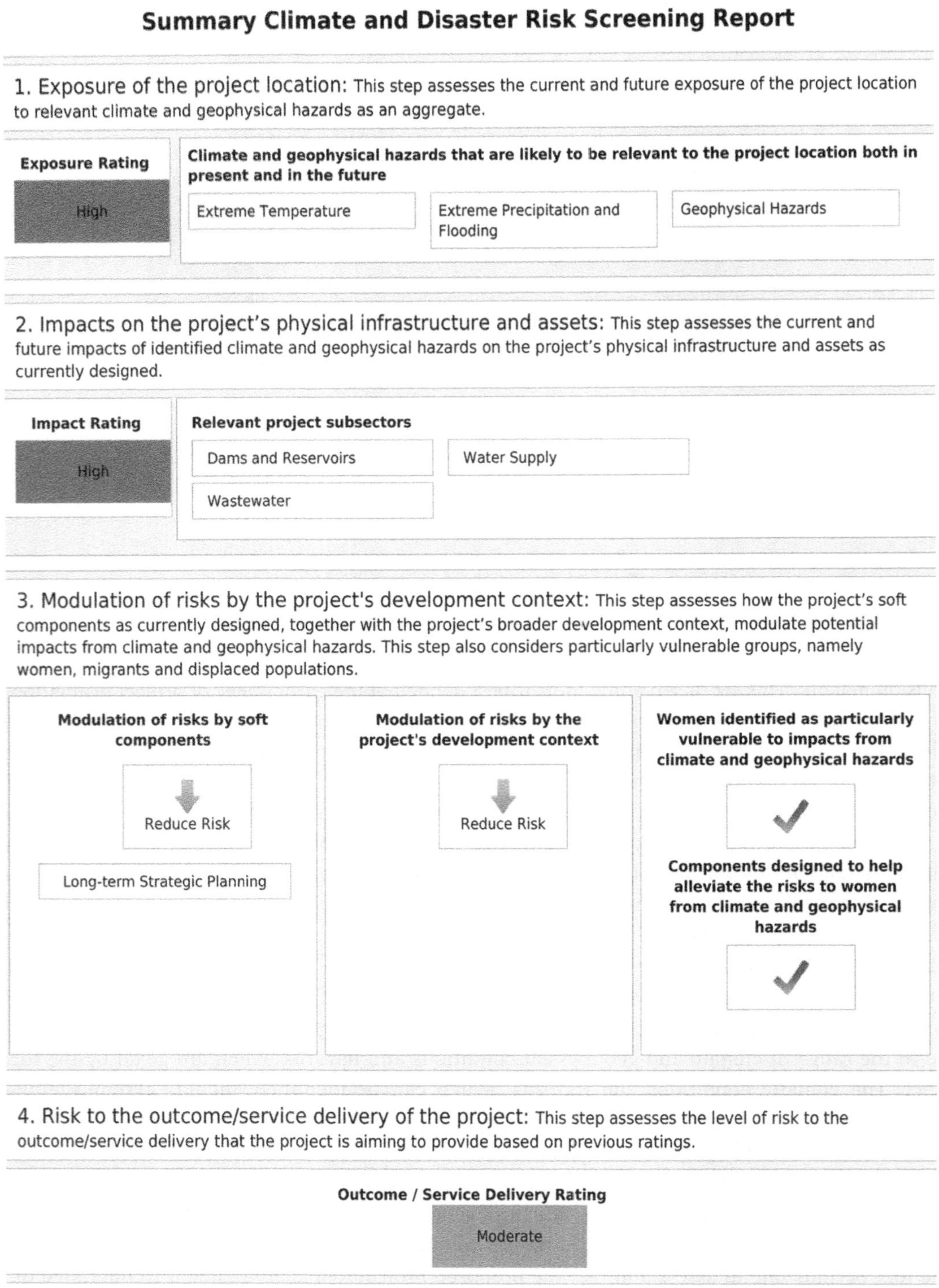

Fig. 2 | Summary climate and disaster risk screening report as output of the Climate and Disaster Risk Screening Tool.

2.3. Decision Tree Framework

This tool goes beyond the CDRST and provides the Project Teams with additional specific questions to ask when screening water-related projects. Its application basically starts after the CDRST (or any other suitable screening method) indicates the project is vulnerable to climate change. If so, it goes into a rapid project scoping exercise to assess the importance of the project's vulnerability to climate change *vis-à-vis* other vulnerabilities. Here, vulnerability thresholds are defined on a project-by-project basis and depend on their design, characteristics, historical tipping points, or others as agreed by stakeholders. If the vulnerability to climate change is deemed significant, it calls for an exhaustive climate risk analysis or stress test, using a bottom-up approach, defined later in this chapter. In the WBG version, it uses the decision-scaling approach (Brown *et al.*, 2012) for robust-based planning of water systems but it can conceivably also use other existent bottom-up approaches. The DTF uses a stress test as a simple way to identify the vulnerabilities of a system, which also permits the evaluation of their response when modifications are performed. If the results of the stress test indicate that the project needs modifications, a further step is needed to verify if its robustness, or other selected performance metrics, can be improved. The results of this analysis will indicate if the project can be adjusted or if more complex methodologies of decision-making under deep uncertainty are needed in search for an adequate performance of the selected metrics, or even if other options should be explored. Section 4 below describes in more detail the characteristics of the DTF.

3. ORIGIN, EVOLUTION, AND POTENTIAL OF THE DTF

3.1. Early initiatives to communicate climate data science

In 2009, the report 'Water and Climate Change: Understanding the Risks and Making Climate-Smart Investment Decisions' (Alavian *et al.*, 2009) was published by the WBG. It was an effort to assist countries with incorporating adaptation to climate variability and change in their work programs. The report also provided the first version of a tool that could be used to assess exposure and screen the investment portfolio on a regional basis.

According to a prevalent trend at the time, the WBG followed traditional approaches that downscaled the results of several GCMs under given IPCC scenarios to project streamflow series as input for project design. These approaches (often referred to as 'Top-Down') rely on the initial understanding and manipulation of data derived from GCMs to infer responses at the local scale. Yet, scientific literature has widely recognized the limitations which may occur when manipulating and making decisions solely with these data. These limitations are typically a result of the large uncertainties which are inherited from the configuration and characteristics of GCMs (Stainforth *et al.*, 2005, 2007; Teng *et al.*, 2012; Hartmann *et al.*, 2013), whose original purpose was to simulate climate conditions at a global scale and not hydrology at a local scale. This led water managers to face a wide range of possible scenarios of future climate conditions. If one scenario was chosen over the rest, it may lead to the project being vulnerable to the occurrence of a combination of other *likely* climatic possibilities and the cumulative uncertainties may not, in fact, be adequately characterized.

Acknowledging these fundamental limitations of climate projections, in 2012 the WBG Independent Evaluation Group in its Report 'Adapting to Climate Change' (IEG World Bank/IFC/MIGA, 2012) identified that 'climate model information has generally been unable to inform quantitative decision making' in various previous water resources studies. As such, the report concluded that climate models 'have been more useful for setting context than for informing investment and policy choices'.

Furthermore, in 2014 the International Development Association (IDA), which is the WBG's fund to fight extreme poverty, acknowledged the need for climate change risk assessment under its Seventeen Replenishment IDA17: 'Maximizing Development Impact, Special Theme 3: Climate Change'. It called for 'mainstreaming climate and disaster risk management in IDA countries' strategies, policies and investments' and to 'screen all new IDA operations for short- and long-term climate change and disaster risks and, where risks exist, integrate appropriate resilience measures' (IDA, 2014). This now applies as well to all World Bank projects.

Also, existing research suggests that changes in population dynamics, economic circumstances, demand patterns, land-use change, and others all have important effects on the future characteristics and outputs of water resources systems, and they are equally uncertain decision variables (Groves *et al.*, 2008; Lambin & Meyfroidt, 2011; Stakhiv, 2011; Buytaert & Bievre, 2012). Failing to incorporate these variables, along with an incorrect representation of climatic outputs and their uncertainties, may in turn importantly affect the description and assessment of the characteristic of design risks and costs in water planning and investment design.

3.2. Incorporating decision-making under uncertainty and risk-based approaches

Both the IDA17 call and the acknowledgment of IEG were congruent with the recommendations of the WBG's 2010 World Development Report. There was a need to develop an accepted process for climate change risk assessment in order to reduce the vulnerability of people, their assets, and livelihood to climate-related risks.

This fueled a paradigm change from the traditional method of Predict, then Act approach, to a Flexible and Robust Decision-Making (RDM) approach. Thus, several applications of the RDM methodology (Lempert *et al.*, 2006) were made by the WBG, among them, to various existing projects under preparation in Africa (Cervigni *et al.*, 2015) and to Peru in the 'Strategy for Implementing Lima's Long-Term Water Resources Master Plan' (Kalra *et al.*, 2015).

A summary of existing decision-making methodologies able to deal with climate-related uncertainty, namely cost–benefit analysis under uncertainty, cost–benefit analysis with real options, robust decision-making, and climate-informed decision analysis, was later compiled by Hallegate *et al.* (2012) to help decision-makers identify which method was more appropriate in a given context, as a function of the project's lifetime, cost, and vulnerability.

Conceivably, some questions from the practitioner's point of view can be posed: would these methodologies be applicable in all the cases? If not, in which cases should they be applied? Were there other simpler analyses that could be applied in some simpler cases and if so, what were their characteristics? Looking to respond to these types of questions, a series of analytical approaches have been developed which tend to look at the intrinsic characteristics of a water resources project or system and describe them in terms of exposure, sensitivity, and their adaptive capacity to withstand stress (García *et al.*, 2014; Ludwig *et al.*, 2014; Rosenberg & Madani, 2014; Brown *et al.*, 2015). These approaches are commonly referred to as 'bottom up' since they first assess the characteristics of the existing system and its sensitivity to a wider range of climate inputs and identify the parts of the system that are most vulnerable, and the specific attributes of those vulnerabilities. From here, these approaches look to examine the response, or sensitivity, of a system to a series of climate projections or other variables that stakeholders consider may affect the expected outcomes of their project.

Thus, bottom-up approaches permit analysts to include potential sources of risk such as change in population dynamics or economic conditions, as previously mentioned. Several methodologies of the bottom-up approach exist in the current scientific literature. They include the scenario-neutral approach (Prudhomme *et al.*, 2010), the information-gap theory (Hine & Hall, 2010), the risk informed decision-making (Ayyub *et al.*, 2010), and the decision scaling approach (also known as the Climate Informed Decision Analyses) (Brown *et al.*, 2012). In these approaches, the level of complexity as well as the resources needed to inform decisions is scaled (adjusted) according to the unique characteristics of a project and other relevant factors, as well as the issues and decisions that may arise in the process that are generated by stakeholders. As such, during the various stages of this process, water resources managers as well as local stakeholders can interact in order to identify not just the potential risks, but also agree on the appropriate level of analyses required to understand the vulnerabilities of the system. So these approaches, instead of beginning with the examination of climate projections and hazard calculations, looks to consult relevant stakeholders, explore data available, and others which lead to an informed evaluation of the system characteristics.

Using these approaches allows not only diagnosis of the potential impacts of climate and other variables in a water project, but also generation of adaptation options, which would reduce the effect of the relevant stressors in a system (Hamlet, 2011). Often these approaches permit water planners to assess the chances of success, or failure, of a specific project under current or plausible future conditions (Rosenberg & Madani, 2014). From here, water managers are in the position to propose modifications to the project in order to test the response of such modification to the identified stressors. At this stage, decision-makers and stakeholders would be able to formulate and design more risk–cost-effective resilient and robust solutions for a project.

Since 2011, the WBG has conducted a series of studies, which included components of these risk-based and/or bottom-up approaches. For instance, Rodríguez-Iturbe & Valdés (2011) presented a simple framework, which aimed to understand and predict floods and monthly flows at the catchment level, in various dry and humid ecosystems in rivers of Venezuela and Kenya, and in Arizona and Florida in the USA. This framework proposed 'no-regret' actions in order to produce positive results regardless of the existence of hydroclimatic changes. At the same time, this framework proposed to treat rainfall as a stochastic process to address random fluctuation in rainfall and its impacts on soil moisture seasonality.

In 2012, the Executive Secretariat of the Niger Basin Authority (NBA) and the WBG implemented jointly a climate risk assessment (CRA) initiative for the Niger River Basin (Grijsen *et al.*, 2012). The proposed approach followed a risk-based framework which identified a series of potential climate hazards by analyzing response sensitivities of selected performance indicators (related to hydro-energy generation, irrigated agriculture, navigation, flooding, and the sustenance of environmental flows) to shifts in runoff. The approach translated shifts in future climate variables into runoff changes following an elasticity approach (e.g., what is the runoff response to a $\pm 10\%$ change in precipitation of the existing system?). The study also looked to assess the ability of this water system to tolerate climate variability in terms of the elasticity of performance indicators – both on its operational characteristics, as well as its overall designed capacities and structural reliability. In this case, the results suggested that the impacts of climate changes on the water resources of the Niger River Basin would be mild to modest. It should be noted that historical streamflow records presented more critical situations than those projected by GCMs, diminishing the usefulness of the projections for exploring system vulnerabilities to climate change in a stress test framework. A similar elasticities approach to examine runoff response to climate variables was later applied to assess the climate vulnerability of the Tina Hydropower Project in the Solomon Islands (Grijsen, 2016).

In a more comprehensive effort in 2013, the WBG completed a study which looked to exhaustively assess the impacts of climate risks and development alternatives on water and agriculture in the Indus Basin of Pakistan (Yu *et al.*, 2013). A multi-modeling framework was developed to understand the interdependencies between climate, water, food, and the economy. The results showed that the most sensitive parameters to climate in this basin are inflow from stream tributaries, crop water requirements, and depth of groundwater. This framework also helped to explore the implications of possible adaptation investments scenarios. The results, for instance, indicated that investments in canal efficiency and crop yield would minimize the impacts of future climate risks and meet the food self-sufficiency objectives of the country. The importance of the effect of non-climate factors, such as mandatory distribution of water volumes among provinces, was also highlighted.

3.3. Structure and process of the DTF

As above-mentioned, a tool capable of addressing climatic and non-climatic uncertainties in water resources planning and project design, while building on state-of-the-art scientific methods and the practical experiences from the studies mentioned above, was needed. But mainly, able to answer what should be an obvious question of a hypothetical Bank Task Team Leader or likewise mid-level decision-maker or water engineer in any WBG Client Country:

> *'I know that my plan or project is vulnerable to the effects of climate change, which might be to some extent uncertain, and I am convinced of the importance of taking them into account in my plans and designs. How best to do it in a manner commensurate with the perceived vulnerability of the investment and the corresponding justifiable effort and cost?'*

Thus, the WBG, in partnership with organizations such as the Alliance for Global Water Adaptation (AGWA), the U.S. Army Corps of Engineers, the University of Massachusetts-Amherst, and the Stockholm International Water Institute (SIWI), among others, started in 2011 to develop an updated framework for analysis – hopefully meeting those characteristics – that in 2015 resulted in what is now called the DTF.

For its development, similar decision support tools used by other institutions were reviewed and many consultations, internally and externally in international technical meetings, were held. The resulting DTF is a bottom-up tool for decision-making under uncertainty that encompasses a step-by-step decision-making approach resembling the attributes of a branch of a tree. It looks to deepen or scale up the level of analysis only when needed by the perceived vulnerability of the project at each step. This characteristic in turn helps decision-makers to make better use of their human and financial resources and make use of known and proven hydrology and water resources methodologies.

To improve the effectiveness of water resources management planning and project design under climate variability and change, it incorporates into the analysis the climate science insights provided by constantly improving models. It then compares its effects *vis-à-vis* other non-climate uncertainty. It also offers transparency that lets the decision-maker or engineer know what is going on at each step and allows them to make the decisions.

In addition, the DTF also looks to maximize the utility of previous WBG diagnostic tools such as the CCKP and the CDRST. Also, while using up-to-date scientific knowledge to assess and manage climatic and non-climatic risks in water planning and projects, it is flexible in that sense, and it can conceivably use many methodologies or models according to the characteristics and needs of the problem at hand and the requirements of the user. The DTF is, therefore, designed in such a way that the process can be followed mainly within the capabilities available to water managers and increase their confidence about the overall assessment of climate risks while increasing their capacity to manage them. A graphic overview of the DTF is presented here in Figure 3.

One of the principal characteristics of the Decision Tree process is that it is hierarchical, meaning that the continuity of the process is assessed in each phase of analysis. The DTF involves four main phases. Phases 1 through 3 correspond to the risk *assessment* stage of analyses. Lastly, Phase 4 provides the elements necessary to *manage* risk. As such, the four hierarchical phases of the DTF process are: Screening (Phase 1), Initial Analysis (Phase 2), Climate Stress Test (Phase 3), and Climate Risk Management (Phase 4). The above steps are described in more detail in Supplementary Material, Information 1, and graphically in Supplementary Material, Information 2.

4. SUMMARIZING THE IMPLEMENTATION OF THE DTF

From its inception, the DTF was designed to be applied to all types of water infrastructure projects at the earliest stages of study (pre-feasibility or feasibility) so that whatever recommendations for risk management could be provided in advance of the more detailed and definite design stages. The DTF was not conceptualized as an independent tool but as an integral component of the pre-feasibility or feasibility studies, consistent with existing planning and evaluation protocols. Therefore, its added level of effort and cost should be commensurate with the

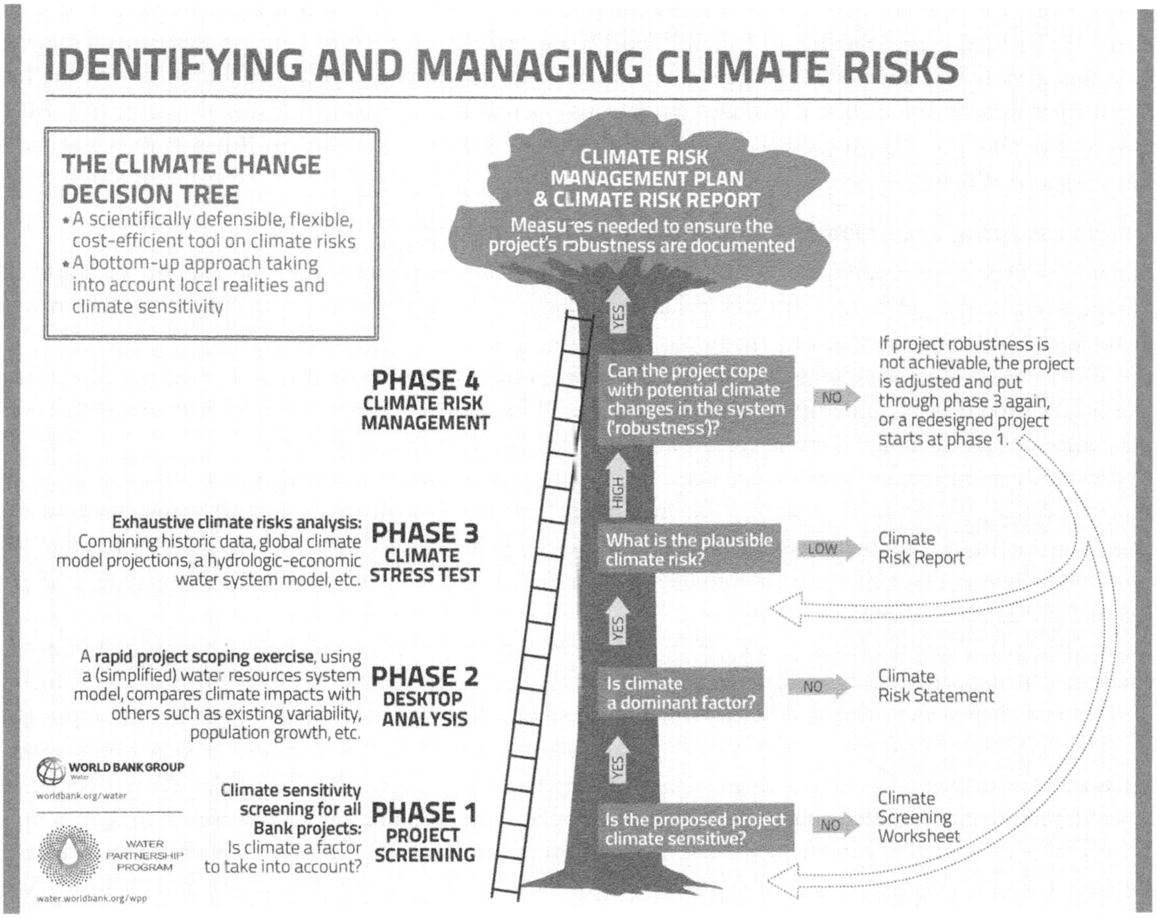

Fig. 3 | Decision Tree Illustration. *Source: Water Partnership Program (WPP), Water Security for All: The Next Wave of Tools. Annual Report 2013/2014.* World Bank Group.

effort and cost of these study stages. For the same reason, the DTF was originally a tool for analysis once the plans and pre-designs have been elaborated.

Since its launch by the WBG in 2015, the DTF has been able to successfully disseminate the climate component of a project *vis-à-vis* other potential risks while providing management strategies. The DTF has been piloted in 10 WBG projects and one more is currently still in implementation. These are: (1) the Upper Arun Hydroelectric Project (UAHP) in Nepal, (2) the Mwache Multipurpose Reservoir in Kenya, (3) the Poko Hydroelectric Project in Indonesia, (4) the Cutzamala Water System (CWS) in Mexico, (5) the Evaluation of the SACMEX Urban Water Resilience in Mexico City, Mexico, (6) the Kabeli-A Run of River Hydroelectric Project in Nepal, (7) the Lower Nzoia Irrigation and Flood Control Project in Kenya, (8) the Multipurpose Water Resources System in the Chancay-Lambayeque Basin in Peru, and (9) the qualitative design of Resilient Strategies for the Malanje and Lubango Water Utilities, Angola. Current activities include concluding stages of the Modeling of Water Security and Resilience for the Valley of Mexico, Mexico.

The DTF has also been the basis for the preparation of WBG sector guidelines such as (1) the Resilience Guidelines for Water Supply and Sanitation Utilities and (2) the Hydropower Sector Climate Resilience Guidelines (World Bank, Mott MacDonald, 2017). The latter is now being tested in the field in collaboration with the International Hydropower Association (IHA) and the European Bank for Reconstruction and Development (EBRD) in three to five pilot projects. At present, the projects that the WBG is targeting are the Kabeli-A in Nepal and the Mpatamanga in Malawi.

4.1. From assessing uncertainties to investment decision-making

As shown in previous sections, the WBG in collaboration with its partners and the support of world-class researchers has made significant advances in facing the practical challenges posed by climate change in water-related plans, investment design, and operations.

The WBG – as do other similar institutions – has specific policies and rules for project financing and is favoring the use of decision-making under uncertainty approaches to incorporate climate- and non-climate-related uncertainties into its decision-making framework, where the DTF is an important resource.

As previously discussed, the WBG's Safeguard Policies for Investment Project Finance operations have evolved with an increasing attention to social and environmental issues (through the ESF) and within these, climate change, and within this, adaptation and resilience resulting in guidelines for designing climate-informed projects around four major climate change commitments, which include climate and disaster risk screening, and adaptation and mitigation co-benefit tracking.

Although the DTF is included as a tool for the climate and disaster risk screening first step, its usefulness goes beyond that. It supports project teams to also assign co-benefits by specifically assessing the project's vulnerabilities to climate change and suggesting linked risk management actions. As mentioned in the Pakistan Indus Basin study (Yu *et al.*, 2013), non-climate factors (such as land-use change) may, in some cases, be equal or more important than those attributed to climate. Beyond the climate change realm, the DTF also helps the project team to assess other non-climate-related vulnerabilities and their relative importance among themselves and, in comparison, with those attributed to climate. It will also help the project team to assess and compare the project behavior in relation to its intended delivery reliability and other indicators such as cost, robustness, and resilience, when subject to uncertain climate and non-climatic stressors.

As such, examples of piloting applications of the DTF as it has evolved from the climate change screening phase to non-climate uncertainty screening and decision-making for project investments and prioritization are reviewed below.

4.2. Climate change and non-climate uncertainties

The UAHP in Nepal (see Supplementary Material, Information 3) was the first pilot where the DTF was applied (Bonzanigo *et al.*, 2015). The objective of the study was to assess the vulnerability of the pre-feasibility 335 MW design of the UAHP to changes in the initial (climate and other) conditions under which the proposed design was based; and to recommend adaptation strategies through iterative CRA and climate risk management. Using visual aids, such as interactive parallel plotting (see Figure 4), proved to be useful to illustrate this type of comparative analysis results for decision-makers.

4.3. Vulnerabilities, trade-offs, and risk management actions

The Mwache Multipurpose Reservoir study in Kenya (HRG UMass, 2017a) aimed to assess the risks of Mwache's Dam 2015 design, associated with climatic and demographic change as well as to evaluate adaptation and risks management options from a water supply perspective. It also examined trade-offs between domestic, irrigation, and environmental uses, and the climate change stress test was expanded to include demographic stressors.

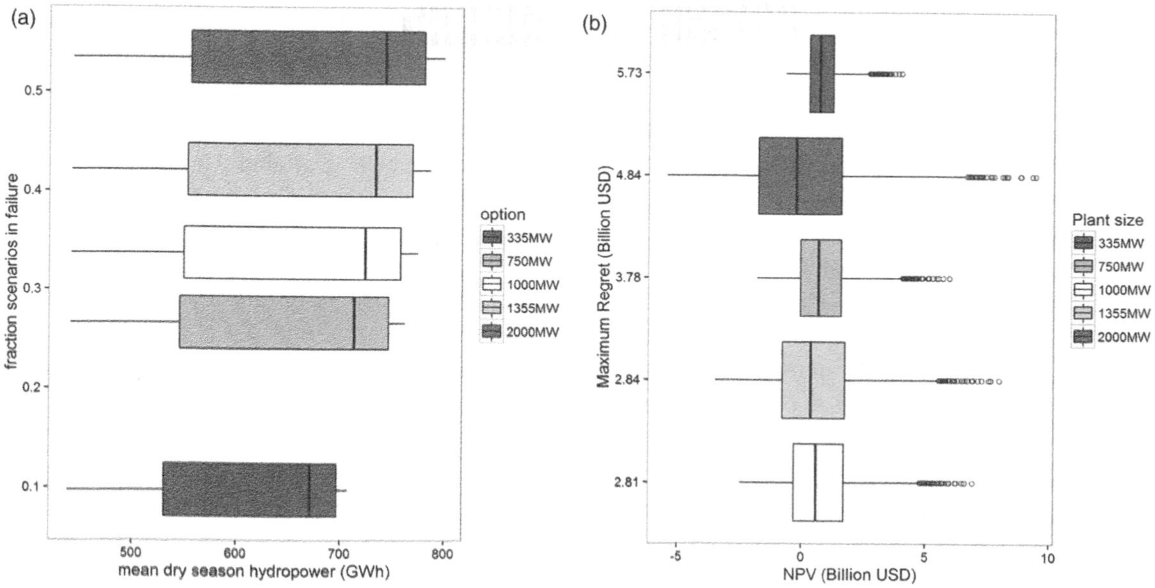

Fig. 4 | Comparative analysis for different plant size scenarios in the UAHP, Nepal: (a) fraction of scenarios in failure is determined by the NPV that is less than 0; (b) maximum regret vs. NPV. *Source:* Ray *et al.* (2018).

In Phase 2 of the analysis, the DTF was able to detect that the project, in its original design, was sensitive to climate uncertainty, hydrological model uncertainty, natural climate variability, and long-term demand changes. For instance, water demand was estimated to increase by up to 200% by 2030 due to demographic expansion. Also, the climate models were found not to agree in the same direction of change when projecting precipitation changes. Therefore, Phase 3 was deemed necessary.

The stress test in this stage was done using a simple rainfall–runoff model (with a storage component accounting for soil moisture and groundwater) and a water resource system model which included the effects of sedimentation. Stakeholders identified safe yield (the amount of water that can be delivered annually with a 95% reliability ratio, or 80 MCM/year) and overall and critical reliability (reliability was defined in this context as the percentage of intervals during the simulation period that the reservoir can meet the prespecified demand; critical reliability was defined here as the percentage of intervals during the simulation period that the reservoir can meet at least 60% of the prespecified demand). This stage later showed that the risk of the Mwache Dam failing to meet the safe yield target was low. Yet, the reliability of the system under various climatic conditions was found to decrease as domestic water demand increased.

Lastly, Phase 4 was triggered to assess the performance of the system under four different reservoir capacities. This stage found that larger designs, despite minimal benefit increases in terms of average safe yield, significantly improved the resilience of the system to drought conditions. Here, resilience was defined as the ability of the system to recover from delivery failures and it was measured by the probability that a successful period will follow a failure period. This phase also found that modifications in the operating rules could importantly help to increase resilience. Changing the allocation rules between domestic, irrigation, and environmental uses could be important to achieve the goals of the individual uses and may increase their reliability by up to 10%.

4.4. Prioritization of investments and trade-offs between objectives

The study of the CWS in Mexico (HRG UMass, 2017b) was developed given the growing concern for the vulnerability of the Mexico City water system to shocks, and the capacity of the system to adjust to the climate and demographic change. The CWS (Supplementary Material, Information 4) is a highly complex inter-basin water transfer to the Mexico City Metropolitan Area (MCMA) which has about 23 million inhabitants. Drinking water supply and irrigation are also provided to local users in the Cutzamala basin. This CWS involves seven reservoirs: El Bosque, Valle de Bravo, Villa Victoria, Tuxpan, Colorines, Ixtapan del Oro, and Chilesdo – the first three providing storage. Pumping stations (with about 1,100 m of elevation difference), open channels, tunnels, pipelines, aqueducts, and a single water supply treatment plant are also part of the system, which delivers around 15 m^3/s to the MCMA.

Built between the late 1970s and the early 1990s, the CWS is operated by the Valley of Mexico River Basin Organization (OCAVM in Spanish) of the National Water Commission of Mexico (CONAGUA in Spanish), delivering bulk water volumes for the city of Toluca, and the Water System of Mexico City (SACMEX). Local uses in the Cutzamala basin have priority. Reservoirs minimize spills and maximize storage to assure the reliability of water supply. Reservoir levels are also influenced by water quality considerations and recreational/tourism use preferences.

Although authorities were confident that the CWS had been reliable up to the time of analysis, supply shortages during the 2009 and 2013 droughts raised concerns about future worsening conditions due to climate change and other uncertainties, so several new investment options had been proposed to strengthen reliability, as shown in Table 1.

With support from the WBG, the development of an integrated water resources management plan to enhance the reliability and resilience of the CWS and enhance its confidence was initiated. The WBG, OCAVM, and the Hydrosystems Research Group of the University of Massachusetts (HRG UMass) applied the DTF (see Supplementary Material, Information 4) to identify vulnerabilities to changes in climate and evaluate the possibility of greater water supply delivery from the system. Using a broad set of performance indicators including yield, cost, robustness, and resilience, the proposed interventions of Table 1 were evaluated both individually and combined as optimized portfolios. Two additional options (the last two rows of Table 1) – a pressurized tunnel at Villa Victoria and a pressurized tunnel plus storage expansion – were also evaluated, as well as a preliminary

Table 1 | New investment options.

Intervention	Brief Description of Intervention
Villa Victoria	Extra storage and canal
Bosque-Colorines Canal	Increased capacity of canal between Bosque and Colorines reservoirs
Temascaltepec reservoir	Additional reservoir with larger storage capacity
Platform at Bosque	Decreased dead storage level at Bosque reservoir
Tuxpan Pump	Pumping groundwater below Tuxpan during dry season
Tuxpan Irrigation	New larger capacity canal for irrigators at Tuxpan
Villa Victoria Pressurized Tunnel	New pressurized tunnel (20 m^3/s) to connect pumps to Villa Victoria and Villa Victoria to Los Berros
Villa Victoria Pressurized Tunnel and Storage Expansion	50 MCM extra storage in Villa Victoria and the pressurized tunnel (above)

Source: HRGU Mass (2017b).

evaluation of optimizing reservoir and canal operations and the effects of deferred maintenance of canals and pumps.

As stated in the (HRG UMass report 2017b), the Temascaltepec reservoir provided the greatest increase in the maximum reliable yield, robustness, and resilience of the system. However, it was the costliest and would take the longest to develop. The best-performing investment portfolio that did not include Temascaltepec consisted of the Villa Victoria, Bosque Platform, and the Tuxpan Pump interventions. However, 'the Villa Victoria Pressurized Tunnel in conjunction with an expansion of the Villa Victoria's current capacity yielded significant improvements in all performance metrics considered' falling just short of the approximate yield increase which would be provided by an investment in Temascaltepec (HRG UMass, 2017b). At substantially less costs per m^3/s, it becomes the best performing intervention that does not include Temascaltepec and it does perform better for a wider range of futures than the current system, thus improving the robustness of the system to future climate change.

The analysis also preliminarily evaluated the effects of optimized reservoir and canal operations (Figure 5). The results indicate that changing the operational rules of the system provided benefits that are comparable or superior to many of the investment portfolios. Specifically, there is a potential to somewhat increase the reliable yield of the system without investment in additional infrastructure. Also, a preliminary analysis also indicated that lack of proper maintenance of canals and pumps will be reflected not only in less delivery capacity and more frequent and longer failures but also in less ability to recover from shocks (HRG UMass, 2017b).

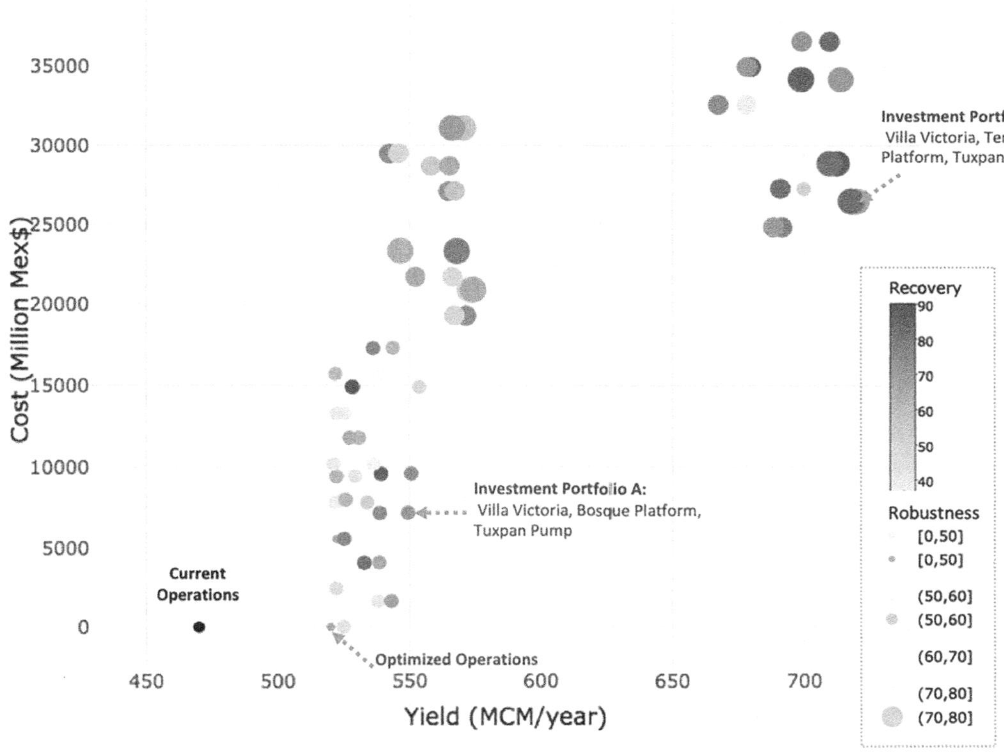

Fig. 5 | Evaluation of investments in the Cutzamala System options against their performance in terms of water yields, cost, resilience (recovery), and robustness levels.

The example highlighted in Figure 5, with the use of visualization tools, is of importance to inform local authorities on how best to select and prioritize new investments considering not only the financial domain but also other metrics such as reliability, resilience, robustness, and performance during drought. The DTF helps to understand trade-offs between yield, robustness, resilience, and economic performance metrics. This is a useful feature in Phase 4. As some of these objectives may be conflicting, the use of visual aids such as spider-plots of trade-offs between key performance indicators like those shown in Supplementary Material, Information 5, for example, may help stakeholders' decisions and evaluate the consequences of such decisions.

5. DESIGNING CLIMATE-INFORMED PROJECTS AROUND WBG'S FOUR MAJOR CLIMATE CHANGE COMMITMENTS

Guidelines drafted to help project teams comply with this mandate, previously described in Section 2, offer the DTF as one of the WBG's tools to 'help task teams allocate climate risk assessment efforts in a way that reflects various projects' sensitivity to climate-related (and other long-term) risks. When more in-depth analyses are needed, methodologies for decision making under uncertainty (DMU) are suggested. These DMU methodologies help identify specific project vulnerabilities and solutions that may help alleviate these, despite the large uncertainty about future climate and other dynamic factors.' This is no different from what the DTF has been used for up to date. The difference is that previously the DTF has been applied once plans or pre-designs have been elaborated and, in this case, the tool would be involved in the preparation process *before* Board approval. That, together with the new emphasis on resilience and co-benefit tracking, poses new stimulating challenges to the application of the DTF that the WBG has started to confront with research partners.

5.1. Incorporating resilience in the decision-making process, work in progress, and next steps

Resilience is a concept that intuitively makes one think about shocks and an ability to recover. While it is a concept that is not uniform and is still under construction, some of its first definitions were introduced many years ago, evolving from the disciplines of psychology and psychiatry in the 1940s (Manyena, 2006). Some of the simplest definitions of resilience in this domain applied to individuals refer to it as 'the capacity of human beings to survive and thrive in the face of adversity'. It involves 'being able to recover from difficulties or change – to function as well as before and then move forward'. The equivalent of 'bouncing back' (Lynde, 2015) is derived from the Latin word *resilio*, meaning 'to jump back' (Manyena, 2006). That is why in water resources it is usually connected to the occurrence of extreme hydrometeorological or seismic events during the life cycle of structures and services.

However, the shock or disturbance may be of such nature that it does not affect the infrastructure *per se*, but its ability to deliver a certain service level for which it was designed. So, a complementary popular conceptual definition of resilience associates it to 'the ability of human communities to withstand external shocks or perturbations to their infrastructure and to recover from such perturbations' (Bocchini *et al.*, 2014). Two key aspects are introduced here. First, resilience is a property not only of structures or infrastructures but mainly of communities. Second, resilience is not only about being able to withstand a certain disturbance, but also about having the institutional framework, resources, and means for a prompt, efficient, and effective recovery (Bocchini *et al.*, 2014). At the same time, recently the concept of resilience has also been enriched by also adding the characteristics of the recover. In fact, resilience could also be seen as a catalyst of change, toward building back better or safer after a shock or disruption event (Wilby, 2020).

As a response, several efforts have been underway at the World Bank toward building resilience to weather-related disasters, as well as to plan and design of water-related infrastructure based on resilience indicators, to guarantee that this infrastructure will keep providing the required level of service in the face of shocks of different nature (World Bank, 2013, 2017b, 2017c; World Bank/GFDRR, 2015). The IPCC or the Rockefeller Foundation definition of resilience has been favored by some sectors, this is 'the capacity of individuals, communities, and systems to survive, adapt, and grow in the face of stress and shocks, and even transform when conditions require it. Building resilience is about making people, communities and systems better prepared to withstand catastrophic events – both natural and manmade – and able to bounce back more quickly and emerge stronger from these shocks and stresses' (Rockefeller Foundation, 2015).

Yet, the overall universal difficulty to define resilience is also translated at the World Bank sphere (World Bank, 2017b). Furthermore, encompassing concepts of resilience including robustness and adaptability have also been suggested (World Bank, Mott MacDonald, 2017). But in most of the pilot DTF applications mentioned in Section 4.4, a simpler concept based on Hashimoto *et al.* (1982) was used. This refers to the concept of recovery, or how fast a system recovers from a failure once failure occurs. This concept facilitated the operationalization during Phase 3 of the DTF according to the services provided by the infrastructure. This went from simpler, single structure–single use cases to more complex multi structure cases such as in the Cutzamala study in Mexico, where resilience was evaluated for the whole system. Nevertheless, the focus has been on the infrastructure and its ability to provide the services.

But the resilience framework for water resources systems is much more complex, as reviewed by Wang & Blackmore (2009). Regarding resilience to climate change, the WBG considers two types of projects: (1) those that when screened show vulnerability to climate change but after modifications the risk is managed and (2) those that are specifically designed to increase robustness and resilience to long-term climate changes (World Bank, 2017b). The first type of projects needs to *be* resilient and the second type needs to *build* resilience. In order to address these aspects, the project resilience rating system was developed as part of the recently launched WBG's Adaptation and Resilience Action Plan mentioned in Section 2, which has a strong focus on people's resilience and sees projects as instruments that contribute to people's resilience ('resilience *through* projects'). These projects, however, must themselves be resilient and able to deliver their services to the population ('resilience *of* projects'). A new resilience rating system has been recently released by the World Bank to support clients scaling up climate adaptation actions both at the infrastructure level and at the user level (World Bank, 2021).

Up to now, the DTF has been used as a tool for type (1) projects. An example is the Cutzamala system study for Mexico City described previously. However, the DTF pilot applications presently in progress in Nepal (testing the application of the Hydropower Sector Climate Resilience Guidelines in a run-of-river hydroelectric development), Kenya (resilience of an irrigation and flood control project in the lower Nzoia river basin), and Peru (resilience of the Chancay-Lambayeque water resources system at the basin level) are a good opportunity to explore interrelated project resilience performance indicators at three levels: infrastructure, beneficiaries of the services, and the river basin community.

To be applied for type (2) projects, modifications and upgrades to the DTF will need to be made. A reference is the pilot work that has continued in the MCMA after the Cutzamala study with the collaboration of the Rockefeller Foundation and the HRG UMass, aiming to transform the way water infrastructure investment programs are designed to obtain 'freshwater resilience' for the city. This concept emphasizes that freshwater ecosystems (think lakes, rivers, and aquifers) can handle changes, particularly climate, and still continue to deliver their essential services. This is the fundamental principle of resilience: being able to respond and adapt to shocks and stresses and to transform when conditions require it.

Using the approach called 'Freshwater Resilience by Design' (Brown, 2017, 2018), the work focuses on multiple objectives (economic/service, social/equity, and environmental) consider multiple futures (climates, population growth and demographic change, economic growth and change, and societal preferences change) and links 'urban investment program and policies with investments and policies in the connected river basins, and the evaluation of benefits and costs in each location' (Brown, 2017, 2018). This approach also incorporates aspects of projects and activities that strengthen the resilience of the communities and users.

Adding resilience and robust design considerations to any project in order to reduce risk and uncertainty typically increases the costs. These vulnerability-reducing design features often conflict with traditional economic project justification and decision rules (e.g., maximize net benefits, maximize the internal rate of return, and benefit–cost ratio greater than one) and trade-offs must be considered, as was done in the study of the Cutzamala system (Supplementary Material, Information 4). Moreover, in urban water investment projects such as water and wastewater, for example, that are complex in nature and depend not only on physical factors but also on social factors, it is difficult to estimate the benefits when focused on increasing resilience. Initial work being discussed with RAND aims to develop a framework for the economic assessment of the resilience benefits. That is, 'the net benefits associated with the absorption of shocks and stressors, the recovery path following a shock, and any co-benefits that accrue from a project, even in the absence of a shock' which may be positive or negative for a given project (Bond *et al.*, 2017).

Regarding multiple objectives and consideration of both climate and non-climate factors, the need for a multidimensional stress test was posed for hydropower investments facing climate, geophysical and financial uncertainty (Ray *et al.*, 2018) and updates to the DTF have been proposed, such as (1) replacing its 'climate change stress test' with a multidimensional stress test and (2) 'the addition of a Bayesian network framework that represents joint probabilistic behaviour of uncertain parameters' as sensitivity factors to aid in the combination of conditions under which a water system fails to meet its performance targets, and better understand the relative magnitudes of the varied risks faced (Ray *et al.*, 2019). The use of a multidimensional stress test is being piloted in the analysis of the Kabeli-A Run of River Hydroelectric Project in Nepal, in collaboration with the IHA and the EBRD.

5.2. Lessons learned and the way forward

The DTF has been operationally piloted in various water projects to both enhance the assessment of climate and non-climatic risks and provide guidelines to manage such risks. Toward this end, these efforts have promoted a highly participatory approach in situations dealing with hydropower, water supply, irrigation, flood control, and river basin planning, both individually and in multi-use schemes, involving both single major hydraulic structures as well as complex multi-structure systems. Co-creation and validation workshops organized with stakeholders in each pilot study as well as training courses were useful to reinforce stakeholders' understanding of their systems, define their performance metrics, and evaluate their risks and risk-management options.

The first pilot study was a faithful application of the methodology described in the WBG's DTF book (Ray & Brown, 2015). From there on, successive pilots needed to include methodological enhancements to cover non-climatic uncertainties in more depth, economic considerations, regard for ground water, nature-based solutions, environmental and other trade-offs, operation rules, maintenance, robustness, resilience of infrastructure (Ray *et al.*, 2018), and prioritization of investment options. Although initially intended as a tool for project screening, the DTF is now being applied to assess multiple-investment basin plans and prioritize investments considering multiple metrics, including robustness and resilience.

Thus, being able to assess and manage risk in a systematic and flexible way, the DTF has been an important contribution to the WBG's efforts to mainstream assessing and managing the risk posed by the climate change and other important uncertainties. Nonetheless, as the pilot studies suggest, there are challenges and improvements that have yet to be addressed.

Looking forward, immediate necessities that need to be explored include those concerning the integration of important components of the hydrological cycle. For instance, the application of the DTF to complex urban systems evidenced that it still needs to improve its capabilities to integrate groundwater processes with surface water. This would require the integration of models which help to include the interaction of groundwater in flood and drought occurrences. In line with this, current efforts in Mexico City look to incorporate these principles within a machine learning approach to simulate infrastructure, physical, and decision processes under a resilient framework. Yet, the DTF also offers the opportunity for its application without the need of application of sophisticated models, particularly in areas where data are scarce, non-existent, or highly uncertain. Indeed, a previous initial experience of applying the DTF in two water utilities in Angola, which met these characteristics, suggests that its adjustments toward a more qualitative type of exercise are also promising.

At the same time, the DTF has not yet been used to assess and manage water quality risks. Also, the consideration of the way future land-use changes, human activities, and social factors may affect the performance of a water resource project still needs to be further deepened. In line with this, the DTF would also require incorporating and addressing issues concerning seawater intrusions and salinization of soils. A similar improvement yet to be considered relates to the inclusion of processes that would enhance the management of risks and resilience building in coastal zones, such as projected sea-level rise or the occurrence of cyclones and tropical storms. Also, modeling in detail the effect of maintenance or lack thereof as well as the definition of adequate indicators for resilience or equity as well as for non-climate stressors has been challenging.

6. CONCLUDING REMARKS

Important benchmarks in the development and consolidation of the DTF are the robustness policy prescription of the WBG's 2010 World Development Report, the IDA 17 climate and disaster risk screening requirement, the Climate Change Corporate Commitments incorporating the Water Global Practice, the development of the CCKP, and the CDRST. The DTF unites them as a practical assessment and managing tool that also provides a way to integrate the use of traditional hydrologic and water resources methodology. At the same time, the DTF allows the addition of various of the latest advances in decision-making under uncertainty and bottom-up approaches, as well as new information provided by the latest climate science modeling.

Although not finished yet, the pilot phase with the incorporation of research and knowledge advances is considered positive. Further insights on the practicality and flexibility of this approach are expected from the pilot studies under process, the project and community resilience screening indicators being developed through the WBG Adaptation and Resilience Action Plan, the results of the freshwater resilience concept application in Mexico City, and the results of the Hydropower Resilience Guidelines testing and validation. In particular, achieving a better understanding of which types of projects (hydropower, irrigation, municipal water supply, and flood control) require a better-defined set of specific analytical steps in order to assure uniformity of outcomes.

The flexibility of the DTF offers the opportunity to incorporate advances from the academic/research community – not only to overcome some of the noted needs but also to use the framework beyond considering the physical dimensions to integrate human systems, such as those aspects of resilience related to economic, social, and institutional-governance. In line with this, incorporating and adjusting various indicators analyzed by Hallegatte *et al.* (2016) and the proposed collaborative work in resilience benefits would enhance the understanding not only of physical infrastructure resilience but also of the socioeconomic resilience of water projects. Similarly, this approach still needs to refine the incorporation of green infrastructure as risk management alternatives. Including a reasonable representation of human and natural systems would enhance the application of this framework for resilient and robust decision-making in natural and socially sensitive areas.

Efforts have been made to deepen understanding of the financial mechanisms and investment strategies of water projects under uncertain conditions. But other issues, like the role of water funds and governmental bodies in financing strategies, and economic evaluation procedures that stem from the DTF need yet to be incorporated.

Similarly, the role of resilient and robust water infrastructure projects to build resilience to sustain city, river basin, and country-level development and water security goals, recently introduced, is still incipient. This task would need to include an integrative assessment and management of risks of additional systems where water systems constitute an important component such as in health, food, commerce, transport, and international trade.

Equally important for institutions is not to forget the need for giving increasing attention to narrowing the gap between the science and its application in the field. In the operational dimension, mainstreaming still has a way to go. The technical level of the methodology is considered high but, even as modifications and additions are being considered and introduced at the upper end, there have been suggestions about lowering the level of its description so that it can be adopted and applied more easily. The methodologies are well sustained scientifically but would benefit from being translated to a practitioner language that makes sense to the technician and the decision-maker in the field, especially in the developing world.

DISCLAIMER

The findings, interpretations, and conclusions expressed in this review paper are those of the authors and do not necessarily reflect the views of the World Bank, its Board of Executive Directors, or the governments they represent.

DATA AVAILABILITY STATEMENT

All relevant data are included in the paper or its Supplementary Information.

REFERENCES

Alavian, V., Maher Qaddumi, H., Dickson, E., Diez, S. M., Danilenko, A. V., Hirji, R. F., Puz, G., Pizarro, C., Jacobsen, M. & Blankespoor, B. (2009). *Water and Climate Change: Understanding the Risks and Making Climate-Smart Investment Decisions*. The World Bank, Water Sector Board, Bank Netherlands Water Partnership Program, and Norwegian Ministry of Foreign Affairs, Washington, DC.

Ayyub, B. M., Prassinos, P. G. & Etherton, J. (2010). Risk-informed decision making. *Mech. Eng. Mag. Sel. Artic. 132*(1), 28–33.

Bocchini, P., Frangopol, D. M., Ummenhofer, T. & Zinke, T. (2014). Resilience and sustainability of civil infrastructure: Toward a unified approach. *Journal of Infrastructure Systems 20* (2), 04014004.

Bond, C. A., Strong, A., Burger, N., Weilant, S., Saya, U. & Chandra, A. (2017). *Resilience Dividend Valuation Model: Framework Development and Initial Case Studies*. RAND Corporation, Santa Monica, CA. Available at: https://www.rand.org/pubs/research_reports/RR2129.html.

Bonzanigo, L., Brown, C., Harou, J., Hurford, A. & Ray, P. (2015). *South Asia Investment Decision Making in hydropower. Decision Tree Case Study of the Upper Arun Hydropower Project and Koshi Basin Hydropower Development in Nepal*. World Bank Report No. AUS11077, June 17, 2015.

Brown, C. (2017). *Freshwater Resilience by Design, Presentation in the World Water Week*. SIWI, Stockholm, Sweden.

Brown, C. (2018). *Freshwater Resilience by Design*. Presentation in the Resilient Water Decisions Workshop, World Bank, Washington, DC.

Brown, C., Ghile, Y., Laverty, M. & Li, K. (2012). Decision scaling: linking bottom-up vulnerability analysis with climate projections in the water sector. *Water Resour. Res. 48*(9).

Brown, C. M., Lund, J. R., Cai, X., Reed, P. M., Zagona, E. A., Ostfeld, A., Hall, J., Characklis, G. W., Yu, W. & Brekke, L. (2015). The future of water resources systems analysis: toward a scientific framework for sustainable water management. *Water Resour. Res. 51*, 6110–6124. doi:10.1002/2015WR017114.

Buytaert, W. & De Bievre, B. (2012). Water for cities: the impact of climate change and demographic growth in the tropical Andes. *Water Resour. Res. 48*(8), 1–13. doi:10.1029/2011WR011755.

Cervigni, R., Liden, R., Neumann, J. E. & Strzepek, K. M. (2015). *Enhancing the Climate Resilience of Africa's Infrastructure: The Power and Water Sectors. Africa Development Forum*. World Bank, Washington, DC. Available at: https://openknowledge.worldbank.org/handle/10986/21875.

García, L. E., Matthews, J. H., Rodriguez, D. J., Wijnen, M., DiFrancesco, K. N. & Ray, P. (2014). *Beyond Downscaling: A Bottom-Up Approach to Climate Adaptation for Water Resources Management*. Available at: https://openknowledge.worldbank.org/handle/10986/21066.

Grijsen, J. C. (2016). *Tina River Hydropower Development Project, Solomon Islands Climate Risk Assessment*. World Bank.

Grijsen, J., Brown, C., Tarhule, A., Ghile, Y. & Taner, M. (2012). Climate risk assessment for water resources development in the Niger River Basin. *Water Partnersh. Progr. (WPP)/TWIWA, HydroPredict 12*.

Groves, D. G., Yates, D. & Tebaldi, C. (2008). Developing and applying uncertain global climate change projections for regional water management planning. *Water Resour. Res. 44*(12). doi:10.1029/2008wr006964.

Hallegate, S., Shah, A., Lempert, R., Brown, C. & Gill, S. (2012). *Investment Decision Making under Deep Uncertainty, Application to Climate Change*. The World Bank Sustainable Development Network, Office of the Chief Economist, Washington, DC.

Hallegatte, S., Bangalore, M. & Vogt-Schilb, A. (2016). *Socioeconomic Resilience: Multi-hazard Estimates in 117 Countries*. The World Bank.

Hamlet, A. F. (2011). Assessing water resources adaptive capacity to climate change impacts in the Pacific Northwest Region of North America. *Hydrol. Earth Syst. Sci. 15*(5), 1427–1443. doi:10.5194/hess-15-1427-2011.

Hartmann, D. L., Tank, A. M. G. K. & Rusticucci, M. (2013). *Climatie Change* 2013: The Physical Science Basis, IPCC Fifth Assessment Report, IPCC, AR5, 31–39.

Hashimoto, T., Stedinger, J. R. & Loucks, D. P. (1982). Reliability, resiliency, and vulnerability criteria for water resource system performance evaluation. *Water Resour. Res. 18*(1), 14–20.

Hine, D. & Hall, J. W. (2010). Information gap analysis of flood model uncertainties and regional frequency analysis. *Water Resour. Res. 46*(1). https://doi.org/10.1029/2008WR007620.

HRGUMass (2017a). *The Decision Tree Initiative: Water Resources Development and Allocation for the Mwache Dam*. Final Report. World Bank.

HRGUMass (2017b). *Including Climate Uncertainty in Water Resources Planning and Project Design-Decision Tree Initiative. Pilot Studies of the Cutzamala Water System*. Mexico-Draft Final Report. World Bank, Washington, DC.

IDA (2014). *Report from the Executive Directors of the International Development Association to the Board of Governors, Additions to IDA Resources: Seventeenth Replenishment, IDA17: Maximizing Development Impact*.

IEG World Bank/IFC/MIGA (2012). *Adapting to Climate Change: Assessing the World Bank Group Experience*. Phase III of the World Bank Group and Climate Change, Washington, DC.

Kalra, N., Groves, D. G., Bonzanigo, L., Molina Perez, E., Ramos, C., Brandon, C. & Rodriguez Cabanillas, I. (2015). *Robust Decision-Making in the Water Sector; A Strategy for Implementing Lima's Long-Term Water Resources Master Plan*. Climate Change Group, Office of the Chief Economist, Washington, DC.

Lambin, E. F. & Meyfroidt, P. (2011). Global land use change, economic globalization, and the looming land scarcity. *Proc. Natl. Acad. Sci. U. S. A. 108*(9), 3465–3472.

Lempert, R. J., Groves, D. G., Popper, S. W. & Bankes, S. C. (2006). A general, analytic method for generating robust strategies and narrative scenarios. *Manage. Sci. 52*(4), 514–528.

Ludwig, F., van Slobbe, E. & Cofino, W. (2014). Climate change adaptation and integrated water resource management in the water sector. *J. Hydrol. 518*(PB), 235–242. doi:10.1016/j.jhydrol.2013.08.010.

Lynde, D. W. (2015). *Resilience Definitions and Concepts*. Available at: https://dmh.mo.gov/docs/mentalillness/resilience.pdf.

Manyena, S. B. (2006). The concept of resilience revisited. *Disasters 30*(4), 433–450.

Merrey, D. & Awulachew, S. B. (2015). Critical roles of water in achieving the proposed SDGs: a nexus perspective (water-energy-food-climate change). In: *Presented at the Workshop on 'Capacity Development in Advancing Water and Sustainable Development'*, February 24–25, 2015, United Nations, New York City, NY.

Prudhomme, C., Wilby, R. L., Crooks, S., Kay, A. L. & Reynard, N. S. (2010). Scenario-neutral approach to climate change impact studies: application to flood risk. *J. Hydrol. 390*(3–4), 198–209.

Ray, P. A. & Brown, C. M. (2015). *Confronting Climate Uncertainty in Water Resources Planning and Project Design: The Decision Tree Framework*. World Bank, Washington, DC. Available at: https://openknowledge.worldbank.org/handle/10986/22544.

Ray, P. A., Bonzanigo, L., Wi, S., Yang, Y. -C. E., Karki, P., García, L. E., Rodriguez, D. J. & Brown, C. M. (2018). Multidimensional stress test for hydropower investments facing climate, geophysical and financial uncertainty. *Global Environ. Change 48*, 168–181.

Ray, P. A., Taner, M. Ü., Schlef, K. E., Wi, S., Khan, H. F., Freeman, S. S. G. & Brown, C. M. (2019). Growth of the Decision Tree: Advances in Bottom-Up Climate Change Risk Management. *JAWRA Journal of the American Water Resources Association* 55(4), 920–937.

Rockefeller Foundation (2015). *Resilience Webpage*. Rockefeller Foundation, New York. Available at: https://www.rockefellerfoundation.org/our-work/topics/resilience/.

Rodríguez, D. J., van den Berg, C. & McMahon, A. (2012). Investing in Water Infrastructure: Capital, Operations and Maintenance. Water Papers, World Bank, *Water Partnership Program*.

Rodríguez-Iturbe, I. & Valdés, J. B. (2011). *Uncertainty and Climate Variability in the Design and Operation of Water Resources Projects: Examples and Case Studies*.

Rosenberg, D. E. & Madani, K. (2014). Water resources systems analysis: a bright past and a challenging but promising future. *J. Water Resour. Plann. Manage. 140*(4), 407–409. doi:10.1061/(ASCE)WR.1943-5452.0000414.

Stainforth, D., Aina, T., Christensen, C., Collins, M., Faull, N., Frame, D. J., Kettleborough, J. A., Knight, S., Martin, A., Murphy, J. M., Piani, C., Sexton, D., Smith, L. A., Spicer, R. A., Thorpe, A. J. & Allen, M. R. (2005). Uncertainty in predictions of the climate response to rising levels of greenhouse gases. *Nature 433*(7024), 403–406. doi:10.1038/nature03301.

Stainforth, D. A., Allen, M. R., Tredger, E. R. & Smith, L. A. (2007). Confidence, uncertainty and decision-support relevance in climate predictions. *Philos. Trans. Royal Soc. A365*(1857), 2145–2161.

Stakhiv, E. Z. (2011). Pragmatic approaches for water management under climate change uncertainty. *J. Am. Water Resour. Assoc. 47*(6), 1183–1196.

Teng, J., Vaze, J., Chiew, F. H. S., Wang, B. & Perraud, J. -M. (2012). Estimating the relative uncertainties sourced from GCMs and hydrological models in modeling climate change impact on runoff. *J. Hydrometeorol. 13*(1), 122–139.

UN (2019). *The Sustainable Development Goals Report 2019*. United Nations, New York.

UN Water/WBG (2018). *Making Every Drop Count: An Agenda for Water Action. High-Level Panel on Water Outcome Document*. Available at: http://www.unwater.org/high-level-panel-on-water-outcome-document/.

Value of Water Campaign (2017). *The Economic Benefits of Investing in Water Infrastructure*. Supported by a diverse group of more than 33 leaders in the US water industry, including American Society of Civil Engineers, Boston Water and Sewer Commission, Water Environment Federation, and others. Available at: http://www.TheValueofWater.org/resources.

Wang, C. -H. & Blackmore, J. (2009). Resilience concepts for water resource systems. *J. Water Resour. Plann. Manage. – ASCE 135*. doi:10.1061/(ASCE)0733-9496(2009)135:6(528).

Wilby, R. L. (2019). A global hydrology research agenda fit for the 2030s. *Hydrol. Res. 50*(6), 1464–1480. https://doi.org/10.2166/nh.2019.100.

Wilby, R. L. (2020). Resilience viewed through the lens of climate change and water management. *Water 12*(9), 2510.

World Bank (2010). *World Development Report 2010: Development and Climate Change*. © World Bank. License: CC BY 3.0 IGO, Washington, DC. Available at: https://openknowledge.worldbank.org/handle/10986/4387.

World Bank (2013). *Building Resilience: Integrating Climate and Disaster Risk into Development: Lessons from World Bank Group Experience*. The World Bank, Washington, DC.

World Bank (2016). *World Bank Group Climate Change Action Plan 2016–2020*. World Bank, Washington, DC.

World Bank (2017a). *The World Bank Environmental and Social Framework*. Washington, DC.

World Bank (2017b). *Operational Guidance for Monitoring and Evaluation (M&E) in Climate and Disaster Resilience-Building Operations*. Washington, DC.

World Bank (2017c). *World Bank Resilience M&E (ReM&E) Good Practice Case Studies*. Washington, DC.

World Bank (2018a). *Annual Report 2018*. Washington, DC.

World Bank (2018b). *Adaptation & Resilience Action Plan, Key Messages*. Washington, DC.

World Bank/GFDRR (2015). *Investing in Urban Resilience, Protecting and Promoting Development in a Changing World*. Washington, DC.

World Bank Group (2021). *Resilience Rating System: A Methodology for Building and Tracking Resilience to Climate Change*. World Bank.

World Bank, Mott MacDonald (2017). *Hydropower Sector Climate Resilience Guidelines*. Final Report (Beta Version). Washington, DC.

Yu, W., Yang, Y. -C., Savitsky, A., Alford, D., Brown, C., Wescoat, J., Debowicz, D. & Robinson, S. (2013). *The Indus Basin of Pakistan: The Impacts of Climate Risks on Water and Agriculture*. The World Bank.

First received 16 November 2021; accepted in revised form 16 November 2021. Available online 1 December 2021

WORLD WATER COUNCIL IWA PUBLISHING

Water Infrastructure Planning, MDCU doi: 10.2166/wp.2021.243

Climate risk-informed decision analysis (CRIDA): 'top-down' vs 'bottom-up' decision making for planning water resources infrastructure

Joe Manous[a] and Eugene Z. Stakhiv[b,c,*]

[a] U.S. Army Institute for Water Resources, Alexandria, VA USA
[b] Johns Hopkins U., Baltimore MD, USA
[c] ICIWaRM, Alexandria, VA, USA
*Corresponding author. E-mail: estakhi1@jhu.edu; eugene.z.stakhiv@usace.army.mil

ABSTRACT

Climate risk-informed decision analysis (CRIDA) is a guidebook that lays out an evaluation framework and decision procedures to deal with climate uncertainties that are consistent with traditional agency water resources planning frameworks. CRIDA guidelines complement existing institutional guidance on recognizing circumstances when more complex risk-based climate analysis may be needed, above those required by standard planning procedures. The procedures are based on the concept of 'decision-scaling' judgments to qualitatively assess levels of future risk and analytical uncertainty stemming from climate change-related uncertainties, and as a guide for choosing specific analytical approaches and appropriate levels of analysis. CRIDA addresses how much detail is appropriate for a given problem setting, depending on infrastructure type and function, whether it is new design or rehabilitation of existing infrastructure, modular design or long-life infrastructure. CRIDA was structured to resolve the contentious issue of deciding under what circumstances a 'top-down' climate scenario-driven analysis ought to be conducted versus a more traditional 'bottom-up' vulnerability assessment, based on conventional agency project feasibility procedures. The procedures for such vulnerability assessments and planning procedures are well-represented in classical approaches, such as those included in the 1983 U.S. Water Resources Council's 'Principles and Guidelines'. These commonly used procedures promote normative evaluation protocols and decision rules that generate alternative solutions which minimize risk-cost outcomes.

Key words: 'Bottom-up' water resources planning frameworks, Decision-scaling, Non-stationary climate analysis, Risk-based decision analysis

INTRODUCTION

For millennia, water resources management has been strongly linked to the economic, social and environmental development of civilizations and, later, nations. Harnessing water resources has transformed variable and often destructive hydrologic flood extremes into reliable, socially desired benefits (municipal water supply, irrigation, hydropower, navigation and environmental flows) and has ameliorated the ruinous effects of droughts. As needs have grown and interventions have increased, water resources managers worldwide have designed more complex water management systems and made more elaborate trade-offs among social, economic and environmental goals. To address these complexities, a wide array of technical, analytical and governance procedures for water management have evolved to keep pace with growing societal demands and the intricacies of public decision-making. However, large uncertainties associated with climate change have added to those planning and management complexities by challenging the foundations of hydrologic and hydraulic analyses associated with the assumptions of a stationary climate (Milly *et al.*, 2008).

The water resources engineering profession has always dealt with risk and uncertainty, as it is inherent in the randomness of natural hydrological phenomena that serve as the basis for design criteria and engineering standards. However, in addition to resorting to empirical relationships based on physical properties, as well as statistics and probabilities to characterize the uncertainties that were known, engineers primarily accounted for the 'known unknowns' by stressing design redundancies and safety factors as a way of dealing with residual risks – i.e., those risks which could not be easily quantified or which represented acceptable or 'tolerable' risks (Institute for Water Resources, 2010).

Hence, while risk and uncertainty analysis are not a new issue for policy makers, engineers or scientists associated with water management, climate change has highlighted the need to reassess the sequence of cascading uncertainties caused by the climate models themselves, together with natural variability, future scenario uncertainties and decision-related uncertainties associated with public needs, objectives and values. Added to these largely intractable climate model uncertainties, is the more fundamental problem of dealing with hydrologic non-stationary.

In a changing world, planning, decision-making and designing water resources management require long-term projections of hydrological time series that include trends due to anthropogenic intervention and climate change (Kjeldsen & Rosbjerg, 2005; Barsugli *et al.*, 2012; Bayazit, 2015). Climate variability is a random variation from a long-run average distribution, whereas climate change is a trend or a shift in the long-run distribution.

Estimating future climate impacts has proven to be particularly contentious and analytically frustrating. In many cases, effective solutions have been largely dependent on the skills of experienced individuals rather than being based on protocols and standards for systematic and reproducible approaches to planning under climate uncertainty. In general, a simple well-understood model is generally preferable to a sophisticated one with large uncertainties (Matalas & Fiering, 1977; Fiering, 1982; Rogers, 1997). For water resources planning studies, stationarity is still a useful concept, as a starting point for stress tests and vulnerability analyses to make reliable predictions for engineering design (Stakhiv, 2011; Brown & Wilby, 2012; Slater *et al.*, 2021). Observations of past patterns and information are key elements for a successful prediction for hydrological processes (Lins & Cohn, 2011; Bayazit, 2015).

Water resources management decisions made today will have long-term effects because the projects have lifetimes in the order of 50–100 years. Therefore, planners must attempt to account for those impacts in the future (Thompson *et al.*, 2013). This may require an adjustment to or adaptation of frequency analysis, concepts like return period and risk, and hydrological design methods to the conditions of a changing world (Salas *et al.*, 2012; Salas & Obeysekera, 2014). Synthetic hydrologic series for specified future time periods can be generated by sampling uncertainty from variability due to climate change in a probabilistic way (Brown *et al.*, 2012). These are used to estimate the frequency distributions of decision variables under alternative water management strategies, showing the extent of a system's ability to meet the design requirements (Brown & Wilby, 2012). Uncertainties of future demand due to population growth and development in river basins can also be considered in a comparable manner. The combination forms the basis of a complete risk-based decision-making process, such as required by the U.S. Water Resources Council's 'Principles and Guidelines' (U.S. Water Resources Council, 1983).

A bewildering array of various climate modeling approaches, analytical techniques for non-stationary hydrologic analysis and planning evaluation approaches were devised to deal with the key issues. These center on climate change impact analyses, mainly for devising greenhouse gas reduction mitigation strategies. In turn, this array of methods has created additional challenges with associated uncertainties (Slater *et al.*, 2021), that include the following:

1. Downscaling a multitude of climate scenario precipitation and temperature outputs and transforming into hydrological outputs (runoff and flood and drought frequencies).

2. Dealing with non-stationary climate signals with standard frequency-based hydrologic analysis is currently the basis for project economic justification and design.
3. Adapting existing planning evaluation approaches that are not readily suited for dealing with large magnitude climate uncertainties.
4. Designing and economically justifying robustness and resilience for water projects under conventional economic project evaluation and justification procedures and criteria.

It should be recognized that many water infrastructure design procedures and standards already encompass and compensate for a considerable portion of climate and weather extremes and hydrologic uncertainties through various design redundancies and 'safety factors' (Matalas & Fiering, 1977; Ayyub, 2018). This historically conservative engineering approach implicitly and explicitly incorporates many, but not all, aspects of 'robust' and 'resilient' design features called for by many advocates of climate vulnerability assessments (Lempert & Groves, 2010; Hallegatte *et al.*, 2012; Lempert, 2019).

TRADITIONAL WATER RESOURCES PLANNING PARADIGM UNDERPINS CLIMATE ADAPTATION ANALYSIS

Most modern water resources planning is based on the tenets of operational hydrology and rational planning introduced by the Harvard Water Program (Maass *et al.*, 1962; Fiering, 1967). The U.S. Water Resources Council's (1983) 'Economic and Environmental Principles and Guidelines for Water and Related Land Resources Planning Implementation Studies (P&G)' stand as a representative example of such traditional and proven water resources planning and evaluation protocols that were derived from the principles and concepts of the Harvard Water Program.

These procedures represent a unique combination of decision theory, public choice theory and benefit–cost theory (Stakhiv, 1986). As such, they encompass most, but not all the basic risk analysis requirements needed to assess infrastructure vulnerabilities that evolve primarily from uncertainties associated with climate change as expressed via hydrologic and hydraulic uncertainties, as well as future socioeconomic drivers. Risk and uncertainty analysis, in all aspects and dimensions of planning – social, environmental, economic, hydrologic and hydraulics, have been a core requirement of U.S. federal planning for the last five decades.

Many leading water resources infrastructure management institutions, such as the U.S Army Corps of Engineers (2000, 2019), World Bank (2020, 2021), Wright (2015), Asian Development Bank (2015, 2021), and the Netherlands Rijkswaterstaat (2016; Van Alphen, 2015), have developed comparable elaborate protocols for evaluating the socioeconomic and environmental feasibility of a variety of water projects that are consistent with the ideas and principles of the P&G. These protocols and guidelines undergo constant revisions and updating to accommodate new planning objectives (e.g., sustainable development goals, gender equity and environmental justice), updated economic and financial decision criteria, as well as new methods and techniques for hydrologic analysis, watershed modeling and climate uncertainty analysis.

Water resources project planning is essentially a three-tiered process. The process begins with a feasibility phase to determine if there is a real need for a project, and then lays out options for further study. This is followed by an advanced planning phase, where options are evaluated according to publicly derived planning objectives and evaluation criteria, and legally prescribed normative decision rules. Finally, if a project option is technically feasible as well as economically, socially and environmentally justified, it is selected and authorized to proceed to the design phase.

Project planning emphasizes procedures and rules that result in economic–social–environmental trade-off analyses that seek to maximize project net benefits. The overall aim is to maximize economic performance and service reliability over the life of a project (Yoe, 2017). On the other hand, once a particular project alternative has been selected, engineering design strives to ensure that a catastrophic project failure is avoided by emphasizing measures that minimize the risk of hydraulic failure. Both the planning function and the subsequent design function require very specific, but different types of information about future climate and hydrologic loadings.

At all levels of planning and evaluation, a great deal of hydro-meteorologic information is required but use of available data is complicated by the fact that global warming is likely to result in a non-stationary climate regime (IPCC, 2021). Much effort has gone into tools and models to forecast future hydro-climatology based on a wide array of climate models and speculations about future hydrologic responses. Unfortunately, this has often led to greater uncertainty and a broadening of ranges of potential future scenarios and associated flood and drought frequency analyses (Koutsoyiannis *et al.*, 2008). Such scenarios are the traditional bases for hydrologic and hydraulic design. Consequently, any choice for a particular subset of future scenarios to plan, design or invest has become an increasingly subjective enterprise, centering on which climate scenarios to consider, and what hydrological analytical tools could be employed to deal with these cascading uncertainties (Slater *et al.*, 2021). Planners are confronted with the basic question of how much investment can be justified to minimize risk or how much is too much or too little investment?

Hence, there is a need for quantitative and pragmatic guidance that builds on and extends conventional planning and design procedures and practices that are internally consistent and can encompass future climate uncertainties and non-stationary. Existing conventional practices, associated with risk-informed decision-making under a stationary climate regime, provide a coherent approach and entry point to problems that require one to consider '*known unknowns*' such as climate change-related consequences. The results of these analyses must then be presented at a level of understanding to decision-makers that is not dependent on a complete, quantitative understanding of future probabilities that are inherently unknowable (Dessai & Hulme, 2004; Pielke & Wilby, 2012; Wilby & Murphy, 2019). Such guidance, however, must also consider the financial, technical and institutional limitations, and capabilities of planners and decision-makers to complete master planning, pre-feasibility and feasibility studies.

'TOP-DOWN' VS 'BOTTOM-UP' CLIMATE IMPACT ANALYSIS

In the context of this discussion, 'bottom-up' impact analyses, vulnerability assessments and 'stress tests' are meant to reflect the traditional manner in which water resources projects are considered. These begin with an assessment of vulnerabilities in existing water infrastructure and problems and anticipated needs at a site-specific or watershed level. Use of historical climate and hydrologic data is typically the first step. This approach contrasts with methods advocated by many climate scientists and resource management agencies who advocate that climate-based analyses begin with a consideration of data generated by climate models, e.g., general circulation models (GCMs) and scenarios for greenhouse gas emissions (Kundzewicz & Stakhiv, 2010; Brown & Wilby, 2012; Nazemi & Wheaton, 2014). The latter approach would be considered 'top-down'.

The 'top-down', or GCM-centered, approach was originally designed for devising climate mitigation policies, i.e., strategies for reducing greenhouse gases and not intended for site-specific adaptation analysis (Anagnostopoulos *et al.*, 2010; Conway *et al.*, 2019). Since the advent of the UN Intergovernmental Panel on Climate Change First Assessment Report (IPCC FAR, 1990), climate science has evolved exponentially and has better-defined pathways of future plausible climate scenarios.

On the other hand, traditional 'bottom-up' planning approaches must factor in multiple drivers of change to produce contextualized information that is relevant for decision-makers. Therefore, the best that can be offered in terms of derivations that certain climate scenarios will materialize is an inherently judgment-based statement of 'plausibility' (Mendoza *et al.*, 2018). Bottom-up approaches focus on water resources problems and how they affect communities, through consideration of potential future risks on various population sectors and the residual risks associated with solutions. In a typical water resources management study, climate change is but one additional risk factor among a myriad array of other vulnerabilities, rather than the principal factor of future changes. Use of traditional planning processes developed over the past century that focus on climate change and its manifestations remain invaluable and should not be discarded (Stakhiv, 2011). This is especially relevant when it has been shown that most climate model simulations provide widely divergent outputs (Koutsoyiannis *et al.*, 2008; Knutti & Sedláček, 2013; Giuntoli *et al.*, 2015; Eisner *et al.*, 2017; Pechlivanidis *et al.*, 2017).

For example, Giuntoli *et al.* (2015), among many others involved with the Inter-Sectoral Impact Model Intercomparison Project's (ISIMIP) multi-model ensemble comparisons, found that results from different impact models simulating the same systems under the same climate change conditions showed considerable variability. Unfortunately, the trend has been for many leading climate scientists and international institutions to advocate the most extreme scenarios, such as the RCP 8.5 (Representative Concentration Pathways) scenario as part of 'robust' planning solutions for major infrastructure (Ritchie & Dowlatabadi, 2017; Hausfather & Peters, 2020).

Hausfather & Peters (2020) called on policy makers to 'stop using the worst-case scenario for climate warming as the most likely outcome – more-realistic baselines make for better policy'. This is an important acknowledgment because these scenarios drive not only global and national climate mitigation and adaptation policies, but virtually all future infrastructure planning. As such, they skew the outcomes towards expensive and ultimately infeasible solutions. These scenarios are required in many national climate change assessments and strategic planning documents, so that economic development scenarios, project planning and regulatory decisions can be developed and compared against a uniform and consistent baseline. The basic question confronting planners in which baseline does one plan for an uncertain future 50–100 years hence.

Though a new set of future scenarios designed for global and national mitigation policy formulation have been developed – the shared socioeconomic pathways – offering a broader view of what a world 'with and without' future climate policy options might look like (O'Neill *et al.*, 2014; Nature Editorial, 2019), they are still far from being useful for water planners at the site-specific, project level (Hausfather & Peters, 2020). Thus, when water resources planners are forced to use highly unrealistic scenarios by their respective governments, as part of comprehensive river basin planning strategies or site-specific project planning, they are placed in a very difficult position of analyzing and planning for implausible and extreme scenarios, resulting in very costly and mostly unrealizable strategies, plans and project designs for future population needs (Shortridge *et al.*, 2017).

A commonly encountered display of future climate uncertainty for major river basins is exhibited in Figure 1. This figure represents the Mekong River Basin annual precipitation derived from three different climate models, based on two high-emission scenarios, RCP6.0 and RCP8.5, resulting in six climate scenarios that had been 'approved' by the Mekong River Commission (MRC, 2018) for comprehensive basin planning purposes. These scenarios represent the typical wide disparity among results in the various climate change scenarios and the models that generate the temperature and precipitation projections based on the scenarios, which are in use today that influence policies, vulnerability assessments and decision-making.

From a project planning or design perspective, when viewing Figure 1 it is quite difficult, if not impossible, to draw any sensible insights, much less conclusions about vulnerabilities and specific water management impacts for a particular project, project site or watershed in the Mekong basin from an ensemble of selected climate scenarios that represent the high end of possible greenhouse gas emissions scenarios. Clearly, there is a wide divergence in projected precipitation among the three climate models for the same scenarios, ranging from a 30% precipitation decrease for the GISS model/RCP8.5 scenario to a +40% precipitation increase for the GFDL/RCP8.5 scenario. Practically speaking, it is hard to imagine any combination of economically feasible

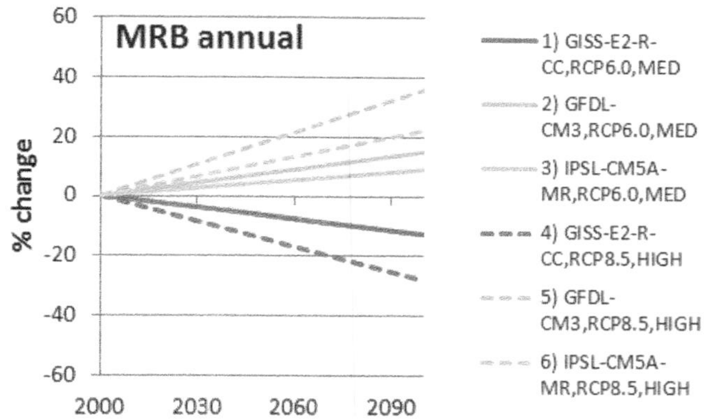

Fig. 1 | Projected change to annual precipitation relative to baseline period (1986–2005) in the whole Mekong River Basin (MRB) under six recommended climate change scenarios (MRC, 2018).

solutions that can resolve water resources problems for the Mekong basin across such a wide range of disparate 'worst-case' scenarios.

This situation is a common one for most analyses relying on a 'top-down' approach, i.e., one beginning with climate models and emissions scenarios provided by the IPCC, and then working to downscale the information to the geographic levels of interest. Though this approach was originally used as the basis for national-level 'climate policy' decisions largely for mitigation purposes (Dessai & Hulme, 2004; Kundzewicz & Stakhiv, 2010), the top-down analytical model was soon extended for many other adaptation purposes, including agriculture, forestry and water resources, inappropriately so, according to many (Koutsoyiannis *et al.*, 2008; Anagnostopoulos *et al.*, 2010; Pielke & Wilby, 2012).

Presently, it is impossible to predict which climate scenario will emerge to be the dominant one in a particular region – especially in the monsoon-driven climate of the Mekong River basin. Similar wide disparities among climate model outputs have been shown for many river basins around the world (Giuntoli *et al.*, 2015; Eisner *et al.*, 2017; Hattermann *et al.*, 2017; Kettner *et al.* 2018).

Recognizing these considerable uncertainties and disparities among various climate scenarios and models, it is quite impossible to design a functional water management project, or system, based on such vague and inconsistent climate information. Nor is it likely that applying 'robust decision-making' (RDM) methods can overcome such widely divergent climate model outputs (Daron, 2015; Herman *et al.*, 2015; Giuliani & Castelletti, 2016). Robust strategies are understood as those which satisfy performance criteria against most sets of future conditions (Bhave *et al.*, 2016; Wilby & Dessai, 2010). Project alternatives that meet 'performance criteria' across many different climate scenarios, i.e., 'robust' solutions, generally cannot meet normative economic efficiency criteria (i.e., benefits > costs) that are used by most water management institutions (Tol, 2003; World Bank, 2010; Stakhiv, 2011; Dennig, 2018).

Hence, there is an incongruity between the aims of 'RDM' that develop 'robust' solutions which cover a wide range of climate scenarios, and those of traditional decision frameworks that aim for economically optimal solutions. Traditional decision frameworks focus on the most likely or most plausible scenarios, both socioeconomic and climate based. This critical step, of assessing plausibility, relies on professional judgment, to a large degree, based on the known trends and circumstances within a particular watershed, area or region (Mendoza *et al.*, 2018). Once discrete projects/solutions are formulated, the decision rules applied are those that maximize net benefits or minimize the risk-cost, typically used by most public engineering institutions.

ALTERNATIVE APPROACHES TO DECISION-MAKING UNDER UNCERTAINTY

There is an additional discontinuity between the detailed risk analysis methods currently required for traditional hydrologic analysis and project benefit–cost analysis (BCA), with the inability to derive legitimate probabilistic information from an ensemble of highly divergent climate scenarios. While vulnerability assessments and bottom-up frameworks are useful in testing the robustness of planning alternatives, this approach requires some sense of the likelihood of selected future scenarios, usually with either uniformly sampled scenarios or ensemble projections treated probabilistically (Shortridge & Zaitchik, 2018).

Dessai & Hulme (2004) covered three important concepts in their paper 'Does climate policy need probabilities?' The first being the notion of whether probabilities can be legitimately assigned to various climate change scenarios and outcomes. The second questioning whether climate change scenarios even belong in the realm of probabilistic computation. The third issue raised was whether 'top-down', climate scenario-driven analysis can be applied to contemporary resource management issues. They proposed a 'bottom-up' variant to deal with typical resource planning and management climate adaptation policies, because they included an array of very complex interacting social, economic and resource capacity issues and vulnerabilities. Figure 2 best encapsulates their 'top-down' and 'bottom-up' dichotomy, which is an appropriate starting point for climate adaptation.

Borgomeo *et al.* (2018) assert that under conditions of climate uncertainty, decision-making is a three-way tradeoff between risk, cost and robustness. The incremental costs of meeting added robustness/reliability criteria can be quantified based on either risk attitudes of the public ('tolerable risk') or existing institutional, legislative or engineering standards and/or community criteria. In fact, most engineering design standards are routinely adjusted to meet contemporary notions of 'tolerable risk' (see Stakhiv paper, this volume).

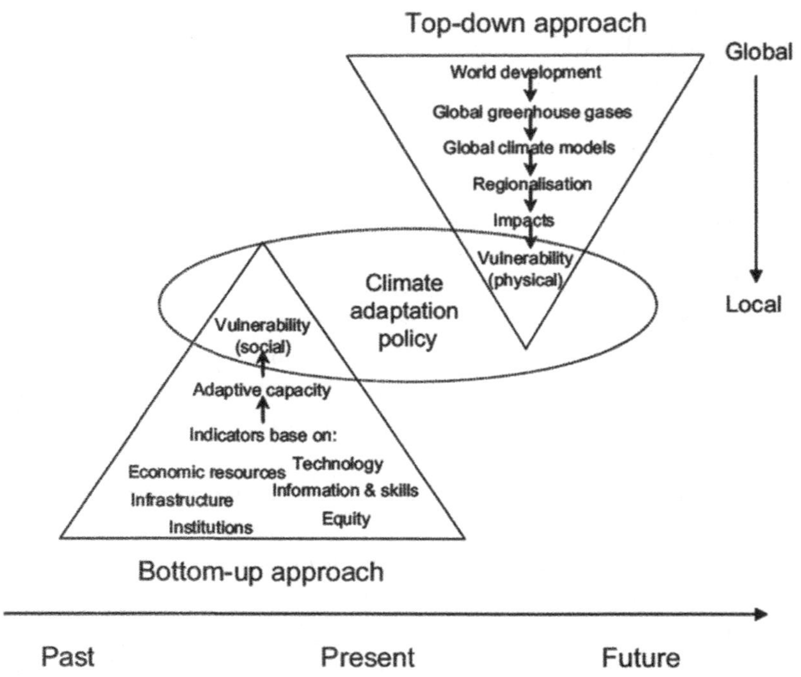

Fig. 2 | 'Top-down' and 'bottom-up' approaches used to inform climate change adaptation (based on Dessai & Hulme, 2004).

For water resources agencies, 'robustness' has two fundamental performance dimensions: reliable service delivery (municipal water supply, irrigation and hydropower) over time and safety – i.e., avoiding catastrophic project failure (e.g., levee breaches, dam failure, pipeline ruptures, etc.). The incremental costs of robustness are those that reduce 'residual risks' in the future, either in performance and/or safety, which can readily be derived using standard approaches for formulating cost-effective solutions based on existing conventional procedures and engineering standards (Stakhiv, 2011).

There are, however, two fundamental but very different and mostly incongruous aspects of climate risk-based decision-making that must be resolved within current practices. The first is that perceived climate non-stationary requires a different or modified suite of hydrological methods and models for conventional project justification approaches required by such evaluation protocols as the P&G and many other agency procedures that require BCA (Read & Vogel, 2015; Salas *et al.*, 2018; Serago & Vogel, 2018). The second is that climate change scenarios cannot be credibly assigned probabilities. The question is whether non-stationary can be treated independently of determining the probabilities of multiple climate scenarios?

Many alternative evaluation approaches that have been proposed, attempt to deal with climate change scenarios as a qualitative, judgment-based 'plausibility' issue, without specifically dealing with either scenario probabilities or explicitly with hydrologic non-stationary (Dessai & Hulme, 2004; Slater *et al.*, 2021). While climate scenario probabilities are not essential to formulating projects, some standardized approaches to hydrologic non-stationary would be useful for existing project evaluation procedures, as described, for example, in Salas *et al.* (2018).

Analyzing the risk of disastrous extreme hydrologic events (floods or droughts) lies at the heart of BCA and project justification. Hydrologic frequency analysis of observed conditions provides the evidence to estimate the probabilities of harmful outcomes, alongside extensive stochastic simulations of water systems to sample spatial and temporal variability (Borgomeo *et al.*, 2018). However, since there is now pressure to evaluate water investments in the context of a moving non-stationary climate framework (Milly *et al.*, 2008), different analytical frameworks have been proposed to overcome the difficulties of dealing with non-stationary and assigning probabilities to climate scenarios (Slater *et al.*, 2021).

However, Kroll *et al.* (2015) noted another underlying problem of trying to quantify non-stationary in flood flows, for it is not just related to climate change, but is shown to result from a variety of anthropogenic processes including changes in land use, climate and water use, with likely interactions among those processes making it very difficult to attribute trends to a particular cause, at a particular site. Serago & Vogel (2018) developed a simple statistical model which can mimic observed flood trends and the frequency of floods in a non-stationary world.

It is argued that decision analysis methods based on probabilistic risk analysis of historical hydrological observations alone may no longer be adequate to identify a water investment capable of meeting a performance or service standard (Borgomeo *et al.*, 2018). Others, such as Lempert (2019) and Dennig (2018) argue that the applicability of BCA, based on probabilistic characterizations of uncertainty, can no longer be relied on as an approach to ranking decision alternatives in the face of large uncertainties, especially those associated with climate change. For BCA to provide decision support under these circumstances, it needs to be reframed to account for the large uncertainties facing many types of decision challenges.

Stakhiv (2011) advocated changes in BCA procedures from a somewhat different standpoint, considering anticipated large climate uncertainties in the analytical process, and the need to formulate more robust water projects and systems. He argued that current economic procedures need to be revised to counter what he termed the 'triple discount dilemma' common to most agency procedures. In other words, current economic evaluation criteria and decision rules are set up in such a way as to discount (minimize) the effects of low probability–high consequence catastrophic events that may be more common during a future non-stationary climate. The socioeconomic effects of uncertain, low probability–high consequence events, such as extreme floods and droughts, can also be discounted by the choices of probability distributions that diminish the importance of low probability events (Haimes, 2009; Pindyck, 2010; Botterill & Cockfield, 2013).

In addition, e.g., high economic discount rates of 7–10% used by the major public engineering organizations, such as the Corps of Engineers and World Bank, coupled with a decision rule that optimizes project outputs only to the level that maximizes net benefits, often resulted in projects that provided protection against floods in the 50–100-year range of return periods. The trend towards optimization, as part of systems analysis, coupled with the built-in 'triple discount dilemma', ultimately led to 'brittle' solutions (Fiering, 1982) – i.e., projects that were neither robust nor resilient (able to quickly return to design project performance levels after a failure).

In the search for more 'robust' decision-making approaches, an expanded approach that employed multi-scenarios based on a multiple models approach, using 'ensemble' climate projections favored by many climate scientists and hydrologists, generated scores of different precipitation and runoff simulations for river basins around the world (Lehner *et al.*, 2020). These, however, have not been as useful in clarifying future states of hydrologic runoff in given river basins, nor in reaching robust conclusions about the utility of these methods (Eisner *et al.*, 2017; Hattermann *et al.*, 2017; Krysanova *et al.*, 2017).

Hence, approaches that advocate 'RDM' which undertake evaluations of these ensembles also propagate a 'top-down' cascade of uncertainties that are difficult to unravel and comprehend by the average water resources planner working in any major water agency or department (Shortridge *et al.*, 2017). Nevertheless, they are part of a suite of newly developed approaches to deal with climate uncertainty as part of either policy setting, strategies for river basin development and investments and project planning (Groves *et al.*, 2014).

Daron (2015), Lehner *et al.* (2020) and Bhave *et al.* (2016) challenge the utility of RDM practices, especially for developing nations, where data are sparse, and there are disagreements over which scenarios should be used for analysis. Conventional forecast-driven approaches to climate change adaptation create a cascade of uncertainties that can overwhelm decision-makers and delay proactive adaptation responses. One of the significant limitations and shortcomings of robust planning frameworks is the tendency to favor static alternatives to be implemented in the near-term, which could result in costly overdesign, particularly in the case of long-life infrastructure (Borgomeo *et al.*, 2018; Herman *et al.*, 2020). Clearly, there exist solutions that are insensitive to a wide array of climate scenarios, but they are either often considered economically infeasible under current benefit–cost criteria, or would likely be either socially or politically infeasible, requiring massive regulatory interventions and widespread behavioral adjustments at the national and local levels.

The added problem is that there are always uncertainties and residual risks associated with technological solutions – particularly non-structural solutions for floodplain management that are based on floodplain delineations and depend on uncertain flood warning triggers and timely evacuations during a flood (Haimes *et al.*, 1995). RDM seeks to identify those areas of residual risk that fall outside the range of stationary climate analysis. Risk 'tolerance' and residual risks are politically determined criteria, via public preferences (Institute for Water Resources, 2010).

Concurrently, several different and more pragmatic approaches called 'adaptation pathways' (Walker *et al.*, 2013), have been developed, which focus on defining a comprehensive set of alternative sequences of decisions and identifying the conditions that could guide decision-makers along the most appropriate sequence based on defining critical performance thresholds. Haasnoot *et al.* (2013) demonstrated this approach for flood control in the Netherlands, which included novel visualizations of different sequences of investments, that they called 'pathways'. The pathways are combinations of different projects or measures that are implemented depending on which critical performance thresholds are exceeded. What is not and developed in this approach are the evaluation criteria which link the analytical procedures that transform future climate scenarios to defining when these acceptable performance thresholds are breached. In other words, clearly identifying the thresholds for changing a course of action.

What the 'adaptation pathways approach' entails is the formulation of a core set of projects/actions that would be required, as a starting point, under any combination of scenarios, if only to keep pace with growing population demands and project rehabilitation requirements. Then, over a period of time, depending on new information and perhaps technological developments that alter future demand patterns (more efficient water saving devices, changes in industrial technologies, etc.), certain climate response pathways may be narrowed down to a better-defined strategy, with well-defined alternative adaptation pathways depending on the improved climate scenario information.

Another related approach, termed 'decision-scaling' (Brown *et al.*, 2012), has been described and applied in the Great Lakes region (Brown *et al.*, 2011) and the Niger River Basin (Ghile *et al.*, 2014). Decision-scaling provides techniques for extensively exploring plausible climate conditions to help identify key climate-related thresholds that distinguish different robust strategies. Both approaches – adaptation pathways and decision-scaling – belong to the subset of methodologies that rely on a 'bottom-up' approach. Decision-scaling is defined by Brown *et al.* (2012) as

> '... the use of a decision analytic framework to reveal the scaling of climate information that is needed to best inform the decision at hand. The methodology first identifies the climate conditions that are *relevant to the decision* and then uses that information to link to what is *credible* in available climate information'.

In other words, Brown *et al.* (2012) suggest that before launching into a 'top-down' multiple ensemble climate scenario robustness analysis, first look at what types of decisions and scales of analyses are required regarding the specific existing circumstances and vulnerabilities of a project or system of projects. These circumstances could vary at many different scales – from a strategic river basin plan, an infrastructure project, major dam spillway rehabilitation project, reservoir reregulation or capacity expansion of an irrigation project or water supply treatment plant. Then, identify those elements that require more detailed analysis simply to meet the projected demands and updated hydrological information since the projects were constructed. Finally, select the type of analysis and degree of specificity most appropriate to the nature of the problem that was identified. In other words, do not automatically begin with top-down, multi-scenario analysis. Brown & Wilby (2012) also propose several different quantitative approaches that link GCM scenarios with stochastic hydrologic ensembles as the basis for joining 'top-down' with 'bottom-up' analysis, if required.

Water resources management is inherently about incremental adaptation, adjusting to changes in climate, technology and population demands. As climate events are more closely monitored and models improve, and the variability, frequency and intensity of certain extreme events become more apparent over the next two decades, suggesting distinctive trends, then the sequence and timing of a suite of water resources projects that were designed for flexibility, robustness and resilience, can be evaluated with greater certainty. Then, the numerous permutations of alternative development pathways will be more readily apparent and can coalesce into a definable adaptation strategy.

CLIMATE RISK-INFORMED DECISION ANALYSIS

A dominant factor in the way 'public engineering' has evolved in the past two decades is linked to the advent of the six reports of the UN Intergovernmental Panel on Climate Change (IPCC, 2021), as well as national greenhouse gas reduction commitments required by the Paris Accords of 2015. These reports created a climate model-driven 'top-down' analytical paradigm which was unsuited for most water resources management problems – particularly those related to site-specific infrastructure design. Because climate models have been so undependable for site-specific project planning, the water resources community banded together in 2010 to develop an alternative pathway for analyzing projects under climate uncertainty. Consequently, a number of practicing water management agencies and lending institutions, such as the World Bank, Corps of Engineers, Rijkswaterstaat, etc. formed an 'Alliance for Global Water Adaptation' (AGWA, 2020) http://alliance4water.org/.

The first outcome of this alliance effort was the World Bank's *Confronting Climate Uncertainty in Water Resources Planning and Project Design: The Decision Tree Framework* (Ray & Brown, 2015), widely known now as the 'decision tree approach'. Shortly thereafter, under the longstanding cooperation between the U.S. Army Corps of Engineers (USACE) and the Rijkswaterstaat (the Dutch national water management authority), the Climate Risk-Informed Decision Analysis (CRIDA) project was initiated (Mendoza *et al.*, 2018). (https://agwaguide.org/docs/CRIDA_Sept_2019.pdf). Meanwhile, the Asian Development Bank was also crafting its own set of guidelines for dealing with climate uncertainties within their planning frameworks (Watkiss *et al.*, 2020).

Both the World Bank 'decision tree framework' and the associated CRIDA planning methodology emphasize a 'bottom-up' analytical perspective to water resources project planning and decision-making, in view of the wide disparities in climate model outcomes. They are complementary analytical frameworks, since CRIDA also incorporates the 'decision-scaling' approach developed for the World Bank.

What differentiates CRIDA from other approaches is the application of a step-by-step approach that marries a conventional water resources planning and evaluation framework with decision-scaling rationale, emphasizing a 'bottom-up' methodology to risk assessment and management for those problems requiring a formal examination of potential climate-related uncertainties to justify the selection of a project alternative. The CRIDA version of decision-scaling focuses more on guiding the types of decisions that are made within a typical planning process, and what credible climate information might be needed for those decisions, and arrays those against the analytical uncertainties of the available information and anticipated future risks of project performance associated with climate change.

In CRIDA, it is the overall formal planning and evaluation process, i.e., the institutionally approved 'evaluation and decision model' that guides the determination of which analytical path is taken in dealing with model uncertainties versus those decision frameworks that are based on climate uncertainties as the introductory and encompassing step. Existing institutional frameworks, like the U.S. Water Resources Council (1983), which tie together national water resources management objectives, public and stakeholder preferences, and socioeconomic decision criteria, guide how agencies should approach particular water problems and the manner in which model and climate uncertainties are to be treated within an existing and politically approved decision-making framework.

At least since the 1970s, risk and uncertainty analysis have been a standard feature of all steps in routine water resources planning processes. The core component, when choosing a particular analytical pathway, has always been asking questions of how important individual uncertainties will impact project performance within an array of possible alternatives. Additionally, in this existing framework, the importance of the societal distribution of risks and uncertainties, how uncertainties influence the design of a project (robustness, resilience and reliability) and impact on costs constitute equally important considerations.

A report by the U.S. National Academy of Sciences (NRC, 2000) noted that '[T]he Corps' [U.S. Army Corps of Engineers] risk analysis techniques and flood damage reduction studies will produce their greatest benefits if these techniques and studies *are executed within a comprehensive planning paradigm and framework designed to make the best social, economic, and environmental uses of the nation's floodplain resources*'. In other words, the NRC recognized that any risk-based vulnerability assessment that is conducted outside a formal and approved societal objectives-based planning framework is incomplete and could lead to skewed or biased outcomes (National Research Council, 2009).

Outcome risk, i.e., uncertainty about either the consequences or severity of those consequences, is routinely reduced by a variety of risk reduction management measures, including both structural and non-structural measures. These measures do not reduce all risk uniformly because they address different parts of the risk management cycle (prevention, preparedness, response and recovery) – some protect property, others are geared towards saving lives at the expense of property. For example, a flood warning and evacuation system have its own highly uncertain components that are very susceptible to a cascading sequence of human error and technological breakdowns (Haimes *et al.*, 1995).

As an example, Figure 3 shows a stepladder hierarchy of typical flood damage risk reduction measures that are routinely employed as part of a strategy of flood risk reduction. At the bottom of the step ladder is 'residual risk', or risks that are unknown or highly uncertain, but are considered as 'tolerable' by society. Figure 3 shows that these residual risks will likely increase in the future, as climate change uncertainties become more apparent.

While economists value water for its abstract economic production values, water engineers measure performance against the specific objectives for which a project or system was designed, a wide array of services which contribute towards economic performance. A number of abstract performance indices were proposed to measure the performance of water resources systems in terms of robustness, reliability, resilience and vulnerability of water resources subjected to climate extremes (e.g., Hashimoto *et al.* 1982a). *Reliability* is the probability that no failure occurs within a fixed period, often taken to be the planning period. *Robustness* describes project designs and operating policies that may be sufficiently flexible to permit their adaptation to a wide range of possible demand conditions at little additional cost. *Vulnerability*, in this context, refers to the likelihood of a project service failure (e.g., due to maximum flood or drought intensity). *Resilience* is often interpreted as a measure of how quickly a system is likely to recover from failure once failure has occurred. Inherently, the ensemble of drought management measures (or flood control measures) that are implemented by any locality or region aims to increase system reliability, resilience and robustness, while decreasing vulnerability (Hashimoto *et al.*, 1982b).

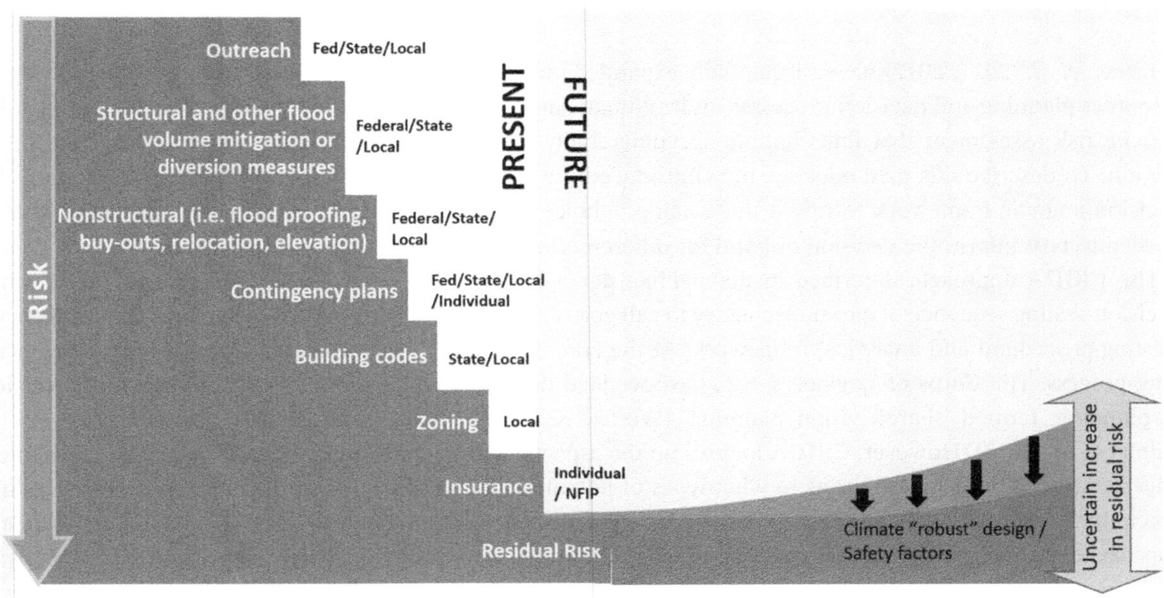

Fig. 3 | Buying down risk: flood risk reduction management measures.

Future climate uncertainty and socioeconomic growth and development trajectories will, almost inevitably, widen the uncertainty bounds of residual risk. Structural measures have proven to be the most reliable of all the measures, and more conservative engineering design factors can significantly reduce future uncertainties, thereby increasing robustness, reliability and resilience.

If certain destructive events occur frequently enough – e.g., a historic flood or drought; an infrastructure failure; or increased demands due to accelerated development programs (new irrigation projects, hydroelectric power or municipal water supply) – institutions will re-examine their portfolio of projects and water management systems, as well as policies and procedures. Such examinations may include operating rules to reassess performance to meet future demands under comparable hydro-meteorological conditions and/or a changing climate. Vulnerability assessments, or 'stress tests', are standard starting points for all inquiries often triggered by a flood or drought; whether expanding the capacity of a dam or navigation lock system due to anticipated growing capacity demands; or major repair and rehabilitation of existing water infrastructure systems that are approaching or have surpassed their design life (levees, hydropower, irrigation systems, dams, etc.).

DECISION-SCALING WITHIN CRIDA

The general concept of 'decision-scaling' as applied to climate-related uncertainties was likely first introduced by Matalas & Fiering (1977), though they did not have a term for the process they described to map the uncertainties associated with future climate scenarios against alternatives of optimal water resources system design. The authors also introduced the notions of robustness, reliability, regret and resilience as key attributes of any system design.

> *'Part of the design problem is to identify the types of climate shift that might be anticipated and to determine if they are sufficiently precipitous with respect to flow characteristics to dictate a change in system design. It is not necessary for this purpose to know or to try to determine whether there is a true climatic shift, this may be an interesting scientific question, important in its own right, but it is virtually meaningless for the design of water resource systems'.*

Brown *et al.* (2011, 2012) have substantially expanded on the concepts of decision-scaling as applied to water resources planning and decision processes under climate uncertainty. They explicitly focused on an approach for climate risk assessment that links bottom-up vulnerability assessment with multiple sources of climate information. To describe this methodology, they introduced the term 'decision-scaling', which refers to the use of a decision analytic framework to reveal the 'scaling' (choice of appropriate level) of climate information that is needed to best inform the decision at hand for different scales of water management problems.

The CRIDA approach, described in manual-like detail in Mendoza *et al.* (2018), is inherently based on a decision-scaling sequence of questions that try to categorize the nature of the problem at hand, within an agency's existing procedural and analytical framework. At the core of the CRIDA approach is a traditional agency planning process. The Corps of Engineers, e.g., have wedded the CRIDA climate decision process with its version of planning, termed 'shared vision planning' (Werick & Palmer, 2008; Bourget, 2011; Brown *et al.*, 2011; Palmer *et al.*, 2013). However, CRIDA focuses on the aspects of project planning that will be affected by climate uncertainties. It poses questions as to what types of potential climate shifts in a particular region may produce precipitous changes in hydrologic flows, and whether they are sufficient to dictate changes in standard approaches to design, so that planners can devise the most appropriate analytical approach to risk analysis under climate uncertainty. It elicits information about how concerned planners are about future risks and analytical uncertainties, depending both on the existing information and on the nature of proposed changes – new construction, reservoir operations management changes, major rehabilitation of existing infrastructure or non-structural regulatory management changes.

There are two ways of viewing the CRIDA decision-scaling approach – both are complementary perspectives. The CRIDA approach is to view the problems more generically, in terms of qualitative 'levels of concern' about 'future risks' from climate uncertainties arrayed against known 'analytical uncertainties'. This appraisal process can be viewed and simplified as four decision quadrants (Figure 4). Another complementary way of looking at the problem is from the more familiar perspective of categories of analysis typically used by water agencies for construction, operations, maintenance and long-range strategic planning (see Figure 5).

In CRIDA, performance metrics are selected to measure a system's ability to reliably deliver services during the life of the project. These are then arrayed and compared against minimum acceptable service levels and thresholds to measure three basic variables: *reliability* (the frequency that system performance meets a target), *resiliency* (the time it takes for a system to return from failed performance to a target) and *robustness* (the range in which a system maintains target performance while facing external stressors) of services under the stress of future change. Compared to traditional economic decision criteria, these three variables form an enhanced level of consideration for decision criteria when high uncertainty and variability are important.

An initial vulnerability assessment, or 'stress test' of how a project or system performs under a variety of selected plausible scenarios or load factors, is always the starting point, and should always begin in Quadrant I (Figure 4), to determine whether the nature of the problem requires moving in a direction that deviates from standard planning procedures and accepted analytical techniques. In other words, explore the range of conditions under which a project will operate as designed, and determine which conditions will cause the project to perform unreliably or fail – either in curtailing delivery of services (water supply, irrigation, hydropower, etc.) or structural failure (spillway failure, overtopping, dam breach, etc.). Such procedures are enumerated in many engineering manuals of the Corps of Engineers or Bureau of Reclamation.

Even if the uncertainties are large enough to warrant additional and more detailed 'stress tests', further probing identified failure points and external loading factors (e.g., an expanded set of plausible climate scenarios), the underlying question remains – is the project robust enough, with its embedded safety factors and redundant design criteria to deal adequately under a foreseeable increase in climate variability? Are existing analytical tools and evaluation protocols sufficient to deal with the added levels of uncertainty?

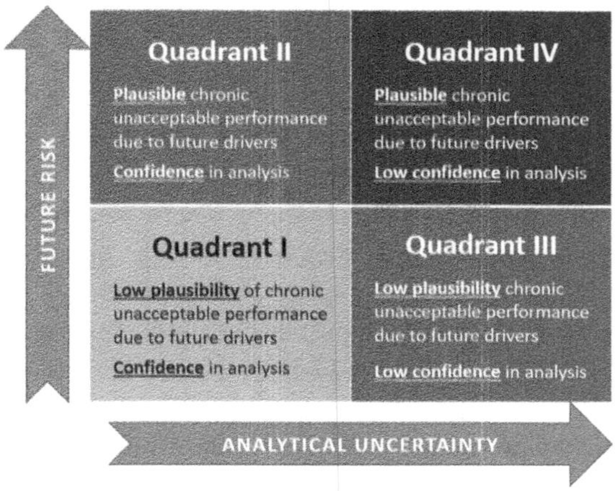

Fig. 4 | Categorizing future risk and analytical uncertainty to establish a 'level of concern' in the planning process (Mendoza *et al.*, 2018).

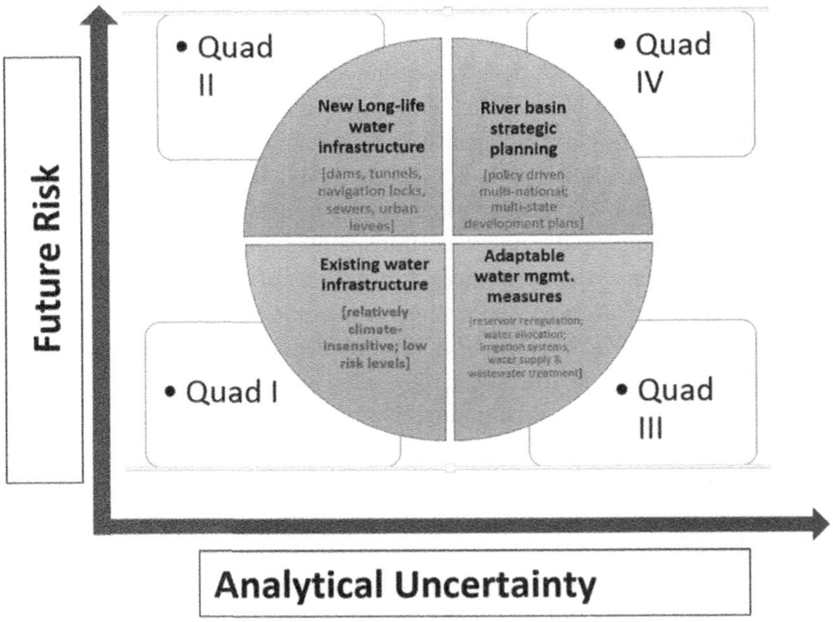

Fig. 5 | Decision-scaling perspectives based on water management functions.

Typically, an analyst is concerned with the consequences of failure from under-designing a project or excessive design costs that cannot be easily justified. An institution's existing practices and design standards often mandate precautionary policies and safety margins. In most cases, the analyst applies standard planning and design guidance as required by existing methodologies, regulations and guidelines (Quadrant I, Figure 4). The stress test should reveal whether there exists a qualitative 'level of concern' about future risks and analytical uncertainties related to either future socioeconomic changes and/or plausible climate scenarios and associated uncertainties. The analysts then determine whether the existing water delivery system or project can address future uncertain but likely plausible stressors, and whether moving beyond standard planning and design approaches is needed and justified.

In Figure 4, 'Future Risk' is the plausibility that a future driver (climate, population growth, demand increase, land use changes, etc.) can be identified that leads to chronic unacceptable failure, i.e., violates one or more selected performance thresholds, with associated adverse consequences. 'Analytical uncertainty' connotes a reduced level of confidence by the analyst in evaluating a project's performance over the life of the project, given the information available to evaluate and to adequately analyze the problem and potential solutions, so that a decision maker can formulate and choose an option with confidence.

The central focus of CRIDA is to determine if the problem encountered can be solved using standard existing 'bottom-up' methods, without necessarily resorting to highly uncertain climate scenarios – meaning whether one can stay within Quadrant I (Figure 4), or whether one needs to enlist other more tailored methods and mechanisms to effectively resolve a range of associated climate uncertainties. Among the decision uncertainties to consider include whether or which future climate scenarios to consider, and/or whether hydrologic 'non-stationary' needs to be factored into the planning and design of a water management strategy or action (e.g., Salas *et al.*, 2018). One of the key underlying questions is whether existing proven analytical methods combined with robust engineering standards are sufficient to encompass the uncertainties inherent in climate variability and future change?

If the 'level of concern' is modest, and there is confidence in the plausibility of selected future scenarios, and the analytical uncertainties are small, then Quadrant I suggests that there is no need to deviate from standard, accepted planning guidance and practices. This default, 'business-as-usual' position is recommended when the stress test reveals minimal impact from future uncertain stressors that are likely to contribute to water resources system failure – either in the form of services or physical infrastructure failure. For most situations, Quadrant I is the starting position for the analyst and decision-makers and may be the endpoint for the CRIDA process if the 'level of concern' is low. This is where the classical 'no regrets' planning options are most sensible and are based on conventional decision rules embodied in agency procedures (Borgomeo *et al.*, 2018).

Based on preliminary stress tests, if the results show significant chronic project or system failures under a number of plausible socioeconomic and/or climate scenarios, then Quadrant II suggests decisive action would be required to build a new project or increase robustness (e.g. raise a dam for added flood storage or raise a levee to protect against higher flood levels) in response to plausible futures. Also, the analytical uncertainty associated with the scenarios is acceptable and can be evaluated using existing risk analysis methods. For example, climate non-stationary is not determined to be as influential as other socioeconomic trends that increase demands which exceed the capacities of the existing storage, treatment or delivery system. The vulnerability assessment may indicate uncertain future climate stressors are credible and have a potentially high impact on system performance. Consequently, the analyst should develop actions or plans that satisfy standard procedural requirements and perform acceptably under potential futures. In this quadrant, the methods advocated by Ray & Brown (2015) are appropriate.

A level of concern that places a situation in Quadrant III suggests that analytical uncertainty may be relatively high, but future scenarios do not indicate a clear increase in risk of impaired performance or system reliability. Hence, this situation may provide evidence and justification to delay major structural project implementation, suggesting consideration of incremental, low-regrets strategies that include monitoring and acquiring better information to address critical futures. In addition to continuing business-as-usual planning procedures, further analysis could suggest incremental adaptive actions be prepared for implementation should a trigger point be reached in a monitoring program.

The level of concern associated with Quadrant IV warrants both decisive and incremental action to increase the robustness of the system. There is plausible evidence and increased risks of future scenarios that will violate one or more critical performance thresholds, together with high analytical uncertainty. Under these conditions, decisive actions, coupled with monitoring to facilitate an incremental and flexible response are justified. Generally, this high level of concern may require substantive institutional adjustments that include changes in planning procedures and decision criteria. A new set of analytical tools, which may include 'top-down' climate scenario analysis (Ray & Brown, 2015; Salas *et al.*, 2018), coupled with an updated evaluation and decision-making paradigm is often required to accommodate this level of concern and address numerous unknowns.

It should be noted that analysis in Quadrants I, II and III essentially rely on standard procedures and analytical methods that can be found in dozens of engineering manuals of the various agencies and ministries that deal with water resources management. It is only in Quadrant IV that future operational risks and both climate and socioeconomic uncertainties are so large that methods specifically tailored for such circumstances should be employed. This is where decisions and analyses that fall within Quadrant IV are best served by the decision framework and methods and procedures found within the 'decision tree framework' proposed by the World Bank (Ray & Brown, 2015). This is also why the 'decision tree framework' and CRIDA are considered to be complementary risk-based planning and evaluation frameworks.

In particular, Quadrant IV, focusing as it does on large-scale strategic river basin planning, is where 'RDM' methods, such as those advocated by Lempert (2019), may be best suited. The Mekong River Basin Management Plan, noted previously, would be a candidate for such analysis, which would fully employ both a combination of 'top-down' and 'bottom-up' decision-scaling methods.

There is, however, a more conventional way of applying the decision-scaling perspective. Figure 5 lays out a comparable and complementary way of viewing the types of typical water management issues that confront water planners from the perspective of familiar water management functions. The two axes are the same as for Figure 4: the vertical axis is future risk and the horizontal axis is analytical uncertainty. For example, the way to overcome analytical uncertainty in Quadrant III is to delay decisions until better information is gathered through monitoring, e.g., and consider a series of incremental actions such as reservoir reregulation or reallocation of storage; incremental expansion of capacity by adding new water treatment units or raising a dam. Quadrant IV is essentially dedicated to large-scale strategic basin planning endeavors that encompass multiple future objectives, increased socioeconomic demands, coupled with a wide range of options and alternative technologies.

When there is high analytical uncertainty but little plausibility of increased future climate risk (Quadrant III), decision-makers face a difficult choice of taking potentially unnecessary (and expensive) actions or taking no action and risk future adverse consequences. In the first case, actions need to be justified based on the limits of analytical knowledge. In the second, the system remains vulnerable, but the risks may be tolerable. With either approach, a flexible adaptive management strategy makes sense – if the relative residual risks are understood and accepted by the stakeholders. Delayed, sequenced or incremental actions can be implemented and actions that expand the lifetime of existing infrastructure or enhance performance may be appropriate. In this instance, an 'adaptation pathways' approach is a reasonable decision framework as a starting point (Haasnoot *et al.*, 2013; Kwakel *et al.*, 2015).

Quadrant III solutions are needed where the future conditions are reasonably well known but there are disagreements on which course of action to pursue because of uncertainties in key decision criteria. Accordingly, the analyst can embark on a series of enabling actions and adaptive management approaches, such as monitoring tipping points and strengthening institutional capacity to detect and implement transitions (Wilby & Murphy, 2019). An example could be the development of new reservoir regulation operating rules or identifying different water allocation quotas for withdrawals and river basin management, keeping pace with an evolving hydro-climatology, as in the case of the Great Lakes (Brown *et al.*, 2011).

CONCLUSIONS

CRIDA (Mendoza *et al.*, 2018) and its companion 'decision tree framework' document that is the basis of World Bank approaches to climate uncertainty (Ray & Brown, 2015), are complementary guidebooks that assist planning practitioners in deciding how to approach a wide range of climate uncertainties for different water resources planning and design functions. CRIDA provides a decision framework for addressing climate uncertainties within an agency's existing institutional guidelines. The 'decision tree framework' focuses more on a decision-scaling process coupled with analytical approaches that tie together 'bottom-up' and 'top-down' climate evaluation perspectives.

For most aspects of water resources planning and management, climate adaptation–type analyses should be conducted from a standard 'bottom-up' risk management perspective because scenario-driven climate models do not provide the specific ground-level information needed for traditional water resources project evaluation and decision-making procedures. In a typical water resources management study or project justification analysis, climate change is but one additional risk factor among a myriad array of social, economic, ecological and behavioral vulnerabilities.

For water resources planning studies, if the analyst determines that future risks due to climate uncertainties and/or analytical uncertainties are fairly well understood, climate stationary is still a useful concept as a starting point for stress tests and vulnerability analyses to make reliable predictions for engineering design (Stakhiv, 2011; Brown & Wilby, 2012; Slater *et al.*, 2021). Observations of past patterns and information are key elements for a successful prediction of hydrological processes (Lins & Cohn, 2011; Bayazit, 2015).

There are four important and major incongruities that water resources planners are asked to reconcile as part of their institutionally mandated project planning risk analysis approaches. First, estimating the likelihood/plausibility of climate scenarios. Second, linking these scenarios to non-stationary watershed hydrology, to quantify the reliability of future project service delivery and the risk of project failure. Third, applying the economic concepts of expectation requires calculations of expected annual damages (or benefits) that employ a selected probability distribution which best reflects a future uncertain climate state. Fourth, almost all benefit–cost decision protocols are based on discount rates and discounting, which further skews the analyses away from robust projects which can withstand more diverse climate uncertainties, by discounting risks and increased variability of distant future climate change, beyond 30–50 years. The sequential application of these largely incongruous steps as part of typical project justification procedures inherently skews the formulation of technically feasible and economically justifiable projects.

CRIDA and its reliance on 'decision-scaling' were devised as a practical way to address and bridge these risk incongruities, in a way that is compatible with existing institutionally acceptable water planning, evaluation and decision protocols. Decision-scaling is essentially a judgmental process that addresses several basic questions related to the degree of climate-related uncertainties that underlie a particular water resources problem. Decision-scaling helps to delineate and separate those aspects of analysis that require advanced methods that ought to be employed to deal with long-term climate uncertainties versus those that require additional analytical rigor using existing methods and analyses.

It turns out that except for mitigation-type policy analysis (prevention and reduction of greenhouse gases) and large-scale strategic river basin planning, the CRIDA approach shows that most existing planning protocols and analytical methods can be suitably modified to deal with hydrologic non-stationaries and climate uncertainties. Furthermore, most existing planning and design approaches are suitable for most water resource management issues because they already incorporate a number of built-in safety factors, redundancies and risk-avoidance measures that inherently are designed to accommodate most future uncertainties associated with climate change.

DATA AVAILABILITY STATEMENT

All relevant data are included in the paper or its Supplementary Information.

REFERENCES

AGWA (2020). *Alliance for Global Water Adaptation, Annual Report 2019–2020.* Available at: https://static1.squarespace.com/static/5e8397698c906c4df39838f5/t/5f3709c38abca4369fca906e/1597442518139/AGWA-Annual-Report-2019-2020.pdf.

Anagnostopoulos, A., Koutsoyiannis, D., Christofides, A., Efstratiadis, A. & Mamassis, N. (2010). A comparison of local and aggregated climate model outputs with observed data. *Hydrological Sciences Journal* 55(7), 1094–1110. doi:10.1080/02626667.2010.513518.

Asian Development Bank (2015). *Water-related Disasters and Disaster Risk Management in the People's Republic of China.* Mandaluyong City, Philippines. Available at: https://www.adb.org/sites/default/files/publication/177695/water-related-disasters-prc.pdf.

Asian Development Bank (2021). *A System-Wide Approach for Infrastructure Resilience. Technical Note.* Mandaluyong City, Philippines. Available at: https://gca.org/wp-content/uploads/2021/01/A-System-wide-Approach-for-Infrastructure-Resilience.pdf.

Ayyub, B. (2018). *Climate-Resilient Infrastructure Adaptive Design and Risk Management.* ASCE Committee on Adaptation to a Changing Climate. ASCE Manuals and Reports on Engineering Practice No. 140. doi:10.1061/9780784415191.

Barsugli, J., Vogel, J., Kaatz, l., Smith, J., Waage, M. & Anderson, C. (2012). Two faces of uncertainty: climate science and water utility planning methods. *Journal of Water Resources Planning and Management 138*(5), 389–395. doi:10.1061/(ASCE) WR.1943-5452.0000188.

Bayazit, M. (2015). Nonstationarity of hydrological records and recent trends in trend analysis: a state-of-the-art review. *Environmental Processes 2*, 527–542. doi:10.1007/s40710-015-0081-7.

Bhave, A., Conway, D., Dessai, S. & Stainforth, D. (2016). Barriers and opportunities for robust decision making approaches to support climate change adaptation in the developing world. *Climate Risk Management 14*, 1–10. doi:10.1016/j.crm.2016.09.004.

Borgomeo, E., Mortazavi-Naeini, M., Hall, J. & Guillod, B. (2018). Risk, robustness and water resources planning under uncertainty. *Earth's Future 6*(3), 468–487. doi:10.1002/2017EF000730.

Botterill, L. & Cockfield, G. (2013). *Drought, Risk Management, and Policy: Decision-Making Under Uncertainty.* Drought and Water Crises CRC Press, Taylor & Francis Group, Boca Raton, FL, USA, pp. 32742–33487.

Bourget, L. (2011). *Converging Waters: Integrating Collaborative Modelling with Participatory Processes to Make Water Resources Decisions. Maass-White Series.* Institute for Water Resources, U.S. Army Corps of Engineers, Alexandria, VA, USA.

Brown, C. & Wilby, R. (2012). An alternate approach to assessing climate risks. *EOS Transactions American Geophysical Union 93*(41), 401–402. doi:10.1029/2012EO410001.

Brown, C., Werick, W., Leger, W. & Fay, D. (2011). A decision-analytic approach to managing climate risks: application to the upper great lakes. *JAWRA Journal of the American Water Resources Association 47*(3), 524–534. doi:10.1111/j.1752-1688.2011.00552.x.

Brown, C., Ghile, Y., Laverty, M. & Li, K. (2012). Decision scaling: linking bottom-up vulnerability analysis with climate projections in the water sector. *Water Resources Research 48*(9). doi:10.1029/2011WR011212.

Conway, D., Nicholls, R., Brown, S., Tebboth, M., Adger, W., Ahmad, B., Biemans, H., Crick, F., Lutz, A., de Campos, R., Said, M., Singh, C., Zaroug, A., Ludi, E., New, M. & Wester, P. (2019). The need for bottom-up assessments of climate risks and adaptation in climate-sensitive regions. *Natre Climate Change 9*, 503–511. doi:10.1038/s41558-019-0502-0.

Daron, J. (2015). Challenges in using a robust decision making approach to guide climate change adaptation in South Africa. *Climatic Change 132*, 459–473. doi:10.1007/s10584-014-1242-9.

Dennig, F. (2018). Climate change and the re-evaluation of cost-benefit analysis. *Climatic Change 151*, 43–54. doi:10.1007/s10584-017-2047-4.

Dessai, S. & Hulme, M. (2004). Does climate policy need probabilities? *Climatic Policy 4*, 107–128.

Eisner, S., Florke, M., Chamorro, A., Daggupti, P., Donnelly, C., Huang, J., Hundecha, Y., Koch, H., Kalugin, A., Krylenko, I., Mishra, V., Piniewski, M., Samaniego, L., Seidou, O., Wallner, M. & Krysanova, V. (2017). An ensemble analysis of climate change impacts on streamflow seasonality across 11 large river basins. *Climatic Change 141*, 401–417. doi:10.1007/s10584-016-1844-5.

Fiering, M. (1967). *Streamflow Synthesis.* Harvard University Press, Cambridge, MA, USA. doi:10.4159/harvard.9780674189287.

Fiering, M. (1982). A screening model to quantify resilience. *Water Resources Research 18*(1), 27–32. doi:10.1029/WR018i001p00027.

Ghile, Y., Taner, M., Brown, C., Grijsen, J. & Talbi, A. (2014). Bottom-up climate risk assessment of infrastructure investment in the Niger River Basin. *Climatic Change 122*(1–2), doi:10.1007/s10584-013-1008.

Giuliani, M. & Castelletti, A. (2016). Is robustness really robust? How different definitions of robustness impact decision-making under climate change. *Climatic Change 135*(3–4), 409–424. doi:10.1007/s10584-015-1586-9.

Giuntoli, I., Vidal, J. -P., Prudhomme, C. & Hannah, D. M. (2015). Future hydrological extremes: the uncertainty from multiple global climate and global hydrological models. *Earth System Dynamics 6*, 267–285. doi:10.5194/esd-6-267-2015.

Groves, D. G., Bloom, E., Lempert, R. J., Fischbach, J. R., Nevills, J. & Goshi, B. (2014). Developing key indicators for adaptive water planning. *Journal of Water Resources Planning and Management*. doi:10.1061/(ASCE)WR.1943- 5452.0000471).

Haasnoot, M., Kwakkel, J., Walker, W. & ter Maat, J. (2013). Dynamic adaptive policy pathways: a method for crafting robust decisions for a deeply uncertain world. *Global Environmental Change 23*(2), 485–498.

Haimes, Y. (2009). *Risk Modelling, Assessment and Management*, 3rd edn. John Wiley and Sons. Available at: https://onlinelibrary.wiley.com/doi/pdf/10.1002/9780470422489.fmatter.

Haimes, Y., Li, D., Tulsiani, V., Lambert, J. & Krzysztofowicz, R. (1995). *Risk-Based Evaluation of Flood Warning and Preparedness Systems*. IWR Report 95-R-6, U.S. Army Institute for water Resources, Fort Belvoir, VA, USA. Available at: https://apps.dtic.mil/sti/pdfs/ADA524938.pdf.

Hallegatte, S., Shah, A., Lempert, R., Brown, C. & Gill, S. (2012). *Investment Decision Making Under Deep Uncertainty – Application to Climate Change*. World Bank Policy Research Working Paper 6193. World Bank, Washington, DC, USA.

Hashimoto, T., Loucks, D. P. & Stedinger, J. R. (1982a). Robustness of water resources systems. *Water Resources Research 18*(1), 21–26. doi:10.1029/WR018i001p00021.

Hashimoto, T., Loucks, D. P. & Stedinger, J. R. (1982b). Reliability, resiliency, and vulnerability criteria for water resource system performance evaluation. *Water Resources Research 18*(1), 14–20. doi:10.1029/WR018i001p00014.

Hattermann, F., Krysanova, S., Gosling, S., Dankers, R., Daggupati, P., Donnelly, C., Florke, M., Huang, S., Motovilov, Y., Buda, S., Yang, T., Muller, C., Leng, G., Tang, Q., Portmann, F., Hagemann, S., Gerten, D., Wada, Y., Masaki, Y., Alemayehu, T., Satoh, Y. & Samaniego, L. (2017). Cross-scale intercomparison of climate change impacts simulated by regional and global hydrological models in eleven large river basins. *Climatic Change 141*, 561–576. doi:10.1007/s10584-016-1829-4.

Hausfather, Z. & Peters, G. (2020). Emissions – the 'business as usual' story is misleading. *Nature 577*, 618–620.

Herman, J. D., Reed, P. M., Zeff, H. B. & Characklis, G. W. (2015). How should robustness be defined for water systems planning under change? *Journal of Water Resources Planning and Management 141*(10). doi:10.1061/(ASCE)WR.1943-5452.0000509.

Herman, J. D., Quinn, J. D., Steinschneider, S., Giuliani, M. & Fletcher, S. (2020). Climate adaptation as a control problem: review and perspectives on dynamic water resources planning under uncertainty. *Water Resources Research 56*, e24389. doi:10.1029/2019WR025502.

Institute for Water Resources (2010). *Exploration of Tolerable Risk Guidelines for the USACE Levee Safety Program*. Proceeding of Workshop, U.S. Army Corps of Engineers, IWR Report 10-R-8. Available at: https://www.iwr.usace.army.mil/Portals/70/docs/iwrreports/10-R-8.pdf.

Intergovernmental Panel on Climate Change (IPCC) (1990). *Climate Change: The IPCC Scientific Assessment (1990)*. Report prepared for IPCC by Working Group. Edited by I. J. T. Houghton, G. J. Jenkins & J. J. Ephraums. Cambridge University Press, Cambridge, UK, New York, NY, USA and Melbourne, Australia, 410 p.

IPCC (2021). *Climate Change 2021: The Physical Science Basis. Contribution of Working Group I to the Sixth Assessment Report of the Intergovernmental Panel on Climate Change*. Edited by Masson-Delmotte V., P. Zhai, A. Pirani, S. L. Connors, C. Péan, S. Berger, N. Caud, Y. Chen, L. Goldfarb, M. I. Gomis, M. Huang, K. Leitzell, E. Lonnoy, J. B. R. Matthews, T. K. Maycock, T. Waterfield, O. Yelekçi, R. Yu & B. Zhou. Cambridge University Press, Cambridge, UK.

Kettner, A., Cohen, S., Overeem, I., Fekete, B., Brakenridge, G. & Syvitski, J. (2018). Estimating change in flooding for the 21st century under a conservative RCP forcing: A global hydrological modeling assessment. In: Schumann, G., Bates, P., Apel, H. & Aronica, G. (eds). *Global Flood Hazard: Applications in Modeling, Mapping and Forecasting*. AGU Geophysical Monograph Series. https://doi.org/10.1002/9781119217886.ch9.

Kjeldsen, T. R. & Rosbjerg, D. (2005). Choice of reliability, resilience, and vulnerability estimators for risk assessments of water resources systems. *Hydrological Sciences Journal 49*(5), 755–767. doi:10.1623/hysj.49.5.755.55136.

Knutti, R. & Sedláček, J. (2013). Robustness and uncertainties in the new CMIP5 climate model projections. *Nature Climate Change 3*(4), 369–373. doi:10.1038/nclimate1716.

Koutsoyiannis, D., Efstratiadias, A., Mamassis, N. & Christofides, A. (2008). On the credibility of climate predictions. *Hydrological Sciences Journal 53*(4), 671–684. doi: 10.1623/hysj.53.4.671.

Kroll, C., Crouteau, K. & Vogel, R. (2015). Hypothesis tests of hydrologic alterations. *Journal of Hydrology 530*, 117–126. doi:10.1016/j.jhydrol.2015.09.057.

Krysanova, V., Vetter, T., Eisner, S., Huang, S., Pechlivanidis, I., Strauch, M., Gelfan, A., Kumar, R., Aich, V., Arheimer, B., Chamorro, A., van Griensven, A., Kundu, D., Lobnova, A., Mishra, V., Plotner, S., Reinhardt, J., Seidou, O., Wang, X., Wortmann, M., Zeng, X. & Hattermann, F. (2017). Intercomparison of regional-scale hydrological models and climate change impacts projected for 12 large river basins worldwide – a synthesis. *Environmental Research Letters 12*(10), 105002.

Kundzewicz, Z. & Stakhiv, E. (2010). Are climate models 'ready for prime time' in water resources applications, or is more research needed? *Hydrological Sciences Journal 55*(7), 1085–1089. doi:10.1080/02626667.2010.513211.

Kwakkel, J., Haasnoot, M. & Walker, W. (2015). Developing dynamic adaptive policy pathways: a computer-assisted approach for developing adaptive strategies for a deeply uncertain world. *Climatic Change 132*(3), 373–386. doi:10.1007/s10584-014-1210-4.

Lehner, F., Deser, C., Maher, N., Marotzke, J., Fischer, E. M., Brunner, L., Knutti, R. & Hawkins, E. (2020). Partitioning climate projection uncertainty with multiple large ensembles and CMIP5/6. *Earth System Dynamics 11*, 491–508. doi:10.5194/esd-11-491-2020.

Lempert, R. (2019). Robust decision making (RDM). In: *Decision Making Under Deep Uncertainty: From Theory to Practice*. Marchau, V. A. W. J., Walker, W. E., Bloemen, P. J. T. M. & Popper, S. W. (eds.). Springer Nature Switzerland AG. doi:10.1007/978-3-030-05252-2.

Lempert, R. & Groves, D. (2010). Identifying and evaluating robust adaptive policy responses to climate change for water management agencies in the American west. *Technological Forecasting and Social Change 77*(6), 960–974. doi:10.1016/j.techfore.2010.04.007.

Lins, H. & Cohn, T. (2011). Stationarity: wanted dead or alive? *JAWRA Journal of the American Water Resources Association 47*(3), 475–480.

Maass, A., Hufschmidt, M., Dorfman, R., Thomas, H., Marglin, S. & Fair, G. (1962). *Design of Water-Resource Systems: New Techniques for Relating Economic Objectives, Engineering Analysis, and Governmental Planning*. Harvard University Press, Cambridge, MA, USA. doi:10.4159/harvard.9780674421042.

Matalas, N. & Fiering, M. (1977). Water resources systems planning. In: *Climate, Climatic Change, and Water Supply (Chap. 6)*. National Research Council, The National Academies Press, Washington, DC, USA. doi:10.17226/185.

Mekong River Commission. (2018). *Basin-Wide Assessment of Climate Change Impacts on Hydropower Production*. Final report. Available at: https://www.mrcmekong.org/assets/Publications/Basin-wide-Assessment-of-Climate-Change-Impacts-on-Hydropower-Production_report-13May19.pdf.

Mendoza, G., Jeuken, A., Matthews, J., Stakhiv, E., Kucharski, J. & Gilroy, K. (2018). *Climate Risk Informed Decision Analysis (CRIDA), Collaborative Water Resources Planning for an Uncertain Future*. UNESCO and ICIWaRM, Paris, France, p. 162. Available at: https://www.agwaguide.org/docs/CRIDA_Nov_2018.pdf.

Milly, P., Betancourt, J., Falkenmark, M., Hirsch, R., Kundzewicz, Z. & Lettenmaier, D. P. (2008). Stationarity is dead: whither water management? *Science 319*(5863), 573–574. doi:10.1126/science.1151915.

National Research Council (2000). *Risk Analysis and Uncertainty in Flood Damage Reduction Studies*. National Academy Press, Washington, DC, USA.

National Research Council (2009). *Informing Decisions in A Changing Climate. Panel on Strategies and Methods for Climate-Related Decision Support*. Committee on the Human Dimensions of Climate Change, Division of Behavioral and Social Sciences and Education, Washington, DC, USA.

Nature Editorial (2019). The CMIP6 landscape. *Nature Climate Change 9*, 727. doi:10.1038/s41558-019-0599-1.

Nazemi, A. & Wheaton, H. (2014). Assessing the vulnerability of water supply to changing streamflow conditions. *EOS 95*(32), 288–288. doi:10.1002/2014EO320007.

O'Neill, B., Kriegler, E., Riahi, K., Ebi, K., Hallegatte, S., Carter, T., Mathur, R. & van Vuuren, D. (2014). A new scenario framework for climate change research: the concept of shared socioeconomic pathways. *Climatic Change 122*, 387–400. doi:10.1007/s10584-013-0905-2.

Palmer, R., Cardwell, H., Lorie, M. & Werick, W. (2013). Disciplined planning, structured participation, and collaborative modeling – applying shared vision planning to water resources. *JAWRA 49*(3), 614–628. doi:10.1111/jawr.12067.

Pechlivanidis, I., Arheimer, B., Donnelly, C., Hundecha, Y., Huang, S., Aich, V., Samaniego, L., Eisner, S. & Shi, P. (2017). Analysis of hydrological extremes at different hydro-climatic regimes under present and future conditions. *Climatic Change 141*(3), 467–481. doi:10.1007/s10584-016-1723-0.

Pielke, R. & Wilby, R. (2012). Regional climate downscaling: what's the point? *EOS Transactions of the American Geophysical Union 93*(5), 52–53. doi:10.1029/2012EO050008.

Pindyck, R. (2010). *Fat Tails, Thin Tails, and Climate Change Policy*. National Bureau of Economic Research. NBER Working Paper No. 16353. Available at: http://www.nber.org/papers/w16353.

Ray, P. & Brown, C. (2015). *Confronting Climate Uncertainty in Water Resources Planning and Project Design: The Decision Tree Framework*. World Bank, Washington, DC, USA. Available at: https://openknowledge.worldbank.org/handle/10986/22544.

Read, L. & Vogel, R. (2015). Reliability, return periods, and risk under nonstationarity. *Water Resoures Research 51*, 6381–6398.

Rijkswaterstaat (2016). *Implementing Risk Based Flood Defence Standards.* Ministry of Infrastructure and Environment, Netherlands, p. 83.

Ritchie, J. & Dowlatabadi, H. (2017). The 1,000 GtC coal question: are cases of vastly expanded future coal combustion still plausible? *Energy Economics.* doi:10.1016/j.eneco.2017.04.01.

Rogers, P. (1997). Engineering design and uncertainties related to climate change. *Climatic Change 37*, 229–242.

Salas, J. & Obeysekera, J. (2014). Revisiting the concepts of return period and risk for nonstationary hydrologic extreme events. *Journal of Hydraulic Engineering 19*(3), 554–568.

Salas, J., Rajagopalan, B., Saito, L. & Brown, C. (2012). Special section on climate change and water resources: climate nonstationarity and water resources management. *Journal of Water Resour Planning and Management 138*, 385–388.

Salas, J., Obeysekera, J. & Vogel, R. (2018). Techniques for assessing water infrastructure for nonstationary extreme events: a review. *Hydrological Sciences Journal 63*, 325–352.

Serago, J. & Vogel, R. (2018). Parsimonious nonstationary flood frequency analysis. *Advances in Water Resources 112*, 1–16. doi:10.1016/j.advwatres.2017.11.026.

Shortridge, J. & Zaitchik, B. (2018). Characterizing climate change risks by linking robust decision frameworks and uncertain probabilistic projections. *Climatic Change 151*(3), 525–539. doi:10.1007/s10584-018-2324-x.

Shortridge, J., Guikema, S. & Zaitchik, B. (2017). Robust decision making in data scarce contexts: addressing data and model limitations for infrastructure planning under transient climate change. *Climatic Change 140*, 323–337. doi:10.1007/s10584-016-1845-4.

Slater, L., Anderson, B., Buechel, M., Dadson, S., Han, S., Harrigan, S., Kelder, T., Kowal, K., Lees, T., Matthews, T., Murphy, C. & Wilby, R. (2021). Nonstationary weather and water extremes: a review of methods for their detection, attribution, and management. *Hydrology and Earth System Sciences 25*, 3897–3935. doi:10.5194/hess-25-3897-2021.

Stakhiv, E. (1986). Achieving social and environmental objectives in water resources planning. In: *Social and Environmental Objectives in Water Resources Planning and Management.* Viessman Jr., W. & Schilling, K. E. (eds.). ASCE, New York, NY, USA, pp. 107–125.

Stakhiv, E. (2011). Pragmatic approaches for water management under climate change uncertainty. *Journal of American Water Resources Association (JAWRA) 47*(6), 1183–1196. doi:10.1111/j.1752-1688.2011.00589.x.

Thompson, S. S., Harman C, M., Srinivasan, V., Hipsey, M., Reed, P., Montanari, A. & Blöschl, E. (2013). Developing predictive insight into changing water systems: use-inspired hydrologic science for the anthropocene. *Hydrology and Earth System Sciences 17*, 5013–5039.

Tol, R. (2003). Is the uncertainty about climate change too large for expected cost-benefit analysis? *Climatic Change 56*(3), 265–289. doi:10.1023/A:1021753906949.

U.S. Army Corps of Engineers (2000). *Planning Guidance Notebook.* ER 1105-2-100. Washington, DC. Available at: https://www.publications.usace.army.mil/Portals/76/Publications/EngineerRegulations/ER_1105-2-100.pdf?ver = OjSERNnMS4-6VFwmhg5kEA%3d%3d.

U.S. Army Corps of Engineers (2019). *Risk Assessment for Flood Risk Management Studies.* ER 1105-2-101. Washington, DC. Available at: https://www.publications.usace.army.mil/Portals/76/Users/182/86/2486/ER%201105-2-101_Clean.pdf.

U.S. Water Resources Council (1983). *Economic and Environmental Principles and Guidelines for Water and Related Land Resources Planning Implementation Studies.* U.S. Government Printing Office, Washington, DC, USA.

Van Alphen, J. (2015). The delta programme and updated flood risk management policies in the Netherlands. *Journal of Flood Risk Management 9*(4), 310–319. doi: 10.1111/jfr3.12183.

Walker, W. E., Haasnoot, M. & Kwakkel, J. H. (2013). Adapt or perish: a review of planning approaches for adaptation under deep uncertainty. *Sustainability 5*(3), 955–979. doi:10.3390/su5030955.

Watkiss, P., Wilby, R. & Rodgers, C. (2020). *Principles of Climate Risk Management for Climate Proofing Projects.* ADB Sustainable Development Working Paper Series No. 69. Manila, Phillipines, p. 34. doi:10.22617/WPS200203-2.

Werick, W. & Palmer, R. (2008). It's time for standards of practice in water resources planning. *Journal of Water Resources Planning and Management 134*(1), 1–2.

Wilby, R. L. & Dessai, S. (2010). Robust adaptation to climate change. *Weather 65*, 180–185.

Wilby, R. & Murphy, C. (2019). Decision-making by water managers despite climate uncertainty. In: W. Pfeffer, J. Smith & K. Ebi (eds). *The Oxford Handbook of Planning for Climate Change Hazards.* Oxford University Press, Oxford.

World Bank (2010). *Cost-Benefit Analysis in World Bank Projects.* Independent Evaluation Group (IEG). World Bank, Washington, DC, USA. Available at: https://ieg.worldbankgroup.org/evaluations/cost-benefit-analysis-world-bank-projects.

World Bank (2020). *Resilient Water Infrastructure Design Brief.* World Bank, Washington, DC, USA. Available at: https://openknowledge.worldbank.org/handle/10986/34448.

World Bank (2021). *Hydrological Risk. Good Practice Note on Dam Safety; Technical Note 1.* World Bank, Washington, DC, USA. Available at: https://openknowledge.worldbank.org/handle/10986/35488.

Wright, D. B. (2015). *Methods in Flood Hazard and Risk Assessment.* World Bank, Washington, DC, USA. Available at: https://openknowledge.worldbank.org/handle/10986/22982.

Yoe, C. (2017). *Principles of Risk Analysis for Water Resources.* U.S. Army Institute for Water Resources, Technical Report TR-2017-R-01, Fort Belvoir, Alexandria, VA, USA.

First received 1 November 2021; accepted in revised form 1 November 2021. Available online 18 November 2021

Water Infrastructure Planning, MDCU doi: 10.2166/wp.2021.287

Evolution of Japan's flood control planning and policy in response to climate change risks and social changes

Toshio Koike

International Centre for Water Hazard and Risk Management (ICHARM), Public Works Research Institute (PWRI), 1-6, Minamihara, Tsukuba 305-8516, Japan
E-mail: koike@icharm.org

ABSTRACT

Building a sustainable society by strengthening disaster resilience is a common goal in the world. It is crucial to promote cooperation between the general public and the science community by sharing data, information, knowledge, experiences, and ideas. Japan has routinely been beset by catastrophic floods caused mainly by destructive typhoons and critically active seasonal fronts. With the turn of the 21st century, changes in climate and society required additional realignment into the standardized procedures that had evolved over the previous half-century. Japan's new policy, 'River Basin Disaster Resilience and Sustainability by All,' takes comprehensive measures, mainly consisting of flood prevention, exposure reduction, and appropriate evacuation, response and recovery, aiming to strengthen disaster resilience and achieve sustainability through concerted efforts among all stakeholders. The policy can play a key role in the achievement of the common global goal.

Key words: Climate change, Disaster resilience, Disaster risk reduction, Flood control planning, Flood control policy, Sustainability

HIGHLIGHTS

- New flood control planning considering climate change is identified in comparison with the conventional planning.
- A new flood control policy considering changes in climate and society is identified in comparison with the conventional policy.
- The roles of science and technology in the new approach are clarified.

INTRODUCTION

Flood control has been a key issue in Japan's policy history due to its unique climatological, meteorological, and hydrological conditions. Japan experienced severe flood damage particularly in a period immediately after World War II. Serious flood disasters claiming more than 1,000 deaths occurred almost every year (Takahasi, 2009). After each disaster, new approaches, methods, and design standards were introduced, including improvements to methods for flood frequency analysis, changes in economic assessment procedures, and various structural measures such as dams, embankment, erosion and sediment control, sewage systems, as well as high-level embankment and comprehensive flood control. As a result, the number of victims and flooded areas decreased drastically in the rapid economic growth period during the 1950 and 1960s.

However, recent climate changes, causing more frequent and severe torrential rainfall events, have increased human losses and economic damage. In addition, Japan's society becomes increasingly vulnerable, as it is reflected in the decreasing ratio of the number of persons offering support to the number of persons needing support in the time of disaster. While such negative changes are becoming apparent, climate studies have been making remarkable progress thanks to high-performance computing and big data science and providing more compelling scientific evidence for improving flood management plans, especially by revising design rainfall, which is statistically calculated using observed data. Furthermore, starting with the adoption of 'The Sendai Framework for Disaster Risk Reduction 2015–2030' at the third United Nations World Conference on DRR in March 2015, the international community agreed on the 'Transforming Our World: the 2030 Agenda for Sustainable Development' in September 2015, and 'The Paris Agreement on Climate Change' in December 2015.

This paper introduces the recent evolution of Japan's flood control planning and policy, focusing on its historical background, the awareness of and response to climate change, recent societal changes, and the roles of science and technology in this evolution.

Evolution of science, technology, and policies to address severe floods after World War II

From the Makurazaki typhoon disaster in September 1945, right after World War II, to the Ise Bay typhoon disaster in September 1959, four-digit casualties were recorded every year. In response, the most advanced science and technology at that time, i.e., the statistical analysis of rainfall observation data and the development of runoff models to rationally calculate flood discharge from rainfall, was introduced into river planning. New approaches were applied to improvement plans for such rivers as Sendaigawa, Shirakawa, and Yodogawa, as well as some others (Nakamura, 2014).

The knowledge and experience from those cases were compiled to draft the Technical Standards for River Engineering and Erosion Control in 1958 (Nakayasu, 1960) for the following three purposes:

1. Clarification of the developmental stages of technology and the presentation of the current technological level;
2. Standardization of the technological level throughout the country; and
3. Integration of technology and management.

The technical standards were published to establish a methodology to rationally promote investment in flood control under a limited national budget, and the publication was a manifestation of the river engineering policy geared toward strengthening preparedness in consideration of national balance while addressing recurring severe flood disasters (Nakamura, 2014).

According to Wakigawa (2007), the basic attitude toward investment in river improvement was also formed during those years. For example, Nakayasu proposed an indicator for economic assessment, arguing that the rate of benefit from disaster prevention (annual benefit rate) must exceed the cost (B/C ratio greater than 1.0). He determined the expected value by calculating the annual mean benefit, which can be produced as a result of construction aimed at the discharge of a certain annual exceedance probability, and obtained a B/C by dividing the expected value by the total construction cost. In 1961, an investigation manual was developed, which became the prototype for the current manual for statistical investigation of flood damage. After that, flood-damage field investigations and flood-damage data investigations were conducted to determine the actual costs of flood damage. Those efforts finally came to fruition as the current manual for statistical investigation of flood damage in 1970. The government also explored ways to investigate the economic impact of flood control projects. It published the guidelines and methods for investigating the economic impact of flood control projects in 1961 and the manual for investigating assets in flood-affected areas in 1962.

In parallel with the establishment of these scientific and technological bases, the necessity of long-term investment in flood control, instead of projects on a single fiscal year budget, was discussed in the wake of a devastating flood in western Japan in 1953. A long-term investment plan, though revised repeatedly, was rejected three times due to opposition from the Ministry of Finance at that time, but it was finally approved by the then cabinet after the Kanogawa typhoon disaster in September 1958 and enacted in 1960 as the Act on Emergency Measures concerning Flood and Erosion Control after the Ise Bay typhoon disaster in 1959. Since then, a long-term flood control project was planned with financial backing and carried out for nine project periods until terminated by the cabinet decision on fiscal structural reform in 2002.

Awareness of and response to climate change

In January 2015, the government devised a new disaster management policy, 'Approaches to disaster prevention and mitigation in response to new stages' (MLIT, 2015a). According to this policy, the Flood Fighting Act was revised in May 2015 with a focus on protecting lives and avoiding catastrophic damage to the country's socio-economy. The revised law stipulates that measures should be designed and implemented to protect the lives of people in preparation for the largest-scale river and inland floods and storm surges. In July 2015, a calculation method was published for estimating the largest scale of a hazard.

However, some concerns were raised for the nationwide application of this approach. The general agreement back then was that research was still halfway to understanding the impact of global warming on natural events of low frequency, such as hazards (river and inland floods) of the calculated largest scale. It was still too early to uniformly use the results of climate change impact assessment in order to estimate the largest scale of a hazard possibly affecting the entire country. Instead, an alternative approach was adopted and has since been in practice. The idea is to divide Japan into 15 climate zones and use the relationship between the rainfall duration and the area obtained from the largest rainfall data observed in each climate zone (MLIT, 2015b). This approach is based on the hypothesis that rainfall of a similar scale can occur at any location even if the location in question is outside the target river basin as long as it is in the climate zone containing the target basin. This is a new attempt to initiate actions to protect the lives of people using not only data observed in the target river basin but also data on extreme events that may occur under the same climate conditions. This can be seen as a step forward from statistical disaster management advocated in the 1958 draft of Technical Standards for River Engineering and Erosion Control.

Immediately after that, a heavy rain disaster occurred in the Kanto and Tohoku regions in September 2015 and in the Hokkaido and Tohoku regions in August 2016. These disasters, having occurred in the areas along rivers managed under national and prefectural supervision, revealed issues related to the issuance of evacuation information, delays in evacuating residents, and catastrophic damage to the local economy caused by extremely severe floods. To address these issues, the national government developed a policy of 'Rebuilding a Risk Conscious and Well-Prepared Society against Water-Related Disasters' (MLIT, 2015c, 2017). In May 2017, the Flood Fighting Act was revised again to mandate that a large-scale flood mitigation council be established in consideration of its area coverage and administrative boundaries. The revision also requires the managers of facilities for people in need of special attention to devise evacuation plans and conduct evacuation drills. Furthermore, the revised act now allows the national government to act on behalf of prefectural governments in carrying out disaster recovery projects and dam redevelopment projects planned under the prefectural jurisdiction.

It was only 2 weeks after the revised law went into effect that Asakura City in Fukuoka Prefecture was hit by a heavy rainfall of 169 mm/h, very close to the maximum hourly rainfall record of 187 mm/h in Japan. A similar severe rainfall event occurred in northern Kyushu in 2012, but Asakura City had created a disaster prevention map, designated local evacuation centers, and conducted evacuation drills. Moreover, evacuation preparation information, advisories and orders were issued as appropriate on the day of the extreme rainfall. Nonetheless, the city suffered severe damage again, with 35 people dead or missing. Consequently, Asakura City was the first case in which the national government took the place of the prefecture in disaster recovery, which was newly stipulated in the revised Flood Fighting Act.

In the 2018 flood disaster in western Japan, the summertime seasonal front was active and remained over the area for an unusually long period; the 48-h rainfall reached a record high at 124 observation points of the Automated Meteorological Data Acquisition System (AMeDAS) throughout western Japan. In the Typhoon No.19 flood disaster, as the typhoon approached Japan while keeping its strength, eastern Japan had massive amounts of rainfall with 120 locations experiencing record 12-h rainfall. In western and eastern Japan together, the record rainfalls were rewritten at about 10% of the 1,300 AMeDAS observation points within only 2 years. The Japan Meteorological Agency (JMA) identified the 2018 July Heavy Rain as an individual event influenced by climate change for the first time, saying, 'Global warming is a contributing factor in this event'.

These severe rainfall events caused massive damage to many areas throughout Japan. In the 2018 flood disaster over western Japan, 245 people were left dead or missing, mainly in the three prefectures of Hiroshima, Okayama, and Ehime. In the 2019 flood disasters due to Typhoon No.19 and the heavy rainfall event in Chiba Prefecture that followed the typhoon, 107 people were either dead or missing. Those two consecutive years saw three-digit casualties for the first time after 1982 and 1983. A closer look at the disasters in 2018 and 2019 finds that sediment-related events occurred at 2,581 locations in the western Japan disaster while levees breached at 142 locations in the Typhoon No.19 disaster. These numbers characterize the nature of each disaster. However, also in the western Japan disaster, levee breaches were found at 27 locations, causing violent inundation in Mabi Town of Kurashiki City in Okayama Prefecture. Likewise, sediment-related events, including landslides and debris flows, also occurred during the Typhoon No.19 disaster at 952 locations, including ones in Marumori Town of Igu County in Miyagi Prefecture. These numbers seem smaller relative to those experienced in much earlier years but clearly indicate that damage was enormous and unheard of in recent years.

Changes in society

Among the confirmed deaths, the elderly aged 65 or older account for 60% in the western Japan disaster in 2018 and 65% in the Typhoon No.19 disaster in 2019. The common feature found among the victims in the recent floods is that most of them were elderly. The ratio of the productive population (15–64 years old) to the aged population (over 65) was 3.9 in 2000 and 2.3 in 2015, and is projected to be 1.4 in 2065 in Japan. These changes and projections suggest that the ratio of people in need of help will increase while that of people who can provide help will decrease.

In the case of the Kinugawa flood disaster in 2015, river administrators kept providing the municipalities at imminent flood risk with flood-related information and informing them of the need for evacuation, but the evacuation order was issued only to residents in limited areas before the levee breach. Many of the residents stayed home instead of evacuating, while watching live coverage of the flooded Kinugawa River, even as it started overflowing at some locations. Then, several thousands of people were stranded due to violent, muddy floodwaters; about 1,300 residents were rescued by helicopter and 3,000 by the ground rescue crew. Furthermore, several nursing-home residents fell victim to a flood when it hit their facility, leaving little time for safe evacuation. Such a situation has been seen more frequently in recent cases. The Kinugawa flood disaster should be viewed as an important lesson that raises an alarm over changes in hazards and vulnerability.

In 2020, the national river-improvement economic investigation manual forecasted significant increases in the number of days for businesses to make a full recovery from 20.6 to 35.6 days for cases with inundation depths of 100–200 cm and 33.6–64.0 days for cases with inundation depths of 200–300 cm. These figures may be another sign suggesting that Japan's society has become more easily affected by hazards.

These analyses indicate that Japan is probably becoming more vulnerable to disasters while having to cope with changes in the intensity of hazards due to climate change. It is essential to take proactive measures to reduce the impact to which such a society may be exposed.

Science and technology for reforming river improvement planning

At the time when methods to calculate the largest scale of a hazard were under discussion, it was generally agreed that it was still too early to uniformly use the results of climate change impact assessment. However, science and technology has become able to provide a solid foundation for planning infrastructure that can cope with climate changes, thanks to collaboration among different research fields of climate change ensemble forecasting and dynamic downscaling using high-performance computing and data integration systems.

The Program for Risk Information on Climate Change (SOUSEI), led by the Ministry of Education, Culture, Sports, Science and Technology of Japan (MEXT), has developed a 'database for policy decision making for future climate change (d4PDF)' (MEXT-SOUSEI, 2019). This is a set of highly accurate model-experiment outputs using high-resolution atmospheric models with a mesh of 60 and 20 km covering the world and the surrounding area of Japan, respectively. Ensemble forecasting outputs for 6,000 years (3,000 years for Japan and its surrounding area) are available for the past climate. Outputs for the future climate have been calculated for 3,240 and 5,400 years, based on the Representative Concentration Pathway (RCP) 2.6 (equivalent to a 2 °C rise) and RCP 8.5 (equivalent to a 4 °C rise), respectively.

Hokkaido University and the Japan Agency for Marine-Earth Science and Technology (JAMSTEC) have created a rainfall dataset by dynamic downscaling the 20 km-mesh RCP 8.5 outputs of d4PDF for Japan and its surrounding area to 5 km-mesh outputs, which are archived in DIAS. This dataset has made it possible to compare the probabilities of extreme events under the present and future climate conditions. The Japanese government recognizes extreme events as one of the principal factors to be addressed in planning river flood control.

Reform directions of flood control policies in response to climate change risks and social changes

Considering the changing climate and society as mentioned above, as much effort as possible should be made for recovery and rehabilitation from recurring severe flood disasters. Still, equally important is to present a basic policy that shows the necessity of investment in flood control for further growth of Japan. Such a policy is the key factor to promote the implementation of essential proactive measures.

The first goal that should be presented in such a policy is the improvement of *disaster resilience*. If proactive measures are taken in planning land use, housing and buildings and evacuation drills, damage should be reduced more effectively by absorbing the impact of a powerful hazard. Moreover, it should be possible to reduce the time-integrated value (disaster risk), which indicates the degree of the loss of social functions, by designing a system to speed up recovery and even achieve 'Build Back Better' by solving other existing issues at once.

The second goal is the introduction of the concept of *sustainability*. Water-related disasters are serious threats to critical infrastructure (public services such as energy, communications, transportation, and water supply), and can cause devastating social and industrial consequences, significantly disrupting sustainability. In the case of Japan, sustainability could be interpreted as maintenance and improvement. For example, comprehensive sediment management has been practiced by monitoring the entire area where sediment transfer takes place from mountains to beaches. Daily maintenance and various conventional flood fighting activities are keys for sustainability of levees. Efforts are also needed to improve disaster prevention capacity by effectively using water-storage facilities available for continuous use.

The third goal is to achieve higher *stakeholder inclusiveness* and community participation, which are essential to enhance disaster resilience and ensure sustainability. In the Typhoon No.19 flood disaster, 142 levees breached, out of which 128 were managed by prefectural governments. In addition, when breaches occurred at the boundaries between the national and prefectural management sections, many of them were on the prefectural side. This fact clearly shows that more comprehensive flood control is urgently needed. For more effective flood control, river improvement policies should collaborate closely with agricultural and other industrial promotion policies and environmental policies, let alone with social infrastructure development policies for urbanization, housing, and transportation.

The three goals of resilience, sustainability and inclusiveness are stated in Japan's Development Cooperation Charter (Ministry of Foreign Affairs, 2015), which was revised in 2015, and the concept of 'Quality Growth' has been announced to the world as the ultimate theme integrating the three goals. In addition to the development of social infrastructure that supported the country's rapid economic growth, it is about time for Japan itself to shift its course toward quality growth in order to cope with increasingly intense hazards and rapidly vulnerable society.

New flood control policies of Japan in response to climate change risks and social changes

The Technical Working Group on Flood Management Planning and the Policy Working Group on Water-related Disaster Risk Reduction Considering Climate Change were established in April 2018 and October 2019, respectively, under the River Subcommittee of the National Land Development Council of Japan to revise flood control planning and develop a new flood management policy.

The Technical Working Group identified RCP 2.6 as the primary scenario for future flood control planning, taking into account that the Paris Agreement in 2015 sets a goal of limiting global warming to 2 °C, or even 1.5 °C, above pre-industrial levels and that the current levels of flood control safety of many rivers are lower than the target levels. On the other hand, it was considered appropriate to use RCP 8.5 as additional references for planning other projects and activities, for example, to check lists of improvement in flood control plans, conduct risk assessment to take disaster mitigation measures, study the use of river management structures for crisis management purposes, and design structures considering future remodeling. The Technical Working Group has also decided to use the outputs of 1951–2010 to define the present climate condition since the current flood control plans are mainly based on post-war data. The future climate condition is, on the other hand, defined using the 2040 outputs since RCP 2.6-based projections indicate that temperature increases after 2040 will stay more or less at the same level.

The Technical Working Group decided to divide the country into 15 climate zones in the same way employed to estimate the largest scale of a hazard. The probabilities of heavy rain events have been compared between the present and future climate conditions, and the rates of change, which are used to multiply the design rainfall statistically derived from the observed data up to 2010, have been calculated to be 1.15 for Hokkaido and 1.1 for the other regions of the country. In determining these rates of change, the following physical properties are taken into account in addition to the numerical comparison of climate change projection data:

1. The rate of increase in temperature rise in higher latitudes.
2. The rate of increase in saturated water vapor pressure in relation to temperature change is greater in the mean summer-temperature zone of Hokkaido than in the other areas of Japan.
3. The rate of increase in sea surface temperature is larger around Hokkaido in any of the various sea surface temperature patterns used in ensemble calculations.

These results were compiled as draft policy recommendations in July 2019, published tentatively in October of the same year, and finalized in April 2021 with the title of 'Policy recommendations on flood control planning under climate change' (MLIT, 2021).

Considering the impacts of climate change, social trends, and science and technology innovation, the Policy Working Group has developed a new flood management policy, 'River Basin Disaster Resilience and Sustainability by All'. Defining a river basin as space consisting of a watershed and floodplain areas, the policy takes

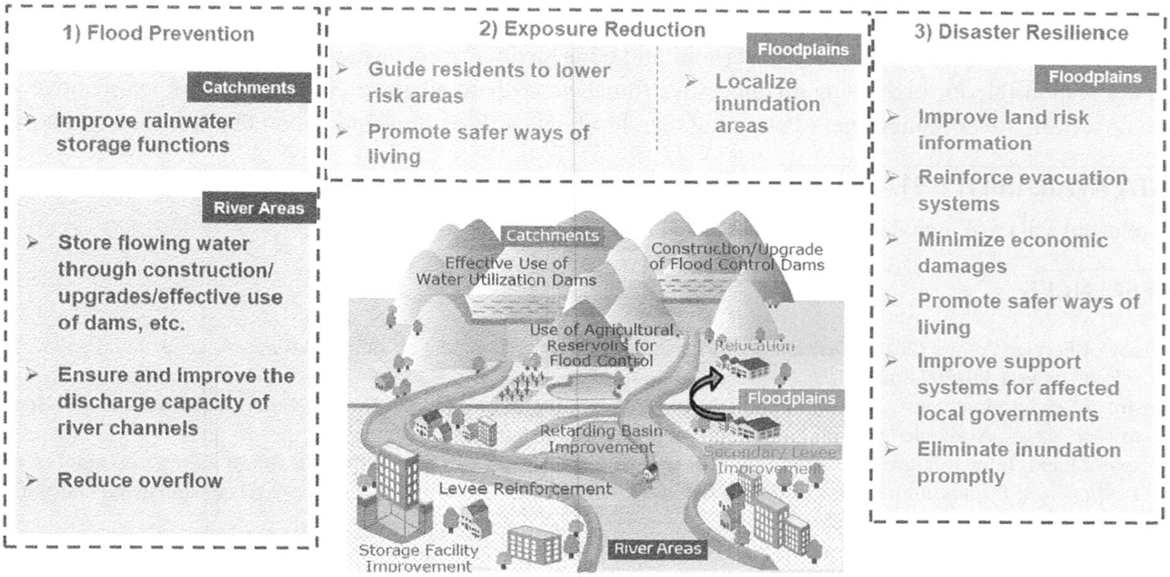

Fig. 1 | Overall structure of 'River Basin Disaster Resilience and Sustainability by All' (MLIT, 2020).

comprehensive, multilayered actions for three main purposes: flood prevention, exposure reduction, and disaster resilience. It calls for all stakeholders in river basins, including the national government, prefectures, municipalities, private enterprises, residents, and water users, to take actions for disaster resilience and sustainability. The Policy Working Group has recommended to the MLIT that it shift from its conventional flood control measures implemented mainly in river areas and floodplains to measures targeted for not only rivers and floodplains but also other areas in the basin. Cooperation with all stakeholders is also recommended toward strengthening disaster resilience and enabling sustainable development. The overall structure of possible measures employed for the new policy is schematically shown in Figure 1 (MLIT, 2020).

After receiving the final council report in July 2020, the Minister of Land, Infrastructure, Transport and Tourism developed a project on River Basin Disaster Resilience and Sustainability by All in 109 basins whose rivers are managed by the national government. The cabinet approved a special budget, about US$140 billion, for the next 5 years. Both the House of Representatives and the House of Councilors of the Japanese Diet unanimously passed the bill for the revision of the nine acts related to the new policy and the amendment of the nine related acts.

CONCLUSION

In Japan, residents in a community traditionally cooperated for water-related disaster risk reduction through various efforts such as improving embankments and building ring dikes or secondary dikes for community protection, flood proofing, and temporary evacuation and participating in flood fighting during flooding. After World War II, measures to improve the safety levels of river and sewage facilities were mainly carried out by administrators. For Japan to cope with an increasing number of hazards intensified by climate change and rapidly vulnerable society, however, there is no room for doubt that additional responses that are both resilient and inclusive are absolutely necessary, as well as continued investment in disaster risk reduction and prevention.

On reconsidering the changing hazards and society, these new challenges can be recognized as an opportunity for sustainable development. Japan's River Law states that a river administrator shall have the authority and duty for river management. Class A rivers are administrated by the Minister of Land, Infrastructure, Transport and Tourism, Class B rivers by prefectural governors, and other smaller rivers by municipal mayors. River administrators are responsible for organizing a cooperative framework where all stakeholders can work cooperatively to develop a Basic River Management Policy and then formulate a River Implementation Plan for each target river.

DATA AVAILABILITY STATEMENT

All relevant data are included in the paper or its Supplementary Information.

REFERENCES

Ministry of Foreign Affairs (2015). *Development Cooperation Charter*. Available at: https://www.mofa.go.jp/mofaj/gaiko/oda/seisaku/taikou_201502.html

Ministry of Land, Infrastructure, Transport and Tourism (2015a). *Approaches to Disaster Prevention and Mitigation in Response to New Stages*. Available at: https://www.mlit.go.jp/saigai/newstage.html

Ministry of Land, Infrastructure, Transport and Tourism (2015b). *Manual for Determining Maximum Expected Hazards for the Prediction of Inundation (River and Inland Floods)*. Available at: https://www.mlit.go.jp/river/shishin_guideline/pdf/shinsuisoutei_honnbun_1507.pdf

Ministry of Land, Infrastructure, Transport and Tourism (2015c). *Flood Control Planning to Reduce Damage in Large-Scale Inundation: Rebuilding a Risk-Conscious and Well-Prepared Society Against Water-Related Disasters by Changing Social Awareness*. Available at: http://www.mlit.go.jp/river/shinngikai_blog/shaseishin/kasenbunkakai/shouiinkai/daikibohanran/pdf/1512_02_toushinhonbun.pdf

Ministry of Land, Infrastructure, Transport and Tourism (2017). *Rebuilding a Risk-Conscious and Well-Prepared Society Against Water-Related Disasters Regarding Small and Medium-Sized Rivers*. Available at: http://www.mlit.go.jp/river/shinngikai_blog/shaseishin/kasenbunkakai/shouiinkai/daikibohanran/pdf/1701_02_toushinhonbun.pdf

Ministry of Land, Infrastructure, Transport and Tourism (2020). River Basin Disaster Resilience and Sustainability by All. https://www.mlit.go.jp/river/shinngikai_blog/shaseishin/kasenbunkakai/shouiinkai/kikouhendou_suigai/index.html.

Ministry of Land, Infrastructure, Transport and Tourism (2021). *Flood Control Planning Under Climate Change*. Available at: http://www.mlit.go.jp/river/shinngikai_blog/chisui_kentoukai/pdf/02_honbun.pdf

Nakamura, S. (2014). *Historical Study on the Standardization of Design Flood*. https://doi.org/10.15083/00007532.

Nakayasu, Y. (1960). Introduction to the Technical Standards for River Engineering and Erosion Control, JSCE Magazine,"Civil Engineering," 44(5), pp.31–32.

Program for Risk Information on Climate Change (SOUSEI) led by the Ministry of Education, Culture, Sports, Science and Technology of Japan (MEXT) (2019). *Database for Policy Decision Making for Future Climate Change (d4PDF)*. Available at: http://www.miroc-gcm.jp/~pub/d4PDF/

Takahasi, Y. (2009). History of water management in Japan from the end of World War II, *International Journal of Water Resources Development* 25 (4). doi.org/10.1080/07900620903274091.

Wakigawa, K. (2007). *A Study on Evaluation Methods for Economic Benefits of Flood Control, Kyoto University Research Information Repository*. https://doi.org/10.14989/doctor.r12136.

First received 9 December 2021; accepted in revised form 9 December 2021. Available online 10 December 2021

Water Infrastructure Planning, MDCU doi: 10.2166/wp.2021.330

Climate change adjustments in engineering design standards: European perspective

Zbigniew W. Kundzewicz[a,b,c,*] and Paweł Licznar[d]

[a] Institute for Agricultural and Forest Environment, Polish Academy of Sciences, Poznan, Poland
[b] Potsdam Institute for Climate Impact Research, Potsdam, Germany
[c] Nanjing University of Information Science and Technology, Nanjing, China
[d] Faculty of Environmental Engineering, Wroclaw University of Science and Technology, Wroclaw, Poland
*Corresponding author. E-mail: kundzewicz@yahoo.com

ABSTRACT

The European Commission Flood Risk Directive review shows that while many nations have embraced the concepts of flood risk management, there is still quite more to do in delineating risk–cost-effective measures and developing cost estimates and financing of those measures. Not mentioned are the necessary changes to existing design standards and protocols which will have to change in order to properly encompass climate change and variability, with associated uncertainties. Adjustments in engineering design standards and changes in hazards are examined, based on trend detection in observational records and projections for the future. Issues of urban and transport (motorways and railways) drainage design are also examined. Furthermore, risk reduction strategies are discussed. Finally, a way of accounting for non-stationarity in determining design precipitation and design floods is tackled. Climate change adjustments in engineering design standards, such as design precipitation and design floods, are reviewed via examples from Europe.

Key words: adaptation, climate change, climate variability, design flood, Europe

HIGHLIGHTS

- Intense precipitation and high river flows have been changing and will change, even if trend detection may not yet consistently show a ubiquitous, and statistically significant, change.
- European nations have embraced the concepts of flood risk management in line with the common Floods Directive, yet existing design standards have to change in order to properly encompass climate change and variability, with associated uncertainties.
- It is necessary to prepare for the existing climate variability, but this is not likely to be sufficient for the future. Current water management practices may be inadequate to reduce the adverse impacts of climate change.
- There is a clear gap between results of scientific studies and needs of practitioners in the domain of climate change adjustments in engineering design. Scientific results are publishable but not necessarily actionable. Both science and practice should try to improve the interface.

1. INTRODUCTION

Though deaths from flood disasters have been reduced markedly over the last century, floods continue to kill thousands of people in an average year, worldwide, and cause material losses in the order of tens of billions of US$. Therefore, risk reduction methods are of considerable practical relevance and scientific interest, virtually in all countries.

The destructive abundance of water can be caused by many different generating mechanisms, such as intense and/or long-lasting precipitation, snowmelt, rain on snow or ice, flow obstruction (e.g. by an ice jam or a landslide), dam failure, storm surge (coastal flooding), and inundations due to inadequate storm sewerage in urbanized areas. Recognizing that flood damages are often caused by improper floodplain management and development, White (1945) famously noted: 'Floods are acts of God – flood damages are acts of man'.

The Directive #2007/60/EC on the Assessment and Management of Flood Risks (Floods Directive, FD) of the European Union (EU) warrants a careful look (EU, 2007). This dedicated legal act was triggered by the observation of increased risk manifested in destructive floods in Europe (e.g. in 1997 and 2002). The reasons for increased risk are (i) likely increase of scale and frequency of floods in the future, as a result of climate change, increased impervious area, inappropriate river management, and construction in flood risk areas and (ii) marked increase in vulnerability due to the increasing number of people and economic assets located in flood risk zones.

The issue of floods and the impact of climate change on flood risks have not been addressed in the earlier Water Framework Directive (WFD) that had introduced the principle of cross-border coordination within river basins. The WFD set the objective of achieving good quality for all waters in the EU, but no objective was related to flood risk management that is now included in the FD. It notes that it is impossible to prevent flooding totally, but it is possible and indeed necessary to reduce and manage the negative impacts of floods on human health, the environment, cultural heritage, and economic activities.

The Directive, which entered into force on 26 November 2007, requires all (27 now, after the Brexit, without the UK) EU Member States to conduct the preliminary flood risk assessment (stage 1), to prepare maps of flood hazard and flood risk (stage 2), and – finally – to develop flood risk management plans (FRMPs; stage 3). The Directive obligated the EU Member States to meet the above three-stage objectives by the end of 2011, 2013, and 2015, respectively, and then to update them by the end of 2018, 2019, and 2021, and thereafter, every 6 years. This gives a possibility to include climate change-related adjustments corresponding to the most recent state of knowledge.

The European Commission released a report (EC, 2019) on the implementation of the WFD and FD. The first FRMPs were prepared and reported by the Member States. The report noted that:

> *'Human choices, historic but also still widespread today, have a significant effect on the occurrence and impacts from flooding and there is evidence that the number of large flood events has increased over the years. Projections are a cause for concern; under a no-adaptation scenario (i.e. assuming continuation of the current protection against river floods up to a current 100-year event), damages in the EU from the combined effect of climate and socioeconomic change are projected to rise from EUR 6.9 billion/year to EUR 20.4 billion/year by the 2020s, EUR 45.9 billion/year by the 2050s, and EUR 97.9 billion/year by the 2080s.'*

Though the EU Member States set a variety of qualitative national flood risk management objectives, and they included associated generic management measures for achieving the objectives, 'not all objectives are sufficiently elaborated to allow for implementation monitoring and not all measures are clearly linked to objectives. Taken together, these deficiencies may pose a challenge for the second cycle (2016–21), when Member States are expected to assess progress' (EC, 2019).

About half of the EU Member States made estimates of the costs of flood measures available, though, in many cases, not covering all FRMPs or measures. In other words, there is quite a bit more to do in clearly identifying the range of measures, and only 11 of 27 Member States used a cost–benefit analysis.

It is important to note that both nonstructural floodplain management measures, as well as structural ones, require a new class of analytical tools that explicitly account for the risks and uncertainties associated with climate change. In all cases, various ministries in each EU country are required to develop standards for regulating development in floodplains as well as for the design of infrastructure. It is only after such regulations and design standards are implemented that the true costs (and benefits) of the FD can be assessed accurately. The assumption of climate 'non-stationarity' in designing new infrastructure is one of the key issues that require resolution. A wide range of, as yet, unresolved analytical issues are associated with this assumption.

2. DESIGN PRECIPITATION AND DESIGN FLOOD NOTION

The assumption of stationarity, i.e. temporal invariance of the N-year annual maximum daily precipitation, R_N, or N-year river discharge, Q_N, with $1/N$ exceedance probability in any 1 year, is commonly used for designing infrastructure: storm sewers, bridges and roads, railways, culverts, as well as structural flood defenses – dikes, embankments, reservoir spillways, and relief channels. The concept of design precipitation and design flood is of crucial importance in natural hazard risk reduction, water management, and climate change adaptation. The engineering design standards (for instance R_{50} or Q_{50}, with 2% exceedance in any 1 year) serve as the basis of both designing the infrastructure and perception of tolerable risk.

However, design events for large N values, such as 500, i.e. R_{500} and Q_{500}, can be problematic even in a stationary case. Observed records extending over a hundred years in the stationary world could allow a reliable estimate of, say, a 10-year flood, Q_{10}, with 10% exceedance probability in any 1 year. However, in some cases, we may have just – say – observation records of 20 years, and the objective is to undertake a rough estimation of a 100-year flood, or even a 500-year or 1000-year flood, requiring excessive extrapolation and increasing uncertainty.

Typically, the broad public fails to understand risk properly. What is even worse, public policy is often driven by misunderstandings regarding risk. Low probabilities (e.g. for the 100-year flood, i.e. flood with 1% exceedance probability) are commonly, and incorrectly, rounded down to zero, and this encourages development in flood-prone areas. As noted by Tullos (2018), homeowners in such areas may be surprised to learn that the odds of their house being flooded over the 30-year lifetime of their mortgage is unexpectedly high – in fact, some 26% chance (i.e. more than one in four) of experiencing a 100-year event.

Yet, in reality, there is a clear and considerable non-stationarity effect (Milly *et al.*, 2008, 2015). The apparent non-stationarity of streamflow is attributable not only to climate change and variability but also to land use and development. Flood peaks increase because of deforestation, urban sprawl, sealing of ground surface, channel constrictions due to flood walls, flood protection embankments, bridge constrictions, and channel straightening (Vogel *et al.*, 2011). Hence, a 100-year flood for a particular location can be dramatically different from a 100-year flood determined for pre-development watersheds. Even more complicated is coping with a watershed where significant changes have been made over a long period of time.

3. CHANGES IN INTENSE PRECIPITATION AND HIGH FLOOD HAZARD – OBSERVATIONS AND PROJECTIONS

Large floods have been recorded since the dawn of human civilization (and, actually, paleo-floods occurred earlier than that), yet information about very old floods is, at best, fragmentary. There is no doubt that costs related to flood damage have been increasing, worldwide, partly due to the increasing exposure of people and assets (Kundzewicz *et al.*, 2014). Kron *et al.* (2019) reported an increase in the number of large flood events and also in economic losses and insured losses, with careful adjustment for inflation (change in the actual value of the currency units used).

The analysis of change in the frequency and intensity and/or magnitude of observed heavy precipitation records showed significant increases in many (but not all) regions of the world. Anthropogenic warming has likely contributed to a global-scale intensification and frequency of heavy precipitation (IPCC, 2013), and this impact is particularly visible in some regions. For winter, a change to wetter conditions and more extreme precipitation was noted in much of Europe, while drier conditions were detected in the south of the European continent, with a slight increase in the occurrence of extreme events.

A large number of studies of changes in flood hazard were examined by Kundzewicz (2012) and Madsen *et al.* (2014). However, there remains only low confidence in the claim that anthropogenic climate change has affected the frequency and magnitude of river floods. There exists a complex interplay between long-term trends, inter-annual and inter-decadal natural variability, as well as nonlinearities and thresholds in the climate system. Pall *et al.* (2011) offer an example of rigorous attribution of observed changes in flood hazard to anthropogenic climate change. In their model-based study, increasing global greenhouse gas concentrations were found to have considerably increased the risk of rainfall-dominated flood occurrence in some river basins in the UK, as observed during the large inundations in the autumn of the year 2000.

In their study of trend detection in maximum river flow records, Kundzewicz *et al.* (2005) did not find the ubiquitous prevailing direction of change, but rather discovered domination of large inter-annual and inter-decadal variability. Typically, eventual trends are not robust and do strongly depend on the start-year and the end-year of analysis. Moreover, trend detection cannot be a meaningful tool for a short-time series (e.g. extending over just a few decades). The occurrence of a single (or several) extraordinary event(s) may disturb the regularity of the trend. Indeed, long-time series of annual maxima of river discharge show no convincing upward trend (e.g. Kundzewicz *et al.*, 2005). There is no conclusive evidence, so far, for ubiquitous and homogeneous, climate-related, increasing trends at larger spatial scale (e.g. regional or national level) in observed extreme streamflow.

It has not been possible to find ubiquitous flood hazard changes in observation records in Europe, so far. There are no clear larger-scale regions in Europe which uniformly exhibit statistically significant increases in flood discharges. However, Kundzewicz *et al.* (2018b) detected an increasing trend in the number of large floods, even if the year-to-year, as well as the decadal, variabilities are strong.

However, for smaller regions in Europe, apparent increases in extreme streamflow have been found. These include, among others, alpine catchments and some maritime-influenced basins (Madsen *et al.*, 2014). In many cases, individual stations in a country or within larger regions may have positive or negative trends (over some intervals) or no evident trend (see a study from Germany by Hattermann *et al.* (2012)). In some areas, where snowmelt is an important flood generation mechanism, decreases in extreme streamflow and earlier spring snowmelt peak flows have been noted, likely caused by increasing temperature. Blöschl *et al.* (2017) detected a clear climate signal in the timing of river floods at the continental scale in Europe. Earlier spring snowmelt floods were found throughout North-Eastern Europe; delayed winter storms associated with polar warming have led to later winter floods around the North Sea and parts of the Mediterranean coast; and earlier soil moisture maxima have led to earlier winter floods in Western Europe.

The study of climate-driven variability in the occurrence of large floods in North America and Europe carried out by Hodgkins *et al.* (2017), only in minimally altered catchments, demonstrated that the number of significant trends was approximately equal to the number expected due to chance alone. Their finding was that natural variability rather than long-term trends drive the variability of occurrence of large floods.

Ivancic & Shaw (2015) demonstrated that large floods do not necessarily depend on the recent precipitation only. The antecedent watershed wetness can play a very important role, and there are ample examples to support this thesis. The precipitation preceding floods in Poland in May and June of 2010 was not at all extreme, yet the capacity of watersheds was filled with water and, as a result, the size of the flood was very large (exceptional for the season). Clearly, trends in heavy precipitation should not be mistaken for trends in high river flow and flood hazard. Sometimes, heavy precipitation does not lead to high river flow and, the other way around, there is high river flow without very heavy precipitation. When examining links between intense precipitation and high river flow, precipitation data should be segregated, based on concurrent soil moisture, resulting from antecedent precipitation or from snow cover. In terms of projections for the future, 'warmer' may mean 'drier'. Hence, the antecedent state of soil moisture in the catchment can considerably modulate the transformation of precipitation into river discharge.

No doubt that hydrological projections for the future are badly needed to inform water management. However, there are considerable differences in flood hazard projections, as demonstrated in a European study by Kundzewicz *et al.* (2017). Since there is a great deal of uncertainty related to the projections of water-related impacts (Kundzewicz *et al.*, 2018a), hence, Kundzewicz & Stakhiv (2010) posed a question 'Are climate models ready for prime time in water resources management applications or is more research needed?'.

Projections for the future can only be based on mathematical models. Typically, their core concepts are physically based, because it is a plausible assumption that laws of physics will not change in the unknown, and largely unknowable but likely warmer future. Yet, models are calibrated on past data, and no account is taken of prospective changes in catchment characteristics. In addition, typically little is done to normalize the measured record to account for evolving changes in the watershed that influence runoff and flooding as a function of time. The assumption that the concepts hold outside the range of calibration is questionable.

An increase in the time horizon and accuracy of flood predictions and projections remains among essential challenges. This includes forecasting the propensity of the occurrence of water abundance synchronized with the rhythm of oscillation of the atmosphere–ocean system (e.g. ENSO: El Niño – Southern Oscillation; NAO: North Atlantic Oscillation; AMO: Atlantic Multi-decadal Oscillation; PDO: Pacific Decadal Oscillation), cf. Kundzewicz *et al.*, 2019, 2020; Norel *et al.*, 2021).

It is advisable to generate river discharge projections for multi-GCMs (General Circulation Models, also known as Global Climate Models) ensembles and multiple realizations of the same model(s). One of the main problems related to GCMs, in the hydrological context, and which is responsible for a major share in total uncertainty, is the large discrepancy between different GCM projections for the same emission scenarios over some regions of the world. Regions, where GCM projections show high uncertainties, warrant more attention from climate modelers.

More understanding is necessary for the impact of bias-correction schemes, parameter uncertainty, and calibration stability within impact (hydrological) models; and uncertainty is related to extreme value estimation. Since there is still low spatial resolution of GCMs, downscaling is often necessary for watershed-scale analyses. Scale-specific assessments are needed – though using GHMs (Global Hydrological Models) may be acceptable for global/continental overviews, it is not so for the regional-scale impact assessment and adaptation actions. Some experts recommended using several available hydrological models in impact studies, but only those were able to mimic the past observations sufficiently well. In global hydrological studies, existing models are often not validated, i.e. they are not evaluated with respect to the quality of their performance in the historical observation period (cf. Krysanova *et al.*, 2018).

In general, there is a puzzling, and distorting, disconnect between a lack of significant increasing trend in the observation record of annual maximum river discharge (or peak-over-threshold series) that could hold ubiquitously or at least for a larger region and the projections for the future. The latter show increase in the frequency of intense precipitation, reflecting, in qualitative terms, the Clausius–Clapeyron law (suggesting links between climate change and flood risk: 1 °C warming should lead to a 6–7% increase of saturation vapor pressure, hence increase of the potential for intense precipitation), as well as regionally organized changes in flood hazard.

In their study of projections for the future, Hirabayashi *et al.* (2013) found that flood hazard is likely to increase in some regions of the globe and to decrease in others. However, flood hazard projections in a region of concern, such as Europe, are largely uncertain, and various approaches reported in the literature may produce dramatically different results (see Kundzewicz *et al.*, 2017, 2018a).

Some experts (e.g. Koutsoyiannis *et al.* 2008, 2009; Beven, 2011, 2018) take the stance that, at the present stage, projections are a waste of time and money, especially since the climate scenarios are unreliable. Sensitivity analyses of Prudhomme *et al.* (2013) show that we do not actually need climate projections to be precautionary about future change. In fact, because of the general unreliability of GCM-based projections for site-specific engineering design, there is a growing body of literature based on the notion of 'bottom-up' analysis, termed 'decision-scaling', that is being promoted by some of the most prestigious institutions such as the World Bank (Ray & Brown, 2015), UNESCO, and the U.S. Army Corps of Engineers (Mendoza *et al.*, 2018). The basic premise of both works is that the upgrading of existing infrastructure or the design of new infrastructure should be conducted through a series of 'bottom-up stress tests' to determine the degree of existing vulnerabilities to climate and service demands, followed by increasingly more rigorous analytical methods based on risk-analytic concepts that address the relative reliability of various risk management measures.

4. URBAN AND TRANSPORT (MOTORWAYS AND RAILWAYS) DRAINAGE DESIGN

Adapting standards for designing urban and transport (motorways and railways) drainage to climate change faces several difficulties, many of which may have a common source being imperfections, inaccuracies, and ambiguities in the toolbox that is used today. Due to the stochastic nature of rainfall, reliable operation of sewer systems cannot be fully accomplished and failures are bound to happen. Therefore, we limit ourselves to designing a drainage system in such a way that it is ready to accept the maximum (projected) stormwater input with the frequency of occurrence equal to the permissible frequency of flooding.

Obviously, in a newly designed urban or road drainage system, we cannot know the variability of the maximum flows. Hence, we have to estimate their maximum design size, reaching for a design rainfall intensity, via the classic IDF (intensity–duration–frequency) or DDF (depth–duration–frequency) models. The design rainfall intensity is determined on the basis of the calculated time of runoff concentration and the *à-priori* assumed level of rainfall occurrence. Unfortunately, the frequency of design rainfall is not the same as the frequency of surface flooding. There is only a certain arbitrary relationship between the frequency of design rainfall and the frequency of flooding, which cannot be generalized in an analytical manner for all catchments. This relationship is empirical and is a result of engineering experience rather than comprehensive hydrological studies.

Looking to a future, changing climate and wishing to translate the predicted changes in the maximum intensity of local rain into changes in the frequency of overflows from drainage systems, it is convenient for an engineer to have a specific conversion table. Unfortunately, the recent changes to the design standards of stormwater drainage systems in Europe seem to dilute rather than consolidate the relationship between these frequencies.

The European standard (EN 752, 2008) limited the permissible frequency of flooding from sewage systems, or the inability to collect stormwater, to rare frequencies of their occurrence, while adapting to four types of spatial development, i.e. on average, once in 10 years for rural areas and up to once every 20, 30, or 50 years for different

Table 1 | Recommended design frequencies (storm – for use with simple design methods, flooding – for use with complex design methods), according to the recent standard (EN 752, 2008).

	Design storm frequency[a]		Design flood frequency[a]	
Location	**Return period (1 in *N* years)**	**Exceedance probability in 1 year (%)**	**Return period (1 in *N* years)**	**Exceedance probability in 1 year (%)**
Rural areas	1 in 1	100	1 in 10	10
Residential areas	1 in 2	50	1 in 20	5
City centers/industrial/commercial areas	1 in 5	20	1 in 30	3
Underground/railway underpasses	1 in 10	10	1 in 50	2

[a]For those design storms or design floods, no surcharge shall occur.

types of urban areas (Table 1). For the design of new or modernization of existing drainage systems, the standard recommended correspondingly lower rates of design rainfall: from once a year for rural areas to once every 2, 5, or 10 years for urban areas (Table 1), with a condition that gravity channels cannot be overloaded, e.g. no pressurized flow occurs. Certainly, Table 1 for the conversion of design storm rate and the design flooding rate was not perfect, yet it was straightforward and clear for engineering applications.

A recent version of this standard (EN 752, 2017) proposed to make the permissible frequency of flooding from sewerage dependent on the seven-degree scale of the impact of the threat on the environment for defined locations (Table 2). It can be presumed that the exemplary criteria for inundation hazards are deliberately formulated rather vaguely and descriptively, e.g. 'roads or open spaces away from buildings: $N = 1$ year', or 'roads or open spaces near buildings: $N = 5$ years'. At the same time, it is stipulated that the values of permissible frequency of flooding hazards given in Table 2, as an example, can be both increased in the case 'where the floodwater is moving faster', and also lowered in the cases of 'undertaking rehabilitation of existing systems and where achieving the same design criteria for a new system would entail excessive cost'. As sewer systems are designed for the lifetime of 50–100 years, the latter possibility (i.e. lowering the permissible frequency of hazards) is debatable in view of the generally expected increase in the frequency of heavy rainfall events in the future climate. Moreover, this standard (EN 752, 2017) entails a caveat that 'criteria may largely vary between countries'.

As a result, it is more difficult for an engineer to use the new standard (EN 752, 2017). To manage the risk of sewer flooding, both the permissible frequency of occurrence and the anticipated and estimated effects of flooding, i.e. damage to property, and in particular the impact on human health and safety, should be taken into account. What is worse, there is no clear conversion table for the frequency of the design storm and the frequency of the design flood. In practice, matters are even more complicated, because the real environmental threats caused by flooding from drainage systems can be determined either during their operation or demonstrated by means of hydrodynamic modeling. In the latter case, which is the only possibility at the design stage, other frequencies have to be used. The key problem is how does one calculate the frequencies of a non-stationary climate? In other words, the new standards should have a method for calculating frequencies for infrastructure that will last 50–100 years.

Unfortunately, the European standard (EN 752, 2017) does not contain any indications as to the permissible frequency of surcharges of the sewerage system manholes up to the ground level. In contrast, such values were

Table 2 | Examples of design sewer flooding criteria for standing floodwater, according to the recent standard (EN 752, 2017).

No.	Impact	Example locations	Return period (years)	Probability of exceedance in any year (%)
1	Very small	Roads or open spaces away from buildings	1	100
2	Low	Agricultural land (depending on land use, e.g. pasture, arable)	2	50
3	Low to medium	Open spaces used for public amenity	3	30
4	Medium	Roads or open spaces adjacent to buildings	5	20
5	Medium to high	Flooding in occupied buildings, excluding basements	10	10
6	High	Deep flooding in occupied basements or road underpasses	30	3
7	Very high	Critical infrastructure	50	2

The return period should be increased (probabilities reduced) where the floodwater is moving faster. When undertaking rehabilitation of existing systems and where achieving the same design criteria for a new system would entail excessive cost, a lower value may be considered.

Table 3 | Acceptable frequencies of drainage surcharges damming up to the ground level for calculations checking the operation of the drainage systems according to DWA-A118 (2006).

Lp.	Location	Frequency of overflows (once in *N* years)
1	Rural areas	2
2	Residential areas	3
3	City centers/industrial/commercial areas	More than 5
4	Underground/railway/road and pedestrian underpasses	More than 10[a]

[a]When local security measures are not applied, the frequency of overflows is 50.

previously established in Germany and included in the technical guideline DWA-A118 already in 2006 (Table 3). These German guidelines find practical application also in other countries of Central and Eastern Europe, where approaches to design and construct urban drainage systems primarily established in Germany have often been followed.

5. RIVER FLOOD RISK REDUCTION STRATEGIES

It is recognized that changes in river flood risk may depend (Kundzewicz & Schellnhuber, 2004) on:

1. changes in socio-economic systems (land-use change, increasing exposure and damage potential related to floodplain development and increasing wealth in flood-prone areas, change in flood risk perception);
2. changes in hydrological/terrestrial systems (land-cover change accompanying land-use change: urbanization, deforestation, elimination of natural inundation areas – wetlands and floodplains; river regulation – riverbed straightening and shortening, constructing embankments; damming rivers; changing conditions of transformation of precipitation into runoff by way of increase of impermeable areas; under urban sprawl; watershed management; and structural flood defenses); and
3. changes in climate and atmospheric systems (water holding capacity of the atmosphere, intense precipitation, changing precipitation phase – solid or liquid, seasonality, snowmelt pattern, ice phenomena, and atmospheric circulation patterns).

Over many decades, the USA played the role of the global leader in river flood preparedness, indeed being the pattern for other countries to follow. The Flood Control Act (FCA), passed in the USA in 1936, and subsequent legislation that authorized federal US engagement in structural flood protection have influenced flood protection policy in many countries that followed the US example. Thirty years after the FCA, the National Flood Insurance Act was passed in the USA in 1968, and the National Flood Insurance Program (NFIP) was established. The idea was to identify the 100-year floodplain (with a probability of being flooded of 1% in an average year), as a high-risk area. The flood insurance program discouraged floodplain development, providing incentives for communities that adopted land-use regulations and prohibited future construction below the 100-year flood elevation. However, local governments had ultimate authority on land-use regulations.

A wide array of risk analysis and risk management methods have been developed for flood management that attempts to provide a more uniform and replicable approach to the analysis and consideration of risk–cost-effective flood risk management options. These methods have proven to be relatively effective over the past century, but have come under scrutiny because of climate change and associated non-stationarity and increased uncertainties of the underlying physical phenomena of the generating mechanisms for precipitation, as projected through a myriad of general circulation models.

One cannot directly influence the variability of precipitation, yet one can do a great deal to reduce flood risk. In the EU (European Union) STARFLOOD project, a roster of strategies useful in flood risk reduction were examined (Driessen *et al.*, 2016; Hegger *et al.*, 2016). The fundamental, and deeply rooted, strategies are flood risk prevention measures (keeping people away from destructive water) and structural flood defense measures (keeping destructive water away from people) (Dieperink *et al.*, 2016). The former aim to decrease the consequences of flooding by decreasing the exposure of people and property, via prohibiting or discouraging development in areas at risk of flooding (e.g. via land-use policy, spatial planning, resettling communities with repetitive losses, expropriation policy, and re-allotment policy). The latter aim to decrease the flood damages through infrastructural works (e.g. dikes, dams, and embankments) that increase the capacity of existing channels for water conveyance or the creation of new water storage spaces. Since our efforts to keep people away from destructive water and to keep destructive water away from people sometimes may fail, we also need strategies of flood risk mitigation, flood preparation, and flood recovery (Kundzewicz *et al.*, 2018c).

Flood governance deals with various aspects of relevance to different sectors: natural hazard (here: flood) risk reduction, water resources management (e.g. implementation of the FD of the European Union), and adaptation to climate change (see Lugeri *et al.* 2010).

6. HOW TO ACCOUNT FOR NON-STATIONARITY IN THE DETERMINATION OF DESIGN PRECIPITATION AND DESIGN FLOOD: EUROPEAN PERSPECTIVE

In a non-stationary world, Q_N determined for the 1980s can be considerably different from Q_N determined for the 2020s, and even more so – for future horizons, e.g. the 2040s or the 2070s. One can use the concept of magnification factor to reflect the change in the design flood (Salas *et al.*, 2018).

However, one has to be very careful when attempting to generalize attribution of changes in flood hazard, because there exist various flood generating mechanisms driven by different processes. Mechanisms of changes in flood hazard and flood risk can be really complex, as demonstrated in Wyzga *et al.* (2018).

Noting that climate change has rendered water management more difficult due to uncertainty in future changes of hydrological conditions, Döll *et al.* (2015) advocated for extending the approach of adaptive Integrated Water Resources Management (IWRM) by considering the risk of climate change. They called for embracing the uncertainty of future climate and its impacts in decision-making, by probabilistic assessment of the future water conditions for different scenarios, and 'developing a portfolio of low-regret solutions that reduce vulnerability and can be implemented and modified progressively as future conditions evolve'. Salas *et al.* (2018) reviewed various metrics that may be used for assessing infrastructure investment decisions, including economic risk-based approaches, and planning under the additional uncertainty imposed by non-stationary conditions. In areas with short periods of observation and significant non-stationarity, the reliance on probabilistic assessments of future water conditions may become quite unreliable, and the engineering profession will have to resort to other concepts for decision-making.

Adding safety margins to design flood indices has been traditionally used to cope with risk and uncertainty. However, the unprecedented global economic growth and the prospect of climate change have resulted in a need to plan and design for future 'unknown unknowns' (Mendoza *et al.*, 2018). Much effort has gone into tools and models to produce projections for the future on the basis of multi-model ensembles of climate and impact (hydrological) models. Using multi-model ensembles broadens the ranges of possible future scenarios and associated flood frequency analyses, which are the traditional bases for establishing design floods. As framed by Mendoza *et al.* (2018), 'choosing a particular subset of future scenarios to plan, design, or invest has become an increasingly subjective enterprise, centering on which climate scenarios to consider and what hydrological analytical tools could be employed to deal with these cascading uncertainties. A range of basic questions emerge: How can we justify a particular decision, given all the uncertainties? How do we plan for an action that is neither too early nor too late? How do we convey the resulting analyses, built on a pyramid of uncertainties, to the stakeholders and to political decision makers?'

It is instructive to provide just two examples from overseas, indicating how these issues are tackled in Japan and in the USA in comparison to how European countries approach these issues. Nakamura & Oki (2018) studied paradigm shifts in flood risk management in Japan, a country, whose 20% of the area is flat land, vulnerable to flooding, yet 80% of the national population and 50% of the property are located there. They identified three eras of the modern history of Japan, labeled as changing society (1910–1935), response to megafloods (1935–1970), and response to economic growth (1970–2010) that have largely influenced flood risk management. They informed of developments of design flood in the Tone River basin, where the Tokyo metropolitan area is located. The design flood discharge has increased from 3,750 m^3/s in 1900 to 22,000 m^3/s at present. The development reflected the occurrence of large floods (triggers) that had to be accommodated in planning.

Nakamura & Oki (2018) analyzed 109 river basins in Japan and found 323 revisions of design flood after 1910. They assessed the frequencies of triggers for design flood revision as follows: 34% due to national policy change (e.g. update to the River Act), 28% due to occurrence of a large flood that caused severe damage, 19% due to economic growth (advancement of development or urbanization in the basin), and 4% due to dam construction in the basin. The design flood in 1958 corresponded to the return period of 80–100 years for class A rivers and 50–80 years for class B rivers. After several 'five-year national flood prevention plans', the return periods increased to present (since 2004) values greater than 200 (for class A rivers) and 100–200 (for class B rivers). Nakamura & Oki (2018) note the possibility of another paradigm shift due to climate change and increasing public interest. This can be also related to campaigns against dam construction and intimation that design floods have been overestimated in order to facilitate the construction of dams.

An example of developments taken place in the USA clearly illustrates the science–policy interface. On 30 January 2015, then President of the USA, Barack Obama issued Executive Order (EO) 13690 on Establishing a Federal Flood Risk Management Standard and a Process for Further Soliciting and Considering Stakeholder Input[1], replacing the old Executive Order 11988 (after 38 years of validity). The new document stated that it is the policy of the USA to improve the resilience of communities and Federal assets against the impacts of flooding and to recognize the risks and losses due to climate change and other threats. As per EO 13690, establishing the floodplain results from using one of the following alternative approaches:

1. 'the elevation and flood hazard area that result from using a climate-informed science approach that uses the best-available, actionable hydrologic and hydraulic data and methods that integrate current and future changes in flooding based on climate science. This approach will also include an emphasis on whether the action is a critical action as one of the factors to be considered when conducting the analysis;
2. the elevation and flood hazard area that result from using the freeboard value, reached by adding an additional 2 feet to the base flood elevation for non-critical actions and by adding an additional 3 feet to the base flood elevation for critical actions;
3. the area subject to flooding by the 0.2 percent annual chance flood (i.e. a flood with 500-year return period); or
4. the elevation and flood hazard area that result from using any other method identified in an update to the FFRMS[2].'

---

[1] https://obamawhitehouse.archives.gov/the-press-office/2015/01/30/executive-order-establishing-federal-flood-risk-management-standard-and-.

[2] Federal Flood Risk Management Standard.

The Executive Order 13690 was based on pure hazard criteria, not risk-based and only probabilistic through the connection to the base flood elevation. It is also only related to federally funded projects.

However, it turned out that the lifetime of EO 13690 was rather short. On 15 August 2017, President Donald Trump revoked Executive Order 13690 and issued Executive Order 13807 on Establishing Discipline and Accountability in the Environmental Review and Permitting Process for Infrastructure Projects. This change was reversed again by President Joe Biden on 20 January 2021 that is on his first day in office. President Biden rescinded EO 13807 and issued EO 13990 on Protecting Public Health and the Environment and Restoring Science to Tackle the Climate Crisis[3]. On 20 May 2021, President Biden issued EO 14030 on Climate-Related Financial Risk, reinstating EO 13690.

6.1. Climate change adjustments in urban and transport (motorways and railways) drainage design standards

Madsen *et al.* (2014) offered a summary of existing guidelines in European countries, referring to climate change adjustment of design rainfall. They reported on national guidelines on design rainfall in Belgium, Denmark, Sweden, and the UK.

Germany has been one of the European leaders in the development of an engineering toolbox for calculating and modeling of drainage systems and one of the most advanced countries in the field of adapting urban and transport drainage systems to the projected climate changes in the continent. For example, Staufer *et al.* (2010) showed that the current intensity of precipitation with statistical repeatability, e.g. once every 5 years ($N = 5$), average, is likely to occur in the future more than twice as often ($N = 2$) due to the expected climate change. On this basis, for the verification of future sewer overflows and flooding for a German federal state (*Bundesland*) North Rhine-Westphalia in the (Merkblatt No. 4.3/3, 2009) guideline, it was recommended to change the design precipitation frequency in the standards applicable at that time (DWA-A118, 2006; EN 752, 2008). These changes are summarized in Table 4.

Germany is also well known for a very pragmatic approach to adapting the drainage systems calculation toolbox through iterative updating of the current intensities of design rainfalls. In this country, the need to develop a specialist rainfall intensity atlas for the design of drainage systems was early recognized. In response to this need, the project of the KOSTRA Atlas (German: *KOordinierte STarkniederschlags-Regionalisierungs-Auswertungen*) (Bartels, 1997; Malitz & Ertel, 2015) was launched in the 1990s. The KOSTRA Atlas is a result of a comprehensive statistical study of rainfall maxima with a wide range of durations (5–4,320 min) from the entire network of rain gauges in Germany. Importantly, it provides information in high spatial resolution, i.e. it can be used to read the values of the intensity of design rainfall for any location corresponding to a specific cell (of an area of 66.83 km^2) of the regular grid, covering the whole country. Thanks to advanced procedures for the statistical processing of the maximum rainfall depths from the rain gauge network, including spatial interpolation with the use of kriging, the values read in the grid cells of KOSTRA Atlas are provided with confidence intervals. The first edition of the KOSTRA Atlas was originally based on 30-year rainfall records (1951–1980) and since then it has been updated. Its most recent version, called KOSTRA-DWD-2010R (Junghänel *et al.*, 2017), is based on 60-year rainfall series (1951–2010). The KOSTRA Atlas is comparable in the quality of the products offered to the engineer with the more recent NOAA Atlas 14 in the USA (Perica *et al.*, 2018) or the Polish PANDa Atlas (*Polski Atlas Natężeń Deszczów*) in Poland (Licznar & Zaleski, 2020).

---

[3] https://www.federalregister.gov/documents/2021/01/25/2021-01765/protecting-public-health-and-the-environment-and-restoring-science-to-tackle-the-climate-crisis.

Table 4 | Amendments to DWA-A118 (2006) and EN 752 (2008) with regard to precipitation frequency, reflecting future sewerage overloads, according to Merkblatt Nr. 4.3/3 (2009).

No.	Land development	Precipitation frequency (once in *N* years)	
		For verification of surcharge	For verification of flooding
1	Rural areas	3 instead of 2	50 instead of 10
2	Residential areas	5 instead of 3	100 instead of 20
3	City centers, service, and industry areas	10 instead of 5	100 instead of 30

Attempts to adapt the toolbox for designing urban and roads/railways drainage systems have been also undertaken in other European countries. For example, in the UK, the Environment Agency (UK), in its guidelines from 2016 (updated in 2020), recommends increasing the maximum intensity of design rainfall for small and urban catchments, due to climate change, by up to 40% (upper end, the 2080s) (Environment Agency, 2016). In Belgium, Willems (2013) studied the impact of climate change projected up to 2100 on the necessary revision of the design principles of urban drainage systems. He showed that in Flemish conditions, the 10-year design storm intensity can increase up to about 50% by the end of 2100, or systems designed for a 20-year return period of flooding, might flood with four times shorter recurrence interval of – in order of magnitude – 5 years by the end of the century.

The cited studies and recommendations are, however, only peculiarities of a very conservative practice of designing urban and roads and railways drainage systems. Engineers do not so much use recommendations and results published in scientific articles, but are usually obliged to use methodologies codified in national standards and technical guidelines. However, these documents are often obsolete, both in terms of the topicality of the recorded sources of information about the intensity of design rainfall, but also in terms of the very idea of the functioning of drainage systems. Especially, the standards and guidelines for roads and railways drainage are still rooted in the old philosophy of 'collect and drain'. To support these theses, it is enough to briefly review the standards and technical guidelines from a few selected countries, such as Germany, Spain, and Poland.

Even in Germany, being an advanced country in terms of adaptation to climate change, the changes to the design guidelines for roads and railways drainage are slow and limited. The German railway standard for drainage *Richtlinie 836 – Erdbauwerke und sonstige geotechnische Bauwerke planen, bauen und instand halten* (Ril-836, 2019) allows the use of a simple delay coefficient method together with the oldish Reinhold (1940) rainfall model to calculate the maximum runoff of rainwater.

The Reinhold (1940) rainfall model responds to the challenges of designing drainage systems according to the principle of 'collect and drain' and combined the intensity of rain with different frequencies of occurrence and duration up to 150 min with the reference intensity $q_{(15.1)}$ for the time of 15 min and return period of 1 year. This model can no longer be used in the calculation of drainage systems, equipped with devices for retention and drainage of rainwater, for which the duration of rainfall may be of the order of a dozen or even several dozens of hours.

The Ril-836 (2019) standard recommends to change the $q_{(15,1)}$ intensity, in relation to the original values after Reinhold (1940), for various locations in Germany. For example, for Berlin, a change in the intensity of $q_{(15,1)}$ from 94 to 127 dm^3 $(s\ ha)^{-1}$ was introduced. Moreover, the Ril-836 (2019) standard indicates that the calculation of the maximum rainwater runoff should be in accordance with the road drainage guideline *Fortschreibung der Richtlinien für die Anlage von Straßen, Teil: Entwässerung* (RAS-Ew, 2005). This guideline indicates the KOSTRA rainfall atlas as the source of reliable rainfall intensity for the dimensioning of drainage systems. For calculating rainwater infiltration systems, Ril-836 (2019) recommends the use of the ATV-DVWK-A 138 (2005) technical guideline, also referring to the KOSTRA Atlas.

In Spain, recent guidelines for calculating roads and railways drainage systems for roads (BOE 60, 2016) and railways (NAP 1-2-0.3, 2021) recommend a consistent approach to adopt design rainfall rates for calculating the maximum rainwater runoff. The precipitation intensity is the product of the corrected average daily precipitation intensity corresponding to the desired frequency C and the so-called storminess coefficient, which – according to the standard for a given rainfall duration – can be calculated on the basis of the storminess index, which expresses the ratio between hourly precipitation intensity and corrected average daily intensity. Both standards (BOE 60, 2016; NAP 1-2-0.3, 2021) contain maps with storminess index values for different areas of Spain. The value of daily precipitation with a frequency of $1/N$ for calculations according to the guidelines should be read from the atlas of daily precipitation (*Máximas lluvias diarias en la España Peninsular*), published in 1999. The current standards of Spanish roads and railways drainage can be justified by the fact that they allow for the assumption of daily precipitation values with a frequency of $1/N$ on the basis of a statistical series of annual maximum daily precipitation (with a series length exceeding 30 years), recorded by rain gauges near the designed drainage systems. The Spanish standards also allow for an alternative calculation of the turbulence coefficient on the basis of a nearby representative IDF curve, if it exists and is accepted by the client. These are certainly options that open the possibility of designing on the basis of more current rainfall rates, but the open question is whether engineers will undertake the burden of additional statistical analysis, instead of using simpler, well-established methods from over 20–30 years ago.

Poland faces considerable conservatism with regard to the modernization of the design toolbox, and especially the updating of the intensity of design rainfall. The tools that can be found in the Polish guidelines for designing the roads and railways drainage system lack competence to adapt to changing climate. Both in the national road drainage standard (PN-S-02204, 1997) and in the guidelines for calculating railway drainage systems published 22 years later (Wytyczne obliczania ilości, 2019), it is recommended to calculate the intensity of the design rain, q, on the basis of an outdated empirical, formula:

$$q = 15.347 \frac{A}{t_m^{0.667}}, \tag{1}$$

where A is a constant value taken on the basis of the average annual precipitation in depth H and the probability p of the design rain (see Table 5), while t_m is the rainfall duration time in seconds.

Table 5 | Values of A coefficient in Equation (1) on design precipitation intensity (PN-S-02204, 1997; Wytyczne obliczania ilości, 2019).

	***H* (mm)**			
***p* (%)**	**Up to 800**	**Up to 1,000**	**Up to 1,200**	**Up to 1,500**
5	1,276	1,290	1,300	1,378
10	1,013	1,083	1,134	1,202
20	804	920	980	1,025
50	592	720	750	796
100	470	572	593	627

Formula (1) is a very simple empirical relationship, the origins of which date back to 1926 when local discontinuous records (from 1837 to 1891 and 1914 to 1925) of intense short-duration rains were analyzed for the purposes of modernizing the sewerage system in Warsaw. These results were generalized after the Second World War by Błaszczyk (1954) to the form recommended for use throughout Poland:

$$q = \frac{6.67 \cdot \sqrt[3]{H^2 \cdot C}}{t^{0.67}} \tag{2}$$

where q is the unit rainfall intensity (dm^3 (s ha)$^{-1}$), H is the mean annual rainfall (mm), C is the period of a single exceedance of a given intensity (years), and t is the duration of rain (min).

The design rain intensity in any location in Poland was assumed to depend on a single and easy-to-determine climatic parameter in the form of normal annual rainfall. Later, to further facilitate the use of formula (2), its simpler form (1) with a tabular summary of the numerator values was proposed. Obviously, the simple formula was willingly accepted by the engineers. However, its development was not preceded by comprehensive studies of the design rainfall on rain gauges other than located in Warsaw, even if Poland is a large country (an area of over 312,000 km^2), with spatially variable climates (the range of mean annual rainfall varies from 500 to over 1,000 mm). What is worse, the structure of formula (2) had its roots in rather obsolete Gorbachev's precipitation model (Dębski, 1966), the starting point of which was the speculative concept of the rain-force parameter (the root of the product of precipitation intensity and depth) as a discriminant of rains from clouds formed under the same meteorological conditions.

Obviously, the design rainfall formula (2), developed on the basis of outdated and incomplete rainfall data, without proper statistical analysis and assuming a prior relationship between the annual rainfall depth and maximum rainfall intensities, cannot provide reliable estimates of design rainfall intensity. Comparative studies of empirically determined and verified design rainfall intensities on a network of 100 stations from all over Poland and the rainfall intensities calculated from formula (2), taking into account the average annual rainfalls in depth from the last 30 years was performed by Licznar *et al.* (2018) at the preliminary stage of the PANDa Atlas project. They found that:

1. The intensity of design rainfall does not strongly correlate with the average annual precipitation depth; therefore, the structure of Błaszczyk formula (2) is incorrect and it is not possible to introduce rational systematic corrections to the model.
2. The use of the Błaszczyk formula leads, in a great majority of cases, to a dangerous underestimation of the real values of design rainfall intensity. This underestimation is about 33% over the whole network of 100 analyzed gauges and can even be observed in the case of Warsaw, i.e. the place where the original formula was developed. However, the intensity of rainfall resulting from formula (1) is still widely used when designing roads and railways drainage systems in Poland. In fact, it is often overused, beyond the conditions for which it was developed (rainfall durations from 5 to 180 min). In fact, formulas (1) and (2) are used for much longer durations, e.g. for estimating the necessary volume of storage capacity.

The most recent railway guideline (Wytyczne obliczania ilości, 2019), based on a flawed and outdated Błaszczyk rainfall model (Błaszczyk, 1954), recommends increasing the average annual historical rainfall depth by 3.5%, due to the projected climate change. This reflects the projection of an increase in the average annual rainfall depth for the area of Poland by approximately 3.5% in the time horizon 2050, in relation to the reference period 1971–2000. However, the Błaszczyk model was not based on the data from 1971 to 2000, and the correlation of maximum rainfall intensities with the average annual rainfall depth is not strong. Hence, it is impossible to introduce a systematic correction of formulas (1) and (2) (Licznar *et al.*, 2018).

Finally, engineers using these formulas usually do not have sufficient information about the average annual precipitation depth H, at the project site. Data obtained from nearby rain gauges for various years may differ by much more than the recommended 3.5% correction. At the same time, the recommended increase in the average annual rainfall depth, assuming the average rainfall total for the whole of Poland, $H = 600$ mm, translates into a 2% increase in the calculated intensity of design rainfall. Such a small correction is not only unreliable in relation to the scale of recommended changes in rainfall intensity in other European countries, but is even an order of magnitude smaller than the range of local rainfall intensity confidence intervals read from contemporary rainfall atlases, including the Polish PANDa.

Summarizing the presented review of climate change adjustments in the area of designing urban as well as roads and railways drainage systems in Europe, it can be stated that developed scientific methodologies are already in existence. We know how to project the design rainfall and the band of uncertainty, but we cannot translate them easily into probabilities.

In many European countries, engineers do not have access to or do not use the current information on rainfall intensity for design. It would be very useful to create a common, pan-European rainfall atlas, similar to the KOSTRA or PANDa atlases. Ideally, this action should be a long-term project of adjusting rainfall statistics to changing climatic conditions. It is important that such an atlas could give the possibility of reading the intensity of design rainfall together with confidence intervals due to the small-scale variability of rainfall maxima. Only such a source should be the starting point for introducing corrections due to the projected climate changes. It also seems very important to order and simplify European guidelines as to the recommended permissible rates of flooding and overflows of drainage systems and their unambiguous connection with the frequencies of rainfall assumed for the design.

6.2. Climate change adjustments in design floods

Due to the likely increase in flood risk, despite the huge uncertainty, practitioners and water managers in some European countries and regions are already considering explicitly how to incorporate the potential effects of global change into policies and specific design guidelines via climate adjustments (Kundzewicz *et al.* 2008, 2017).

Interesting development of design floods, to reflect climate change, have been undertaken in the basin of the Rhine, a large international river in Europe (Strategy for the IRBD Rhine, 2015).

Effects of climate change modify the discharge pattern of the Rhine and its tributaries: higher flows are likely to become more frequent. Orientation guidance for flood sensitivity values is given in Strategy for the IRBD Rhine, (2015) for various gauges on the Rhine for a range of characteristic discharges, such as MHQ, HQ10, HQ100, and HQ extreme until 2050. The bandwidths for changes in HQ extreme that may serve as a basis for discussions on possible adaptation measures read: from −20 to +35% at Basel and Maxau, from −15 to +30% for Worms, and from −5 to +20% at Kaub, Cologne (Köln), and Lobith. Even if the width of positive change bands exceeds the width of negative change bands, the uncertainty range is high, especially in upstream gauges on the Rhine.

As a consequence of the great floods of the Rhine in 1993 and 1995, the International Commission for the Protection of the Rhine (ICPR) adopted the 'Action Plan on Floods'[4] for the Rhine that aims at improving flood protection as well as extending and enhancing the floodplains of the Rhine. The plan, consisting of numerous measures, was conceived in phases and was implemented by all riparian countries until 2020, entailing expenses of 12 billion euros.

---

[4] https://www.iksr.org/en/international-cooperation/rhine-2020/action-plan-on-floods/.

Objectives for 2020 included the following:

1. Damage risks were to be reduced by 25%, compared to the reference year 1995.
2. Extreme flood stages downstream the impounded sections were to be reduced by up to 70 cm (60 cm due to water retention along the River Rhine and approximately 10 cm due to water retention in the Rhine basin).
3. The population living in the vicinity of the Rhine was to be made aware of the flood risk via maps of flood hazard and risk, indicating the areas at risk. To increase flood awareness, flood risk maps were drafted and spread for the flood hazard areas.
4. Lags of flood forecasting should be distinctly prolonged.

In some European countries and regions, flood design values have been increased by correction factors (safety margins), possibly based on climate change impact scenarios. Madsen *et al.* (2014) offered a summary of existing guidelines in European countries, referring to climate change adjustment of the design flood. They reported on national guidelines on design floods in Norway and the UK. In Belgium and Germany, the existence of regional guidelines on design flood was reported in the province of Flanders (Belgium) as well as in some German federal states (*Bundesländer*), such as Bavaria and Baden-Württemberg (Germany). The pattern of the construction of these guidelines is quite common: the guideline foresees an *N*% increase, with a future horizon of concern either explicitly specified or not. The specific numbers may depend on the location/region and the return period. In some guidelines, design floods depend also on the prevailing flood season and catchment size. The amplitude of climate change allowance in a particular country or region usually corresponds to the increase of peak river flow, resulting from climatic projections for the future (e.g. end of the 21st century).

In the UK, the Environment Agency (2016) avoided estimating probabilities by providing climate change 'allowances', that is predictions of anticipated change (increase) for peak river flow that local planning authorities, developers, and their agents should use in flood risk assessments. In particular, this guidance aids in preparing strategic flood risk assessments, or flood risk assessments for planning applications, or development consent orders for nationally significant infrastructure projects. As stated by the Environment Agency (2016), there may be circumstances where local evidence supports using other data or allowances, e.g. if the impact of climate change on peak river flow is not the same in all river catchments in a river basin district. However, then the Environment Agency may want to check how and why other data were used in plans.

Peak river flow allowances listed in Environment Agency (2016) show the anticipated increases to peak flow by river basin district. The range of allowances is based on percentiles, e.g. the upper-end allowance is based on the 90th percentile (being exceeded by 10% of the projections). The future horizons considered are the 2020s (2015–2039), the 2050s (2040–2069), and the 2080s (2070–2115). The higher the percentile and the more remote the future horizon, the higher the potential increase of the index, above the 1961–1990 baseline. For the upper-end allowance based on the 90th percentile, there is a projected increase by 105 and 85%, respectively, for Southeast and Southwest river basin districts. Even higher allowances are given for extreme flows (H++), being 120, 105, 95, 90, and 80%, respectively, for Southeast, Southwest, Northwest, Severn, and Thames river basin districts.

7. CONCLUDING REMARKS

There is little doubt that high river discharges and stages have been changing, even if trend detection in flood hazards does not yet consistently show a ubiquitous, and statistically significant, change. Flood hazard impacts may increase, and they might come more rapidly than model projections based on climate scenarios suggest. Hence, the rational expectation is that flood frequencies will be changing (mostly increasing), with associated economic and social consequences. With all these uncertainties, society needs to begin thinking about adaptation investments as a precautionary measure now, as in the case of the Netherlands, where flood safety standards have been very high, because the message should be that we have to prepare without precisely knowing the future flood hazard.

The Dutch have created an incremental adaptive strategy for coping with climate change through a combination of scenarios (different levels of both water regime and socio-cultural conditions) and systematic modeling to define sensitivities of current infrastructure systems to levels of change. The incremental adaptation strategy includes multiple responses that supplement each other, precluding the need to speculate on how much change and when for risk mitigation efforts. This strategy is somewhat unique in that it does not focus on probabilities of change but the physical levels of change. It may well be a window into the future if non-stationarity makes probabilistic analysis much more uncertain.

The spread of projections of hazards and damages associated with destructive water abundance has to be interpreted with caution by decision-makers in charge of natural risk reduction, climate change adaptation, and water resources management. Since it would be naïve to expect the availability of trustworthy quantitative projections of future flood hazards, in order to reduce the risk, one should focus attention on the identification of risk and vulnerability hotspots and improve the situation by reducing risk in such hotspot areas.

Decision-making under uncertainty requires identifying and quantifying the uncertainty involved and then improving a suitable framework for decision-making, including consideration of the risk of action vs. the risk of inaction. The lack of agreement in projections between studies can be interpreted and understood by scientists, but less so by stakeholders in general and practitioners in particular. Despite the caveats typically accompanying large-scale studies, stakeholders in regions where no local hazard projections are readily available view large-scale maps from different sources (e.g. maps in scientific publications) that may strongly diverge in their area of interest. Since they tend to take reported flood hazard projections at face value, they may become confused by noted disagreements.

There is no doubt that it is necessary to prepare for the existing variability of water abundance (to which mankind is not adequately prepared today), but this is not likely to be sufficient for future changes. Current water management practices may simply be inadequate to reduce the adverse impacts of climate change (Kundzewicz *et al.*, 2014). Nevertheless, good adaptation to existing climate and its variability augurs better for adaptation to the future, changed, climate. Robust adaptation procedures, which do not rely on crisp and precise projections of future changes (that can never be obtained), need to be developed. Economic risk-based decision-making is necessary, i.e. search for appropriate levels of infrastructure based on the expected damages avoided vs. the cost of the infrastructure. Under strong and irreducible uncertainty, two alternative courses of action can be envisaged – the precautionary principle (a variation of the min–max concept – to choose the approach minimizing the worst outcome) and/or an iterative and sequential adaptive management approach, based on ensembles and multi-model probabilistic approaches (Kundzewicz *et al.*, 2018a, 2018b).

It is also important to note that engineering design standards serve as the legal basis for infrastructure design, construction, and maintenance. Engineering design standards undergo rigorous and extensive peer review by professional engineering societies. Though they are based on the 'best' peer-reviewed scientific literature, engineering design standards represent a practical subset of a vast body of hydrologic sciences and engineering literature. They are meant to reflect 'best management practices', not necessarily 'best scientific papers'. Hence, looking ahead, it should be recognized that a wide range of new hydrologic analytical techniques that have evolved to deal with the uncertainties of non-stationary climate must be further translated into acceptable and replicable engineering design standards to replace the existing ones that are largely based on empirical methods, such as the 'design precipitation', 'design flood', 'probable maximum precipitation', or 'probable maximum flood'. Complex techniques used for climate adaptation of design rainfalls and most of their application results have to be translated into much simpler rules that could be understood by average engineers, who most probably had never studied climate change. These rules have to be implemented at a wider scale, since climate change is a global phenomenon. Finally, if we insist on implementing adaptation to climate change as a correction, most often increase of design rainfall, we have first to take a careful look at current local rainfall statistics used by engineers in everyday practice that typically need an update. Projection uncertainty of climatic models mixed with an outdated starting point could lead us to nowhere.

The basic problem is that 'risk analysis' requires probabilities. Flood frequencies must be calculated for a non-stationary climate. There are many methods, but no agreement on which to use – comparable to the issue that the USA uses LP3, while most of Europe and Japan use GEV for flood frequency analysis. All new engineering standards need a replicable method for calculating flood frequencies. The engineering profession should standardize computations of flood and drought probabilities under an anticipated non-stationary climate. Several promising methods have been developed by research hydrologists. These must be integrated into standard 'best engineering practices'.

There is a clear gap between results of scientific studies and needs of practitioners in the domain of climate change adjustments in engineering design. Scientific results are publishable but not necessarily actionable. They may provide responses to some scientific questions but may not offer useful answers to burning practical questions. Perhaps both science and practice should try to improve the interface, perhaps each side could take a step backward and listen to what the other party has to say (even if practitioners may not feel particularly interested in non-actionable scientific findings, as long as guidelines, e.g. standards justifying their *modus operandi* are in existence). It would be convenient for a practitioner to get a crisp, scientifically sound number for design, but that is not realistic. One has to deal with a range of values (sometimes a broad range). This calls for the preparation of several variants of plans. Adaptation in stages (with the possibility of add-on effect) is encouraged. Both scientists and practitioners have to live (and cope) with uncertainties, non-stationarities, and non-homogeneities.

ACKNOWLEDGEMENTS

Z.W.K. wishes to acknowledge financial support from the project 'Interpretation of Change in Flood-Related Indices based on Climate Variability' (FloVar) funded by the National Science Centre of Poland (project no. 2017/27/B/ST10/00924). The authors gratefully acknowledge advice of Dr Eugene Z. Stakhiv and constructive recommendations of two reviewers.

DATA AVAILABILITY STATEMENT

All data used in this paper stem from sources that are rigorously referred to.

REFERENCES

Arbeitsblatt DWA-A 138 (2005). *Planung, Bau und Betrieb von Anlagen zur Versickerung von Niederschlagswasser.* Deutsche Vereinigung für Wasserwirtschaft, Abwasser und Abfall e. V., Hennef.

Bartels, H. (1997). *Starkniederschlagshöhen für Deutschland. KOSTRA.* Deutscher Wetterdienst, Offenbach, Germany.

Beven, K. J. (2011). I believe in climate change but how precautionary do we need to be in planning for the future? *Hydrological Processes (Hptoday) 25*, 1517–1520. doi:10.1002/hyp.7939.

Beven, K. J. (2018). On hypothesis testing in hydrology: why falsification of models is still a really good idea. *WIREs Water 5*, e1278. doi:10.1002/wat2.1278.

Błaszczyk, W. (1954). Spływy deszczowe w sieci kanalizacyjnej (wytyczne do normatywu). *Gaz, Woda i Technika Sanitarna 9*, 262–271.

Blöschl, G., Hall, J., Parajka, J., Perdigão, R. A. P., Merz, B., Arheimer, B., Aronica, G.T., Bilibashi, A., Bonacci, O., Borga, M., Čanjevac, I., Castellarin, A., Chirico, G. B., Claps, P., Fiala, K., Frolova, N., Gorbachova, L., Gül, A., Hannaford, J., Harrigan, S., Kireeva, M., Kiss, K., Kjeldsen, T. R., Kohnová, S., Koskela, J. J., Ledvinka, O., Macdonald, N., Mavrova-Guirguinova, M., Mediero, L., Merz, R., Molnar, P., Montanari, A., Murphy, C., Osuch, M., Ovcharuk, V., Radevski, I., Rogger, M., Salinas, J. L., Sauquet, E., Šraj, M., Szolgay, J., Viglione, A., Volpi, E., Wilson, W., Zaim, K. & Živković, N. (2017). Changing climate shifts timing of European floods. *Science 357*(6351), 588–590. https://doi.org/10.1126/science.aan2506.

BOE (Boletín Oficial del Estado) (2016) núm. 60, Jueves 10 de marzo de 2016.

Dębski, K. (1966). *Continental Hydrology: Physics of Water, Atmospheric Precipitation and Evaporation.* Scientific Publications Foreign Cooperation Center of the Central Institute for Scientific, Technical, and Economic Information. Translated from Polish, published for the U. S. National Science Foundation, Washington DC, pp. 315–477.

Dieperink, C., Hegger, D. L. T., Bakker, M. H. N., Kundzewicz, Z. W., Green, C. & Driessen, P. P. J. (2016). Recurrent governance challenges in the implementation and alignment of flood risk management strategies: a review. *Water Resources Management 30*(13), 4467–4481. doi:10.1007/s11269-016-1491-7.

Döll, P., Jiménez-Cisneros, B., Oki, T., Arnell, N. W., Benito, G., Cogley, J. G., Jiang, T., Kundzewicz, Z. W., Mwakalila, S. & Nishijima, A. (2015). Integrating risks of climate change into water management. *Hydrological Sciences Journal 60*(1), 4–13. doi:10.1080/02626667.2014.967250.

Driessen, P. P. J., Hegger, D. L. T., Bakker, M. H. N., van Rijswick, H. F. M. W. & Kundzewicz, Z. W. (2016). Toward more resilient flood risk governance. *Ecology and Society 21*(4), 53. https://doi.org/10.5751/ES-08921-210453.

DWA-A118. (2006). *Hydraulische Bemessung und Nachweis von Entwässerungs-Systemen.* DWA, Hennef.

EC (European Commission) (2019). *Commission Report on the Implementation of the Water Framework Directive (Assessment of the Second River Basin Management Plans) and the Floods Directive (First Flood Risk Management Plans).* Report from the Commission to the European Parliament and the Council, Brussels, 26.2.2019 COM(2019) 95 final. Available at: https://ec.europa.eu/info/sites/info/files/com_report_wfd_fd_2019_en_1.pdf.

EN 752 (2008). *Drain and Sewer Systems Outside Buildings.*

EN 752 (2017). *Drain and Sewer Systems Outside Buildings – Sewer System Management.*

Environment Agency (UK) (2016) *Guidance – Flood Risk Assessments: Climate Change Allowances.* Available at: https://www.gov.uk/guidance/flood-risk-assessments-climate-change-allowances (accessed 22 July 2020).

EU (European Union) (2007). Directive 2007/60/EC of the European Parliament and of the Council of 23 October 2007 on the Assessment and Management of Flood Risks. Official Journal of the European Union L288, 27–34.

Hattermann, F. F., Kundzewicz, Z. W., Huang, S., Vetter, T., Kron, W., Burghoff, O., Merz, B., Bronstert, A., Krysanova, V., Gerstengarbe, F. -W., Werner, P. & Hauf, Y. (2012). Flood risk in holistic perspective – observed changes in Germany. In: *Changes in Flood Risk in Europe.* Kundzewicz, Z. W. (ed.). Special Publication No. 10, IAHS Press, Wallingford, Oxfordshire, UK, Chap. 11, pp. 212–237.

Hegger, D. L. T., Driessen, P. P. J., Wiering, M., van Rijswick, H. F. M. W., Kundzewicz, Z. W., Matczak, P., Crabbé, A., Raadgever, G. T., Bakker, M. H. N., Priest, S. J., Larrue, C. & Ek, K. (2016). Toward more flood resilience: is a diversification of flood risk management strategies the way forward? *Ecology and Society 21*(4), 52. https://doi.org/10.5751/ES-08854-210452.

Hirabayashi, Y., Mahendran, R., Koirala, S., Konoshima, L., Yamazaki, D., Watanabe, S., Kim, H. & Kanae, S. (2013). Global flood risk under climate change. *Nature Climate Change* 3, 816–821. doi:10.1038/nclimate1911.

Hodgkins, G. A., Whitfield, P. H., Burn, D. H., Hannaford, J., Renard, B., Stahl, K., Fleig, A. K., Madsen, H., Mediero, L., Korhonen, J., Murphy, C. & Wilson, D. (2017). Climate-driven variability in the occurrence of major floods across North America and Europe. *Journal of Hydrology* 552, 704–717. doi:10.1016/ j.jhydrol.2017.07.027.

IPCC. (2013). *Climate Change 2013: The Physical Science Basis: Contribution of Working Group I to the Fifth Assessment Report of the Intergovernmental Panel on Climate Change.* Stocker, T. F., Qin, D., Plattner, G.-K., Tignor, M., Allen, S. K., Boschung, J., Nauels, A., Xia, Y., Bex, V. & Midgley, P. M. (eds). Cambridge University Press, Cambridge, UK, NY, USA, p. 1535.

Ivancic, T. J. & Shaw, S. B. (2015). Examining why trends in very heavy precipitation should not be mistaken for trends in very high river discharge. *Climatic Change 133*, 681–693. doi:10.1007/s10584-015-1476-1.

Junghänel, T., Ertel, H. & Deutschländer, T. (2017). *KOSTRA-DWD-2010R. Bericht zur Revision der koordinierten Starkregenregionalisierung und -auswertung des Deutschen Wetterdienstes in der Version 2010.* Deutscher Wetterdienst, Offenbach.

Koutsoyiannis, D., Efstratiadis, A., Mamassis, A. & & Christofides, N., A. (2008). On the credibility of climate predictions. *Hydrological Sciences Journal 53*(4), 671–684. doi:10.1623/hysj.53.4.671.

Koutsoyiannis, D., Montanari, A., Lins, H. F & Cohn, T. A. (2009). Climate, hydrology and freshwater: towards an interactive incorporation of hydrological experience into climate research. Discussion of 'The implications of projected climate change for freshwater resources and their management' by Kundzewicz et al. (2008) *Hydrological Sciences Journal 54*(2), 394–405. doi:10.1623/hysj.54.2.394.Kron, W., Eichner, J. & Kundzewicz, Z. W. (2019). Reduction of flood risk in Europe – reflections from a reinsurance perspective. *Journal of Hydrology 576*, 197–209.

Krysanova, V., Donnelly, Ch., Gelfan, A., Gerten, D., Arheimer, B., Hattermann, F. & Kundzewicz, Z. W. (2018). How the performance of hydrological models relates to credibility of projections under climate change. *Hydrological Sciences Journal 63*(5), 696–720. doi:10.1080/02626667.2018.1446214.

Kundzewicz, Z. W. (2012). *Changes in Flood Risk in Europe.* Special Publication No. 10, IAHS Press, Wallingford, Oxfordshire, UK, p. 516+xvi.

Kundzewicz, Z. W. & Schellnhuber, H. -J. (2004). Floods in the IPCC TAR perspective. *Natural Hazards 31*, 111–128.

Kundzewicz, Z. W. & Stakhiv, E. (2010). Are climate models 'ready for prime time' in water resources management applications or is more research needed? *Hydrological Sciences Journal 55*(7), 1085–1089.

Kundzewicz, Z. W., Graczyk, D., Maurer, T., Pińskwar, I., Radziejewski, M., Svensson, C. & Szwed, M. (2005). Trend detection in river flow series: 1. Annual maximum flow. *Hydrological Sciences Journal 50*(5), 797–810.

Kundzewicz, Z. W., Mata, L. M., Arnell, N. W., Döll, P., Jimenez, B., Miller, K., Oki, T., Şen, Z. & Shiklomanov, I. (2008). The implications of projected climate change for freshwater resources and their management. *Hydrological Sciences Journal 53*(1), 3–10. doi:10.1623/hysj.53.1.3.

Kundzewicz, Z. W., Kanae, S., Seneviratne, S. I., Handmer, J., Nicholls, N., Peduzzi, P., Mechler, R., Bouwer, L. M., Arnell, N., Mach, K., Muir-Wood, R., Brakenridge, G. R., Kron, W., Benito, G., Honda, Y., Takahashi, K. & Sherstyukov, B. (2014). Flood risk and climate change: global and regional perspectives. *Hydrological Sciences Journal 59*(1), 1–28.

Kundzewicz, Z. W., Krysanova, V., Dankers, R., Hirabayashi, Y., Kanae, S., Hattermann, F. F., Huang, S., Milly, P. C. D., Stoffel, M., Driessen, P. P. J., Matczak, P., Quevauviller, P. & Schellnhuber, H.-J. (2017). Differences in flood hazard projections in Europe – their causes and consequences for decision making. *Hydrological Sciences Journal 62*(1), 1–14.

Kundzewicz, Z. W., Krysanova, V., Benestad, R. E., Hov, Ø., Piniewski, M. & Otto, I. M. (2018a). Uncertainty in climate change impacts on water resources. *Environmental Science & Policy 79*, 1–8.

Kundzewicz, Z. W., Pińskwar, I. & Brakenridge, G. R. (2018b). Changes in river flood hazard in Europe: a review. *Hydrology Research 49*(2), 294–302. doi:10.2166/nh.2017.016280.

Kundzewicz, Z. W., Hegger, D. L. T., Matczak, P. & Driessen, P. P. J. (2018c). Flood risk reduction: structural measures and diverse strategies. *Proceedings of the National Academy of Sciences 115*(49), 12321–12325. doi:10.1073/pnas.1818227115.

Kundzewicz, Z. W., Szwed, M. & Pinskwar, I. (2019). Climate variability and floods: a global review. *Water 11*(7), 1399.

Kundzewicz, Z. W., Huang, J., Pinskwar, I., Su, B., Szwed, M. & Jiang, T. (2020). Climate variability and floods in China – a review. *Earth-Science Reviews, 103434*, https://doi.org/10.1016/j.earscirev.2020.103434.

Licznar, P. & Zaleski, J. (2020). *Metodyka Opracowania Polskiego Atlasu Natężeń Deszczów (PANDa).* Instytut Meteorologii i Gospodarki Wodnej. Państwowy Instytut Badawczy, Warsaw, p. 143.

Licznar, P., Siekanowicz-Grochowina, K., Oktawiec, M., Kotowski, A. & Bursztа-Adamiak, E. (2018). Empiryczna weryfikacja formuły Błaszczyka do obliczania wartości natężenia deszczu miarodajnego. *Ochrona Środowiska 40*(2), 17–22.

Lugeri, N., Kundzewicz, Z. W., Genovese, E., Hochrainer, S. & Radziejewski, M. (2010). River flood risk and adaptation in Europe – assessment of the present status. *Mitigation and Adaptation Strategies for Global Change 15*(7), 621–639. doi:10.1007/s11027-009-9211-8.

Madsen, H., Lawrence, D., Lang, M., Martinkova, M. & Kjeldsen, T. R. (2014). Review of trend analysis and climate change projections of extreme precipitation and floods in Europe. *Journal of Hydrology 519*, 3634–3650. doi:10.1016/j.jhydrol.2014.11.003.

Malitz, G. & Ertel, H. (2015). *KOSTRA-DWD-2010 – Starkniederschlagshöhen für Deutschland (Bezugszeitraum 1951 bis 2010) – Abschlussbericht, Offenbach am Main.*

Mendoza, G., Jeuken, A., Matthews, J. H., Stakhiv, E., Kucharski, J. & Gilroy, K. (2018). *Climate Risk Informed Decision Analysis (CRIDA): Collaborative Water Resources Planning for an Uncertain Future.* UNESCO and ICIWaRM Press, Paris and Alexandria, VA, ISBN: 9789231002878.

Merkblatt Nr. 4.3/3 (2009). *Bemessung von Misch-und Regenwasserkanälen. Teil 1: Klimawandel und Möglicher Anpassungsbedarf. Bayerischen Landesamtes für Umwelt (Referat 66).*

Milly, P. C. D., Betancourt, J., Falkenmark, M., Hirsch, R. M., Kundzewicz, Z. W., Lettenmaier, D. P. & Stouffer, R. J. (2008). Stationarity is dead: whither water management? *Science 319*, 573–574.

Milly, P. C. D., Betancourt, J., Falkenmark, M., Hirsch, R. M., Kundzewicz, Z. W., Lettenmaier, D. P., Stouffer, R. J., Dettinger, M. D., Krysanova, V. & Stouffer, R. J. (2015). On critiques of 'Stationarity is dead: whither water management?' *Water Resources Research 51*(9), 7785–7789.

Nakamura, S. & Oki, T. (2018). Paradigm shifts on flood risk management in Japan: detecting triggers of design flood revisions in the modern era. *Water Resources Research*. https://doi.org/10.1029/2017WR022509.

NAP 1-2-0.3 (2021) Norma Adif plataforma. Climatología, hidrología y drenaje, 2a Edición.

Norel, M., Krawiec, K. & Kundzewicz, Z. W. (2021). Machine-learning modeling of climate variability impact on river runoff. *Water 13*, 1177. https://doi.org/10.3390/w13091177.

Pall, P., Aina, T., Stone, D. A., Stott, P. A., Nozawa, T., Hilberts, A. G. J., Lohmann, D. & Allen, M. R. (2011). Anthropogenic greenhouse gas contribution to flood risk in England and Wales in autumn 2000. *Nature 470*(7334), 382–385. doi:10.1038/nature09762.

Perica, S., Pavlovic, S., Laurent, M. S., Trypaluk, C., Unruh, D. & Wilhite, O. (2018) NOAA Atlas 14. Precipitation-frequency atlas of the United States, Volume 11, Version 2.0: Texas. Available at: https://www.nws.noaa.gov/oh/hdsc/PF_documents/Atlas14_Volume11.pdf.

PN-S-02204 (1997) Drogi samochodowe. Odwodnienie dróg, Warsaw, PKN.

Prudhomme, C., Kay, A. L., Crooks, S. & Reynard, N. (2013). Climate change and river flooding: Part 2 sensitivity characterisation for British catchments and example vulnerability assessments. *Climatic Change 119*(3–4), 949–964.

RAS-Ew (2005) *Fortschreibung der Richtlinien für die Anlage von Straßen, Teil: Entwässerung*. FGSV Verlag GmbH.

Ray, P. A. & Brown, C. M. (2015). *Confronting Climate Uncertainty in Water Resources Planning and Project Design: The Decision Tree Framework*. World Bank, Washington, DC. Available at: https://openknowledge.worldbank.org/handle/10986/22544.

Reinhold, F. (1940). Regenspenden in Deutschland. *Archiv für Wasserwirtschaft 56*, pp. 71, Berlin, Germany.

Richtlinie 836 (1999) Erdbauwerke und sonstige geotechnische Bauwerke planen, bauen und instand halten, 1999, DB Netz AG.

Salas, J. D., Obeysekera, J. & Vogel, R. M. (2018). Techniques for assessing water infrastructure for nonstationary extreme events: a review. *Hydrological Sciences Journal 63*(3), 325–352. doi:10.1080/02626667.2018.1426858.

Staufer, P., Leckenbusch, G. & Pinnekamp, J. (2010). Die Ermittlung der relevanten Niederschlags-charakteristik für die Siedlungsentwässerung im Klimawandel. *KA Korrespondenz Abwasser, Abfall 57*, 12.

Strategy for the IRBD Rhine for Adapting to Climate Change (2015). *Report No. 19. International Commission for the Protection of the Rhine (ICPR)*. Koblenz, Germany. ISBN: 978-3-941994-78-2.

Tullos, D. (2018). How to achieve better flood-risk governance in the United States. *Proceedings of the National Academy of Sciences 115*(15), 3731–3734. doi:10.1073/pnas.1722412115.

Vogel, R. M., Yaindl, Ch. & Walter, M. (2011). Nonstationarity: flood magnification and recurrence reduction factors in the United States. *Journal of the American Water Resources Association (JAWRA) 47*(3), 464–474. doi:10.1111/j.1752-1688.2011.00541.x.

White, G. F. (1945). *Human Adjustment to Floods: Department of Geography Research Paper No. 29*. University of Chicago, Chicago.

Willems, P. (2013). Revision of urban drainage design rules after assessment of climate change impacts on precipitation extremes at Uccle, Belgium. *Journal of Hydrology 496*, 166–177.

Wytyczne obliczania ilości wód opadowych i roztopowych na obszarze kolejowym Is-2 (2019) *PKP Polskie Linie Kolejowe S.A.*, Warsaw.

Wyżga, B., Kundzewicz, Z. W., Konieczny, R., Piniewski, M., Zawiejska, J. & Radecki-Pawlik, A. (2018). Comprehensive approach to the reduction of river flood risk: case study of the upper Vistula basin. *Science of the Total Environment 631–632*, 1251–1267. https://doi.org/10.1016/j.scitotenv.2018.03.015.

First received 11 October 2021; accepted in revised form 11 October 2021. Available online 28 October 2021

 Water Infrastructure Planning, MDCU doi: 10.2166/wp.2021.345

The centrality of engineering codes and risk-based design standards in climate adaptation strategies

Eugene Z. Stakhiv[a,b]

[a] Environmental Science Studies Department, Johns Hopkins University, Washington DC, USA
[b] UNESCO International Center for Integrated Water resources Management (ICIWaRM), Alexandria, VA, USA
E-mail: estakhi1@jhu.edu; eugene.z.stakhiv@usace.army.mil

ABSTRACT

Engineering codes, design standards and analytical criteria for hydraulic structures are the final determinative specifications for designing and constructing a water resources project. As such, they are the authoritative and legally accepted standards for project design and construction. Engineering codes and standards are developed to optimize public safety and performance by focusing on structural reliability, which includes a wide range of extreme conditions that encompass most contemporary climate uncertainties, and which are likely to overlap some portion of future climate non-stationary conditions. Current practices of risk-based planning and design standards have evolved incrementally, responding to each catastrophic natural disaster, whether it is geotechnical, floods, droughts or hurricanes. Design standards and building codes encompass an accumulation of changes that progressively reflect changing climate conditions, most notably because they focus on climate extremes. Design standards and embedded 'safety factors' that are based on extremes are likely to encompass a good deal of an anticipated non-stationary climate regime and its associated uncertainties. Modern risk analysis methods and risk-based standards, codes and methods comprise an important part of a progressive autonomous adaptation to climate change. They represent an essential component of 'no regrets' climate adaptation.

Key words: Engineering codes, 'No regrets' climate adaptation, Risk-based design standards, Risk tolerance, Safety factors

INTRODUCTION

New and ageing infrastructure will require huge investments in public infrastructure over the next 30 years because of growing populations, increased trade, prosperity and adaptation to climate change. Global estimates for water-related infrastructure investment requirements are nearly $200 billion/yr. (Oxford Economics, 2017). These investments can serve as the foundation of a 'no regrets' climate adaptation strategy. 'No regrets' adaptation connotes that the benefits of a selected option are justified regardless of whether the projected impacts of future climate change materialize and can be economically justified by existing benefit–cost procedures (Prabhakar *et al.* 2009, OECD, 2018a, 2018b).

Whatever actions ultimately lead to the decarbonization of the global energy system, it will be many decades before they have a discernible effect on the climate (Pielke *et al.*, 2007). Historical emissions dictate that climate change is unavoidable, even under the most favorable assumptions. Hence, the anticipated adverse impacts of climate change require a focus on adaptation that will require innovative technological solutions that increase robustness, resilience and reliability of coastal protection, flood protection, irrigation, water supply and hydropower projects and systems.

Even the most optimistic emissions projections show global greenhouse gas concentrations rising for the foreseeable future (IPCC, 2021). Virtually, every climate impact projected to result from increasing greenhouse gas concentrations – from rising storm damage to declining biodiversity – already exists as a major concern. Existing engineering methods that deal with the extremes of contemporary (stationary) climate variability can serve as a platform for a new class of risk-based methods that will more effectively deal with a non-stationary and more uncertain climate (Stakhiv, 2011; Read & Vogel, 2015; Read & Vogel, 2016; François *et al.*, 2019; Hecht & Vogel, 2020; Vogel & Castellerin, 2017).

Public engineering practices, i.e., that engineering conducted under the auspices of sanctioned public institutions, are bound by numerous overlapping laws and regulations that are conservative and generally risk-averse. Large public works projects (dams, flood control systems, urban drainage systems, wastewater treatment, water supply and irrigation systems) are often legally required to perform and provide services with a high degree of reliability – as close to 'fail-safe' as possible, since communities depend on service reliability and public safety of these water management systems. Public engineering is always accompanied by a dense dendritic network of laws, administrative regulations and public review procedures that guide project planning and design procedures. These engineering planning and design practices are translated into stringent building codes, design standards and analytical criteria (Retief *et al.*, 2013; Ayyub, 2018, 2019).

According to the U.S. Office of Management and Budget, Circular No. A-119 (OMB, 1998), a standard consists of technical definitions, procedures and/or guidelines that specify minimum requirements or instructions for design of infrastructure or equipment. This can be done by specifying either the methods or the results; the latter is termed 'performance specifying'. Most importantly, a standard provides standardization or agreement within an industry or water management sector, which translates to a commonly understood frame of reference among planners, design engineers, manufacturers and contractors.

A code is a standard that has been enacted into law by a local, regional or national authority having jurisdiction, so that the engineer or contractor is legally obligated to comply with the code. Noncompliance can result in being prosecuted. The code may be an industry, government or voluntary professional consensus-based standard. Codes are generally accepted sets of rules that tell engineers what they need to do. Standards provide the 'how to' instructions for implementing codes.

Engineering codes and standards are constantly evolving, as knowledge increases, albeit slowly because of legal liability implications and extensive peer-review and testing processes. Hence, new knowledge and rapid advances in technology are not readily assimilated into engineering practice. Designing infrastructure for future climate uncertainties using novel and untested analytical methods further complicates this matter. Determining an engineer's liability in any future design failure becomes ever more difficult, because of the uncertainties associated with widely varying sources of data generated by climate models and climate scenarios, or unproven design theories for structures at the cutting edge of the profession.

Professional legal liability for infrastructure failure is an even more important reason for carefully vetted and institutionally accepted design procedures and standards (Peck & Hoch, 1985; Hatem, 1989). Averting failure is at the heart of engineering design and explains, to a large degree, why engineering standards and codes that institutions designed to protect the populace from natural and man-made hazards are fundamentally conservative (Petroski, 2006).

Design engineers are legally liable for infrastructure failure if they diverge significantly from acceptable practices and standards (Seemann, 1991). Hence, any changes, such as those associated with incorporating the largely uncertain effects of global warming and climate non-stationarity, require a considerable degree of testing, peer review and formal acceptance by professional engineering societies, federal agencies, ministries and multilateral lending institutions, such as the World Bank and Asian Development Bank (Ittmann, 2020). In general, contemporary laws protect structural engineers who have designed at the accepted professional level of competence but not necessarily at the state of the art (Peck & Hoch, 1985).

It is important to recognize that technologies devised to solve practical water management problems preceded the development of scientific knowledge and associated predictive techniques. Hydraulic engineering (water infrastructure) achievements preceded the science of hydrology by several millennia and began with water development of ancient civilizations in Mesopotamia, Egypt, India, China and Greece aimed at controlling floods and the flow of water, initially for agricultural needs (irrigation) and later for communal water supply (Koutsoyiannis, 2014). Hence, conservative safety factors embedded within design standards that are in use today, are the result of millennia of trial and error and adjustments after many failures (Petroski, 1985, 2006). During these millennia (3000BC–present), climate varied considerably, encompassing higher global temperatures, and lengthy periods of floods and droughts.

Accounting for engineering liability has been true since the times of Hammurabi (1754 BCE), whose code consisted of 282 laws, a number of which dealt harshly with design failure and engineering liability, following the biblical 'eye for an eye' compensation standard: https://avalon.law.yale.edu/ancient/hamframe.asp:

- 227. If a builder builds a house for a man and does not make its construction firm, and the house which he has built collapses and causes the death of the owner of the house, that builder shall be put to death.
- 228. If it causes the death of the son of the owner of the house, they shall put to death a son of that builder.
- 229. If it causes the death of a slave of the owner of the house, he shall give to the owner of the house a slave of equal value.

Adaptation of infrastructure to climate change calls for upgraded risk-based engineering design standards and codes, along with new approaches and frameworks for water management planning, since climate change directly affects the hydrologic basis of planning and design. Climate change introduces many new uncertainties that will stretch the limits of knowledge and risk-based design. Public acceptance of risk and quantifying their degree of risk tolerance is an important component of devising new risk-based standards. Risk acceptance and tolerance by the public is a social construct requiring the acquiescence of, and affirmation by the public, via legislation and approval of professional engineering societies (Institute for Water Resources, 2010; Olsen, 2015; WFEO, 2015; Engineers Canada, 2018).

Design standards, codes and derivative regulations that govern infrastructure design are scientifically well-grounded and often finely negotiated in concert with ministries, agencies and professional societies and are legally balanced, making them slower to adapt (Vrouwenwelder, 2008; Ayyub, 2018). Project planning guidelines, such as embodied in the U.S. Water Resources Council's 'Principles and Standards' (US Water Resources Council, 1983), which precede project design, incorporate specific legislated rules and procedures and, in that sense, can be considered as both codes and standards, as they are legally required for federally funded projects in the U.S. (Major & Stakhiv, 2020).

It should be recognized that many water infrastructure design procedures and standards, already encompass and compensate for a considerable portion of climate and weather extremes and hydrologic uncertainties through various design redundancies and 'safety factors' (Matalas & Fiering, 1977; Auld, 2008; Ayyub, 2018). In reality, the combination of standards, codes and built-in engineering safety factors that aim to deal with a wide range of uncertainties can be viewed as a collective foundation of a 'no regrets', precautionary approach to building more resilient infrastructure (OECD, 2018a, 2018b). Therefore, it can reasonably be posited that this historically conservative engineering approach implicitly and explicitly incorporates many, but not all, aspects of 'robust' and 'resilient' design features called for by many advocates of climate vulnerability assessments and 'robust decision making' (Hallegatte *et al.*, 2012; Lempert, 2019).

UNIFORMITY OF ENGINEERING STANDARDS AND PROCEDURES

Averting failure is the key underlying theme of engineering practice – and is at the heart of engineering design, which is why engineering design standards are conservative by nature, building in redundancies and safety factors to deal with uncertainties and unknowns. Engineering standards are at the nexus of rapid change and orderly processes for adapting to those changes. As Petroski (2006) states: '[S]uccessful design is not the achievement of perfection but the minimization and accommodation of imperfection' and '[E]ngineering is achieving function while avoiding failure'. Acknowledging and estimating the imperfections of our knowledge (uncertainty) and the risks of failure are at the core of hydrologic and hydraulic engineering. Continuous testing and upgrades are embodied in the most recent ensemble of engineering design standards – all of which are subject to constant improvements and updating of techniques, based mainly on failures due to an unending series of highly publicized catastrophic events.

Planning and designing public works infrastructure within the context of any modern public decision-making setting require a high degree of safety, reliability, procedural uniformity, analytical replicability and legal accountability to the public, as part of a lengthy series of steps in a typically cumbersome and complicated publicly conducted planning and review process. In any publicly financed planning endeavor, the first level of accountability rests with meeting publicly stated needs and demands for water-related services, incorporating public inputs into the formulation and selection of options for meeting those demands. Planners must honor those publicly derived choices.

The second level of accountability is following established procedures that are uniformly applied, transparent and analytically replicable. The third level of accountability rests with legally and administratively established rules for the proper operation and maintenance and reliable service delivery and performance of a selected project. The fourth, and most critical level of accountability to the public, is ensuring the reliability and safety of the project, which requires explicit definition and modeling of potential pathways of failure and public acceptance of 'tolerable' risks, residual risks and consequences of failure.

It is important to note, also, that water resources-based infrastructure systems are not stand-alone systems. They are usually part of a larger network of interdependent systems, and thus considered as 'critical infrastructure' – which if it fails, can result in a domino effect of other dependent infrastructure failures. Modern society is highly dependent on many kinds and levels of networks – many of which interact with one another and are highly interdependent. Critical infrastructure systems are the lifelines for our health, safety and prosperity. They are our nerve systems facilitating communications, transportation, trade and financial transactions.

Once any of these critical components fail – say the energy grid, as it did during Hurricane Maria in Puerto Rico on July 20, 2017, there often follows a series of devastating cascading effects on cities and communities. Hospitals, water treatment plants, commercial centers, ports, waterways, transportation and industry are all adversely affected. More than three years after the disaster, Puerto Rico had not fully recovered, and its economy and social welfare systems will take years to recover to pre-Maria conditions (GAO, 2021). The potential failure of 'critical infrastructure' has a heightened level of accountability for the engineering profession, which is why engineering codes and standards go through such rigorous peer review and testing before they become embedded as 'best management practices'.

The core advantages of engineering codes and standards are central to 'best management practices' – i.e., uniformity of approach, replicability of analyses and consistency in outcomes of engineering design and throughout the globe, wherever hydraulic structures may be constructed. Such codified standards and practices serve as the basis for enhanced transparency and public trust and accountability.

INTERNATIONAL PRINCIPLES AND GUIDELINES FOR RISK-BASED DESIGN STANDARDS

Principles for risk analysis and risk-based structural reliability analysis have evolved for over 100 years, are now encoded in a series of guidelines developed by the International Organization for Standardization (ISO, 2015). Most of the ISO standards are derived from and represent the experience and best management practices of leading public engineering agencies, as they are formulated by select committees of practicing professionals. ISO standards exist for environmental management, health and safety, food safety, energy management, etc. https://www.iso.org/standards.html.

There is a great deal of overlap between the uniformity within ISO standards with the more detailed engineering professional standards and agency design regulations. For example, while new ISO standards are being formulated for flood risk assessment (Yunika *et al.*, 2017), they are based on the experience and existing guidelines of national agencies such as the Dutch Rijkswaterstaat, the Corps of Engineers or Japan's Ministry of Land, Infrastructure and Transportation. A similar situation exists with the European Flood Directive of 2007, which set in motion a cascading suite of new rules, regulations, prescribed analytical procedures and design standards (Tsakiris, 2015; Yannopoulos *et al.*, 2015).

In 2009, ISO published a standard for risk management (ISO 31000:2009). The criteria for risk assessment proposed in ISO 31000 largely spring from a probabilistic logic, considering many aspects of standard risk analysis techniques (Purdy, 2010; Lalonde & Boiral, 2012). ISO also published a series of guidelines for adaptation to climate change (ISO 14090:2019). This specifies principles, requirements and guidelines for adaptation to climate change, which include the integration of adaptation within and across organizations, understanding impacts and uncertainties and how these can be used to inform decisions.

There are many professional engineering societies in each country, setting specific codes and standards for planning and design, as there are international bodies such as the International Commission on Large Dams (ICOLD, 2020). ICOLD leads the profession in setting standards and guidelines to ensure that dams are built and operated safely, efficiently, economically and are environmentally sustainable and socially equitable. ICOLD sets the standards for dam design construction and safety for most of the world's engineering agencies.

Of equal interest are the engineering profession's approaches to planning and design under climate uncertainty. In particular, the World Federation of Engineering Organizations (WFEO, 2015) issued a set of 'Principles for Engineers in Climate Adaptation'. These principles were further embellished and encoded by 'Engineers Canada' (2018). Engineers Canada is the national organization of the provincial and territorial associations that regulate and license the practice of engineering in Canada. Similar policy statements and standards are being developed in the U.S. by the American Society of Civil Engineers (ASCE). ASCE Policy 360 deals with climate change (https://www.asce.org/advocacy/policy-statements/ps360–climate-change/). ASCE also published a 'Manual of Practice' for 'Climate-Resilient Infrastructure: Adaptive Design and Risk Management' (Ayyub, 2018).

PLANNING VS DESIGN CONSIDERATIONS

Planning water resources projects, by its nature, encompasses a great deal of information and projections about future conditions, including population, economic growth, land use, etc. Each projection carries varying degrees of uncertainty. Uncertainty is an inherent feature of any future-oriented planning effort. Risk and uncertainty analysis are about quantifying, to the extent possible, the major uncertainties that influence key decisions for selection of the project alternative best suited to solve a particular problem; i.e., how large should the project be (reservoir capacity, levee height, pipeline diameter, channel depth, etc.), and whether it is economically feasible and environmentally acceptable.

Existing engineering standards and criteria, on the other hand, generally represent a combination of, and tradeoff between costs and socially established risk tolerance norms, such as the well-accepted 100-year flood protection standard used for flood insurance purposes and flood protection. There are many other quantified standards to protect against the extremes of natural hazards phenomena, such as earthquakes, floods, hurricanes and droughts, such as the 'probable maximum flood (PMF)', 'probable maximum hurricane', 'design flood' or 'probable maximum precipitation' (Auld, 2008; Ayyub, 2018). More recently, because of climate change uncertainties, risk analysis and management has evolved and broadened to incorporate such broad concepts as 'resilience' and 'robustness' as project planning objectives, in addition to existing and long-held engineering criteria of 'safety' and 'reliability'.

It is no longer acceptable to merely protect against a defined natural hazard event using proven engineering standards. Risk management strategies and solutions should also ensure that communities are able to withstand unforeseen events that are outside the established 'norms' (robust solutions), but also are able to rebound quickly by restoring critical infrastructure services and resuming commercial and economic activities after a devastating event (resilient solutions). In the past, the focus on 'reliability' was a substitute for robustness and resilience in that hydraulic structures not only had to withstand 'loads' that surpassed the historical extremes (robust designs), but also provided a high degree of performance, in terms of delivery of reliable services, whether it be water supply, hydropower, flood control, irrigation or wastewater management under a variety of extreme events.

Risk-based design for planning and designing future water resources projects (hydraulic structures) typically has three important components:

- Probabilistic analysis of future demands on the services of the structure – i.e. services required under a variety of future conditions of growth and development.
- Probabilistic analysis of the reliability of services of the structure provides under various conditions (flood protection, water supply, environmental flows, hydropower, etc.).
- Probabilistic reliability analysis of the structure itself in determining how the structure may withstand different environmental loads.

The first two, future demands and reliability of service delivery, are typically conducted as part of planning processes that are an integral part of project justification. Is the project economically feasible? Does it address ecological and social objectives? Does it meet all the legal and community constraints? Project design methods, on the other hand, have been developed to achieve a reasonable balance between functionality, performance, safety and economic feasibility of such structures. The reliability-based design aims to achieve a consistent and acceptable probability of structural failure across a range of conditions. The risk-based design considers the expected consequences of failure as a principal basis for decision-making and design (Retief *et al.*, 2013; Diamantidis *et al.*, 2016; Sykora *et al.*, 2017).

There is a conceptual divergence, however, between the risk-based analytical approaches used as part of the project planning, with optimization of associated economic justification processes, and that of subsequent engineering design of the selected 'optimized' option. Because of the differences in the many institutional and legislated requirements underlying water resources management, for the same problem set, different planning processes of various agency or ministry project selection frameworks would not necessarily lead to the same outcomes in terms of project capacity, performance, costs or basic design. However, because of the greater uniformity in engineer design procedures, the actual design of any selected project would likely be more consistent in many different settings because of greater uniformity within engineering design practices.

At the core of the differences in analytical procedures and decision rules between project feasibility planning and engineering design is the importance of societal or community 'safety'.

Project feasibility planning centers on economic justification analyses and decision rules, e.g., maximize net economic benefits, while project design focuses on minimizing failure risk and loss of life (Yoe, 2017). An example of this duality is to be found in the most recent Corps of Engineers planning and design guidelines.

The Corps' most recent engineering and design guidance (U.S. Army Corps of Engineers, 2019a) lays out the following principles.

A risk-informed approach will be used for all dam and levee designs for new projects, modifications, improvements, rehabilitation or repairs. Risk assessments are the cornerstone for the application of a risk-informed decision-making approach. The following are the overarching principles to follow:

(a) *Hold life safety paramount.* While seeking to manage risk to people, property and the environment, USACE will consider risk to life safety as priority. Tolerable risk guidelines will be used as the risk-informed decision goal for life safety.
(b) *Risk-informed decision-making.* Decisions for risk management actions will be commensurate with the level of risk that exists when a dam or levee is present to ensure wise federal investments.
(c) *Ensure open and transparent engagement.* USACE will engage project sponsors in all design activities related to their dams and/or levees.
(d) *Learn and adapt.* Risk assessments will be used to evaluate if designs must be up-scaled (i.e., use more stringent design criteria) or downscaled (i.e., use less stringent design criteria) using a risk-informed approach as compared to solely considering traditional standards.
(e) *Do no harm.* Risk-informed designs should not increase the risk to the population and property above the risk the population currently experiences.

Indeed, the (Corps' 2019a) most recent guidance on risk analysis and risk tolerance ('Interim Approach for Risk-Informed Designs for Dam and Levee Projects'), while advocating risk-informed decision-making that balances multiple factors, still prioritizes life safety as paramount. This contrasts with the associated planning guidance that was published by the Corps of Engineers (U.S. Army Corps of Engineers, 2019b) for how to conduct risk assessments for flood risk management (FRM) studies. That planning guidance (ER 1105-2-101) emphasizes selecting a project scale that maximizes net economic benefits, subject to environmental constraints.

The two complementary guidance documents essentially disaggregate the notion of 'risk and tolerable risk' into two distinct components: the design phase which emphasizes safety of lives associated with failure of the designed structures and the planning phase which emphasizes project feasibility based on economic damages associated with the causative flood event. The engineering design guidance is much more risk-averse by highlighting risk analytic procedures that address the reliability of the structure to ensure that it does not fail under a wide range of extreme hydrologic loading conditions.

Under economic optimization procedures, adding resilience and robustness elements to a project or a system that would encompass a broader range of future climate uncertainties, as a way of preventing damages from more frequent and/or more damaging events/hazards would typically add additional project costs (Liu & Son, 2020). For such design entails either expanding the size/capacity of a project and/or additional engineering safety factors to account for extreme future climate uncertainties, which inevitably increase project costs – often beyond the level of initially optimized financial or economic feasibility.

Other important differences between planning and design lie in the numerous economic evaluation and decision criteria. In project planning and justification, the use of expected annual damages (EADs) in a traditional benefit–cost procedure essentially has a two-fold effect in minimizing large or extreme events. The EADs approach required by U.S. law, and commonly used by agencies such as the Corps of Engineers, minimizes or suppresses the costs of and avoidance benefits of large events, whereas the use of a high discount rate deflates the impact of future large events, beyond a 15–25-yr planning horizon (Stakhiv, 2011).

Thus, there is always a tradeoff between risk reduction and cost increases, while minimizing residual risks. One of the underlying goals of project planning, then, becomes the identification and reduction of 'residual risks' of both service failure and/or hydraulic failure through a variety of mitigative actions, which also impose added risks, uncertainties and costs. Also, from an equity standpoint, one must identify the sectors of the population that will bear those residual risks, and what array of mitigation and/or compensation measures are provided as part of the plan.

The U.S. Army Corps of Engineers (USACE) and Bureau of Reclamation (BoR), two of the foremost U.S. federal agencies that develop standards for water-related infrastructure, have accelerated their revisions of many planning and design standards, to rely more on risk-based analysis to accommodate climate uncertainties and design more robust and resilient projects (BoR, 2019). The U.S. General Accountability Office (GAO, 2013), the performance monitoring arm of the federal government, evaluated the USACE and BoR climate change adaptation programs and found that both agencies 'have assessed water resource and infrastructure vulnerabilities and taken steps to develop guidance and strategies to adapt to the effects of climate change'.

The U.S. Department of Transportation is another very important standard-setting agency. They have also tackled the issue of climate uncertainties in revising their numerous standards for roads, bridges and culvert designs. The Transportation Research Board (TRB) of the National Research Council prepared a 'Guide for Design Practices' (TRB, 2019) for hydrologic and coastal infrastructure to deal with climate uncertainties.

ENGINEERING DESIGN AND HYDROLOGIC NON-STATIONARITY

A changing climate is likely to lead to a suite of non-stationary and historically unforeseen conditions that serve as the basis for altering not only project planning procedures, but also engineering design criteria and standards (Milly *et al.*, 2008; IPCC, 2021). According to the most recent IPCC assessment (2021), the frequency, intensity and duration of extreme events (rainfall, runoff, temperature, snow, ice, etc.) are likely to change over time, though it is highly uncertain as to which direction these changes may occur, and at what locations or what magnitudes. Numerous studies have diverged on the frequency, duration and extremity of floods and drought – both regionally and globally depending on which the general circulation model was used, which development scenario was selected and in which continent or region the changes occurred (Pechlivanidis *et al.*, 2017; Kettner *et al.*, 2018).

Projecting climate impacts on hydrological processes is unavoidably prone to a concatenation of uncertainties stemming from selection greenhouse gas emission scenarios, climate models and their parameterization, downscaling techniques, bias correction methods, hydrological model structure and parameters. The uncertainties propagate in very complex and sometimes unknown ways, effectively rendering the outcomes useless for long-term water project design decisions (Wilby & Dessai, 2010; Pielke & Wilby, 2012; Pechlivanidis *et al.*, 2017).

The suite of periodic IPCC reports, beginning in 1990 through 2021 created a climate model, scenario-driven 'top-down' analytical paradigm. This was deemed by many engineering practitioners as unsuitable for most water resources management problems – particularly those related to site-specific infrastructure design (Kundzewicz & Stakhiv, 2010; Stakhiv, 2011; Retief *et al.*, 2013; Sykora *et al.*, 2017; Sykora *et al.*, 2018; Underwood *et al.*, 2020).

Hence, several significant adjustments are required in both the planning and design paradigms currently in use by engineering institutions, to deal with site-specific project analysis under non-stationary climate conditions. These include:

- Top-down vs bottom-up climate downscaling/decision-scaling: bridging the gap between climate-driven scenario analysis and hydrologic non-stationarity at the project level.
- Designing 'robust' structures and resilient systems under existing economic optimization and benefit–cost procedures.
- Factoring in non-stationary hydrology – non-stationary extremes for the design, non-stationary mean and variance for planning.
- Bridging gap between the decision criteria associated with the planning paradigm and design paradigm: minimizing risk-cost vs minimizing hydraulic failure and maximizing safety.

Recent methods that aid planning of water infrastructure to bridge some of these gaps, and to enhance project robustness, resilience and reliability have been developed by such leading institutions as the World Bank, Asia Development Bank, Corps of Engineers and UNESCO (Ray & Brown, 2015; Mendoza *et al.*, 2018; Watkiss *et al.* 2020). These methods are extensions of existing institutionally accepted planning protocols, that focus on the added uncertainties of climate change and non-stationarity, under the rubric of 'climate risk-informed decision-making'. The basic purpose is to decide ('decision-scaling') under what circumstances progressively more detailed levels of climate analysis should be used, or whether existing analytical methods, planning protocols and design standards are sufficient for the problem at hand.

The key for planning and engineering design is the abstraction of an empirical hydrologic data set that is the basis for site-specific project-level analysis and converting it to a functional distributional form or type of empirical non-parametric distribution (Beven, 2016). The next step is to transform that selected probability distribution, reflecting stationary climate conditions, into a non-stationary distribution of extremes, used for design (Salas *et al.*, 2018; Serago & Vogel, 2018). Unfortunately, there are still many epistemic (lack of knowledge) and scientific uncertainties in each step of the process, that need to be resolved. Engineers have found practical ways of circumventing these uncertainties, even as they undertake various permutations of accepted statistical analyses to better inform their decisions.

It becomes a matter of choice and judgment as to how to treat those uncertainties, including the selection of formal probabilistic and statistical frameworks to deal with climate non-stationarity (Beven, 2016). As Mondal & Mujumdar (2017) and Salas *et al.* (2018) have shown, there are many choices available to the practicing engineer, but agency procedures require a uniform, standardized approach to these issues, which have not yet been formalized to any significant degree.

As a partial practical substitute, however, there is some comfort in recognizing that current design standards are based on the analysis of continually updated historical records under the hypothesis of stationarity (Kunkel *et al.*, 2013; Olsen, 2015; Ayyub, 2018; USACE, 2021). This has been a practical and effective approach to deal with repeatedly large but unpredictable events, because simultaneously, many federal and state design manuals, as well as the underlying precipitation and runoff data sets are periodically upgraded. The practical implications of these progressive upgrades are that the risk level of infrastructure that is being designed with today's upgraded standards and data may not necessarily change much in the future (Serinaldi & Kilsby, 2015).

Additionally, Serinaldi & Kilsby (2015) dispute the notion that it is important for engineering design to explicitly incorporate climate non-stationarity into contemporary design. They argue that non-stationary models introduce additional sources of uncertainty; that misspecification of non-stationary models can lead to physically inconsistent results and non-stationary models can provide no practical enhancement of the credibility of results. They advocate an approach that emphasizes the 'risk of failure' during the design life of a project as a more realistic measure of the risk for design purposes, and a more understandable one for decision-making purposes. Read & Vogel (2015, 2016) essentially arrive at the same conclusion.

The key substantive and operational distinction, however, is that hydrologic and hydraulic engineers continuously make important design decisions that affect the populace of future generations. Critical decisions cannot wait for better data about future climate conditions from improved computational technologies. Hence, their decisions are based on a body of time-tested 'best management practices' that have undergone not only periodic and intensive practical peer review by their respective professional societies, along with progressive updating of design criteria and standards, but also the most important consequential and pragmatic reviews of all – the failure or non-failure of their designs and creations.

As an example, the reconstruction of infrastructure that occurred after such catastrophic events as Hurricane Katrina in New Orleans, 2005, or Hurricane Maria in Puerto Rico, 2017, could not wait for a better explication of climate uncertainties. Design decisions were made according to prevailing standards, new data and better models, based on best engineering judgment, given the new set of physical circumstances and societal risk tolerance criteria. The new criteria were determined in cooperation between the U.S. Congress, the engineering profession via the Hurricane Katrina Interagency Performance Evaluation Task Force (Link & Harris, 2007) and academic inputs, such as the recommendations of National Academy of Sciences (National Research Council, 2009). Collectively, those standards and analytical methods represented a major advancement of the state-of-art in engineering design for climate adaptation.

RISK ANALYSIS, RESIDUAL RISK AND TOLERABLE RISK

Engineering standards, criteria and regulations reflect each society's accepted levels of 'risk tolerance'. 'Tolerable level of risk' is a phrase describing a level of risk that has been associated with a national, corporate, regulatory or otherwise socially determined level of acceptable risk. A 'tolerable risk' is a previously unacceptable risk whose severity has been reduced to a point where it is tolerated by way of a social consensus. In effect, it is a 'residual risk' after all acceptable and cost-effective risk management measures have been implemented for a particular situation (Institute for Water Resources, 2010, 2017).

For example, the U.S. Congress could, hypothetically, establish the 0.2% annual exceedance frequency as the tolerable level of risk for FRM in vulnerable urban areas. After all, the 1.0% annual exceedance frequency (100-year flood) is the legal standard for flood insurance purposes and floodplain management in the U.S. Then all urban FRM projects must protect against floods with a recurrence interval of 500 years, or less, as a matter of national policy (Institute for Water Resources, 2017). The problem remains of how to quantify a '500-year flood' under non-stationary climate conditions – i.e., what will be the actual flood volume or discharge 50 years hence? Generally speaking, studies have shown that stationary models tend to underestimate extreme floods relative to non-stationary models (Šraj *et al.*, 2016; Mondal & Mujumdar, 2017; Salas *et al.* 2018).

Different nations, like Netherlands or Japan, set their societal risk tolerance levels using different premises and methods to justify their standards for flood control or earthquake management. Both nations are very risk-averse due to their physiographic circumstances which subject the populace to a wide range of natural hazards and other vulnerabilities. A good part of the Netherlands is below sea level. Hence, their risk tolerance level is very low, manifesting itself as very stringent engineering design standards for sea walls, levees and other flood control infrastructure (François *et al.*, 2019). In the Netherlands, standards for levee and dike design, range anywhere from the 300- to 10,000-year flood depending on location. Japan is a flood-prone mountainous nation that is also subjected to typhoons, tsunamis and earthquakes, so their design standards must offer a higher degree of safety and structural reliability to protect against infrastructure failure.

Most of the 'residual risk', i.e., that risk which cannot be ameliorated through structures, zoning policies or insurance, is largely borne by individuals, homeowners and small businesses. Yet the degree of 'residual risk' is rarely quantified or explained to the public in terms they could understand. Residual risk is almost always underestimated because it is difficult to quantify a cascading series of highly interdependent measures, each of which has its own reliability characteristics and risk of failure, including human error (Perrow, 1984). For example, a local flood protection system is a collection of fragmented measures, including disparate building codes, zoning ordinances and structural measures that are implemented over a long period of time and loosely coordinated by multiple authorities. That combination of measures comprises the definition of a 'brittle system' (Fiering, 1982), whose risks of failure are hard to quantify (Haimes *et al.*, 1995).

ADAPTING CONTEMPORARY ENGINEERING RISK MANAGEMENT PRACTICES TO CLIMATE UNCERTAINTIES

It is understandable that forecasting climate change is a highly uncertain technical endeavor. Yet, these uncertainties need to be managed in some uniform and replicable manner, to ensure planning and design transparency for the public, despite difficulties in identifying and quantifying many of these uncertainties. It is even more difficult to convey this uncertain information to the public and decision-makers. Recognizing that there is a great deal of conjecture about uncertainty estimation in the hydrologic sciences, the engineering profession has devised different ways of dealing with such analytically thorny and largely intractable theoretical issues through a series of risk management practices. These practices have evolved to provide more practical and indirect means for resolving and addressing the accumulation of uncertainties through a variety of best management practices and risk management measures.

There are several parallel paths that are currently being undertaken by the water resources engineering profession as part of climate-related risk management practices (Auld, 2008):

1. Update and revise engineering codes and standards that are directly affected by climate-induced changes in hydro-climatological design criteria (e.g., 1% chance flood; 'design flood'; PMF; 'probable maximum hurricane', etc.).
2. Increase safety factors commonly used in design to account for known climate uncertainties.
3. Periodically revise planning and engineering guidance and regulations to account for new information, analytical techniques, design methods and technologies.
4. Develop new analytical techniques that can effectively deal with climate non-stationarity.

All of these upgrades of standards and other pathways are being vigorously pursued by many federal agencies in both Europe and the U.S. Safety factors and design redundancies are another direct way that engineers have always used to deal with uncertainties of natural hazards, modeling and data collection. Since uncertainty analysis is a well-accepted part of engineering practice that encompasses codes, standards and associated regulatory processes, it should be possible to deal with the growing uncertainty of future climate-related design values by identifying and increasing safety factors when evidence points towards the likelihood of increased climate risks over the lifespan of a given structure (Yen & Tung, 1993; Tung, Yen & Melching, 2006; Sykora *et al.*, 2017; Sykora *et al.*, 2018). These safety factors inherently account for increased project and system robustness, used to minimize the risk of a major catastrophe from occurring.

A primary example of current risk-based analytical methods that can be adapted readily to analyze climate non-stationarity are the Corps of Engineers procedures for flood damage analysis, based on a Monte-Carlo simulation of multiple combinations of flood discharges and stages (Kalyanapu *et al.*, 2011). The Flood Damage Reduction Analysis (HEC-FDA) software developed by the Corps' Hydrologic Engineering Center provides the capability to perform an integrated hydrologic engineering and economic analysis during the formulation and evaluation of FRM plans (U.S. Army Corps of Engineers, 2021). The software computes EAD and equivalent annual damages, and computes the annual exceedance probability (AEP) and conditional non-exceedance probability as required for levee certification, and implements the risk analysis procedures described in EM 1110-2-1619 (https://www.hec.usace.army.mil/software/hec-fda/).

In the Corps' HEC-FDA procedures, risk analysis is applied to both economic performance measures (project net benefits and benefit–cost ratio) and to engineering performance measures (probability of flooding). Figure 1 provides a graphical view of the Monte-Carlo sampling procedure that provides a distribution of damages accounting for multiple sources of uncertainty. The U.S. National Research Council (NRC, 2000), in a review of Corps' risk analysis procedures, found that *'while the measures of economic performance in the method are generally practical and informative, there are too many types of engineering performance measures to be clearly understood by most citizens. Standardization to one or two key measures of engineering performance measures would represent an improvement'*.

The NRC committee also recommended that the Corps use AEP as the performance measure of engineering risk. This is a measure of the likelihood that people will be flooded (including the probability of failure of flood damage reduction structures, such as levees) in any given year, considering the full range of floods that can occur and all sources of uncertainty. In effect, Read & Vogel (2015) recommended a similar approach noting that even for stationary processes, the common application of an average return period is problematic because it does not account for a planning horizon and is an average value that may not be representative of the time to the next flood. Read and Vogel introduce 'reliability' as a substitute for 'return period', as a more useful term to explain the risk of failure during the life of a project. Reliability is defined as the probability that a system will remain in a satisfactory state during its lifetime, i.e., that an exceedance event will not occur within a project life of *n* years.

Design standards and computational methods based on extremes represent a progressively incremental 'autonomous adaptation' to climate change and associated uncertainties. A segment of the engineering profession suggests that traditional risk-based design methods may currently be more useful than non-stationary approaches because it is much too difficult to realistically accommodate all the uncertainties associated with climate change. In fact, many authors argue that factoring in these uncertainties, many of them 'unknown unknowns', simply makes the analyses more complex and bewildering for the public and political decision-makers (Serinaldi & Kilsby, 2015; Underwood *et al.*, 2020).

For instance, 'freeboard' is included in the hydrologic design as a typical safety factor to improve the reliability of levees and embankments, which might fail to deliver the intended level of safety in a changing climate. 'Freeboard' is defined by the Federal Emergency Management Agency (FEMA, 2020) as a factor of safety usually

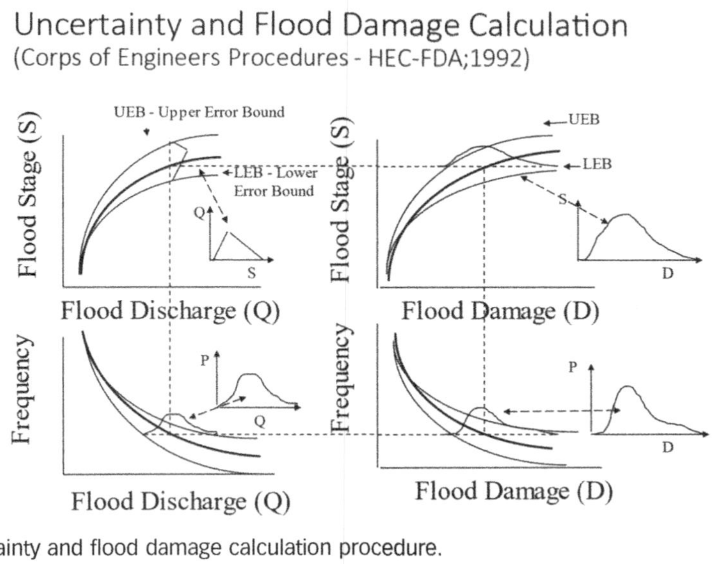

Fig. 1 | HEC-FDA uncertainty and flood damage calculation procedure.

expressed in feet above a flood level for purposes of floodplain management. 'Freeboard' tends to compensate for the many unknown factors which could contribute to flood heights greater than the height calculated for a selected design flood and floodway conditions, such as wave action, bridge and channel constrictions, or the evolving hydrological effects of urbanization of a watershed.

Figure 2 shows how freeboard (typically 3 feet) compensates for the uncertainties and unknowns of engineering analysis – both the hydrologic and hydraulic computations. Freeboard assures that the 100-year design flood can be contained within the 95% confidence bounds. It also can be viewed as potentially protecting against a flood of up to a 500-year return period, under stationary climate assumptions.

Another way of adding a 'safety factor' within the current stationary climate 'design flood' paradigm, is by selecting a 'suitable' flood frequency distribution for analyzing a potential project. The choice of an appropriate probability distribution and parameter estimation method plays a vital role in at-site frequency analysis. U.S. federal agencies are required by law to use the Log-Pearson type 3 (LP3) distribution (England *et al.*, 2018), while the U.K. and Japan recommend the generalized extreme value (GEV) distribution (Kazama *et al.*, 2010). Castellarin *et al.* (2012) reviewed applied statistical methods for flood frequency analysis in Europe. According to their review, the GEV was among the recommended distributions used by Austria, Italy, Germany and Spain.

The selection of an appropriate probability distribution is essential for accurately delineating floodplains and capitalizing national flood insurance funds, as well as selecting the appropriate scale of flood protection infrastructure. Correctly characterizing the flood frequency distribution has the greatest direct influence on the choice of a particular flood management strategy – i.e., deciding whether structural or non-structural flood protection is appropriate for a particular location.

Stakhiv (2011) shows how, for a given set of annual flood peak data in a particular watershed, the selection of a probability distribution can dramatically alter the underlying risk analysis and risk tolerance assumptions for any particular location or choice of an alternative project design. Figure 3 shows a hypothetical comparison between the GEV and LP3 probabilities for flood frequency analysis for the same set of real data of a watershed. It shows that for a given discharge of 25,000 cubic feet per second, that the LP3 shows a recurrence interval of 100 years,

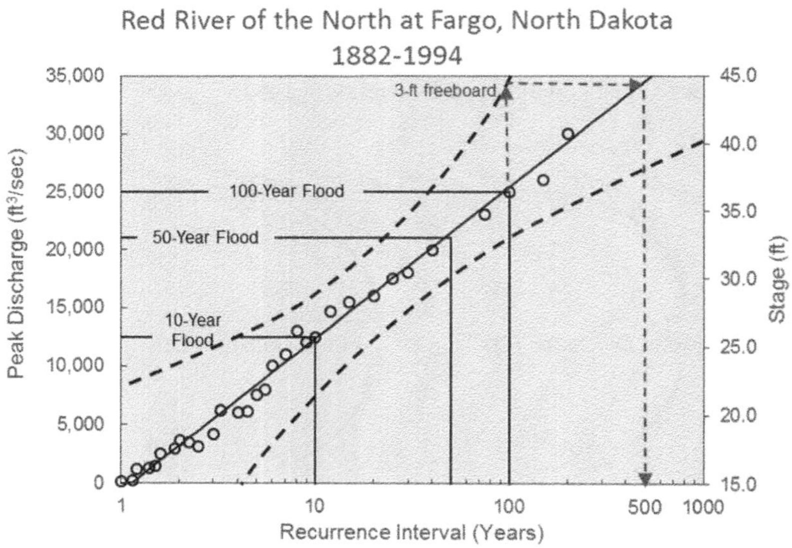

Fig. 2 | The role of 'freeboard' as a safety factor in design of flood protection.

whereas for the same discharge, the GEV shows approximately a 47-year recurrence interval. Conversely, a 100-year discharge computed by GEV equates to a 225-year discharge with LP3.

In this instance, selecting the GEV distribution for levee design or flood plain delineation would partly address concerns about non-stationarity in the future by halving the frequency of a 100-year event computed by the LP3 method. This outcome is consistent with the expectation of increased flood frequencies under many climate scenarios generated by the IPCC (2021). When freeboard is added to the design as an additional safety factor, this would be the equivalent of indirectly adding robustness and reliability to a selected flood control option, solely based on upgrading current codes and standards, without having to resort to inherently difficult and somewhat technically dubious climate scenario analyses (Anagnostopoulos *et al.*, 2010; Kundzewicz & Stakhiv, 2010).

Another way of partially addressing the uncertainties of climate non-stationarity, via engineering codes and standards, equivalent to adding 'freeboard', is the concept of 'flood allowances' employed by many European nations. A 'flood allowance' is defined as 'the maximum change in peak flows to allow for future impacts of climate change' (COST, 2013). Table 1 shows how various European nations use flood allowances to adjust for climate non-stationarity. Several procedures exist to adjust design flow estimates for the combined influence of climate change and land use. Considering the effect of climate change on design flood estimates, a safety margin of 20% is applied in the U.K., as recommended by DEFRA (2009), to compensate for climate change with a time horizon until 2085.

Vogel *et al.* (2011) computed the equivalent of flood allowance factors, in the form of flood magnification and recurrence reduction factors for the U.S. They developed a simple statistical model that could both mimic observed flood trends and the frequency of floods in a non-stationary world. Using historical flood data across the U.S., they calculated flood magnification factors in excess of 2–5 for many regions of the U.S., particularly urban areas with higher population densities. Similarly, they computed recurrence reduction factors which showed that what was once considered the 100-year flood may become much more common in many watersheds.

However, trends in flood peaks do not necessarily imply climate change, as they can arise from many other anthropogenic factors such as land use alterations, urbanization or riverine structures (bridges, constrictions and channel alterations). Vogel *et al.* (2011) found that large increases were generally observed in regions with higher population densities subject to land use changes both in urban areas and upstream watersheds. In particular, they found that '[N]onstationarity in floods can result from a variety of anthropogenic processes including changes in land use, climate, and water use, with likely interactions among those processes making it very difficult to attribute trends to a particular cause'.

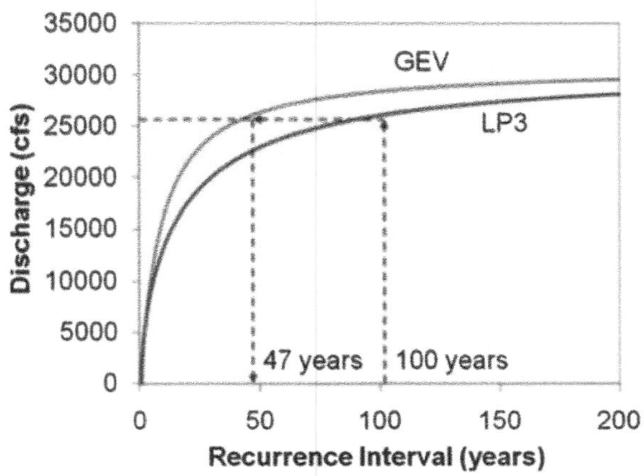

Fig. 3 | Comparison of GEV and LP3 distributions on flood return periods (Stakhiv, 2011).

Table 1 | Summary of existing European guidelines on climate change adjustment factors on design floods and design rainfall (COST, 2013).

Country	Region	Variable	Guideline	Reference
Belgium	Flanders	Design floods	30% increase	Boukhris & Willems (2008)
Belgium	National	Design rainfall	30% increase	Willems (2011)
Denmark	National	Design rainfall	20, 30 and 40% increase for return periods 2, 10 and 100 years	Arnbjerg-Nielsen (2008)
Germany	Bavaria	Design flood with 100-year return period	15% increase	Hennegriff *et al.* (2006)
Germany	Baden-Württemberg	Design floods	Increase between 0 and 75% depending on location and return period	Hennegriff *et al.* (2006)
Norway	National	Design floods	0, 20 and 40% increase based on region, prevailing flood season and catchment size	Lawrence & Hisdal (2011)
Sweden	National	Design rainfall	Increase between 5 and 30% depending on location	SWWA (2011)
United Kingdom	National	Design floods	20% increase for 2085	DEFRA (2006)
United Kingdom	National	Design rainfall	10, 20 and 30% increase for 2055, 2085 and 2115	DEFRA (2006)

As another example of codes and standards, that are periodically adjusted, is the critically important concept of 'probable maximum precipitation' (PMP). The PMP is theoretically the greatest depth of precipitation for a given duration that is physically possible over a given size storm area at a particular geographical location. It is used to develop inflow flood hydrographs, known as PMF, as the design standard for high-risk flood-hazard structures, such as dams and nuclear power plants (England *et al.*, 2014; Wright *et al.*, 2019). The procedures to estimate PMP were standardized by the World Meteorological Organization (WMO) in 1973 and revised in 1986 and 2009 (WMO, 2009). In the U.S., PMP estimates are published in a series of Hydrometeorological Reports (e.g., HMR55A, HMR57 and HMR58/59) by the National Weather Service since the 1950s (Kunkel *et al.*, 2013).

The examples above show that design-related codes and standards that assume a stationary climate are periodically adjusted and upgraded as information becomes available reflecting, to a certain extent, existing climate trends and non-stationary processes. A reasonable case can be made that since the information is inherently focused on climate extremes that are experienced regularly, this updating factor alone could reasonably be considered to account for a significant portion of future uncertainties arising from climate non-stationarity. When additional precautionary measures such as flood allowances, freeboard and other safety factors are used in design to account for other uncertainties linked to hydrologic and hydraulic models and urbanization, it is also reasonable to assume that most of the significant climate uncertainties associated with hydrologic non-stationarity could be accommodated in project design through periodic adjustment of existing design codes and standards.

Further following Auld's (2008) paths to updating engineering risk management strategies, there is yet another, even more fundamental approach that implicitly deals with future uncertainties. Water management functions can be partitioned into five basic categories, each addressing a different set of uncertainties and each offering different ways of dealing with climate variability and uncertainties within their respective evaluation and decision-making processes and engineering standards. Each has corresponding well-documented standardized decision frameworks and associated analytical procedures in the form of engineering regulations, manuals and other guidance. Different water management strategies employ various combinations of the categories listed below (Rogers, 1997; Stakhiv, 2011):

- *Planning new investments* for capacity expansion (reservoirs, irrigation systems, levees, water supply, wastewater treatment and interconnecting systems).
- *Operation and adaptation* of existing systems to accommodate new uses increased demands (e.g., reservoir reregulation and water allocation for ecology, population).
- *Maintenance and major rehabilitation* of existing systems (e.g., dams, barrages, irrigation systems, canals, pumps, etc.).
- *Modifications in processes and demands* (water conservation, pricing, regulation, policies and legislation) for existing systems and water users.
- *Introducing new efficient technologies* (desalting, biotechnology, drip irrigation, wastewater reuse, recycling, solar energy, materials and pumps).
- *Monitoring and regulation of existing systems* to improve performance, increase efficiencies and monitor critical thresholds.

Project maintenance and major repair and rehabilitation offer a good example of a practical engineering approach that obviates the need for a 'top-down' climate scenario approach, taking advantage of the life-cycle properties of infrastructure, and the recognition that water management functions are not static (Rogers, 1997). Rogers demonstrates that, because major rehabilitation of long-lived infrastructure projects normally occurs every 20–30 years, these are the opportunities to integrate new data, technologies, different designs, upgrade materials and analytical techniques as well as adjusting operating rules that can adjust to emerging demands and updated hydrologic data.

More importantly, Rogers (1997) noted that climate uncertainty was not the significant factor in projects whose capacity could be incrementally expanded (e.g., water and wastewater treatment plants, irrigation systems, pumping stations and water conveyance pipelines), finding that the economic discount rate was the more decisive factor in such analyses. Hence, projects that require major rehabilitation (dam spillways, levees and aqueducts) and those that may require incremental capacity expansion because of increased demands can essentially employ conventional analytical procedures that are periodically updated, incorporating the latest data, science and technological upgrades and design standards (Ehsani *et al.*, 2017).

Roger's findings also buttress the conceptual basis of an 'adaptation pathways' approach, which is based on incremental adjustments to infrastructure expansion, built on a form of 'threshold analysis' (Haasnoot *et al.*, 2013; Walker *et al.*, 2013; Mendoza *et al.*, 2018). Performance threshold-based design is essentially standard engineering practice – a 'bottom-up' process where the known or reasonably suspected performance vulnerabilities of existing or proposed infrastructure are identified so that potential conditions that expose those vulnerabilities can be quantified and addressed.

Similarly, Stakhiv (2011) demonstrated that many adaptive management adjustments can be made to existing infrastructure, in terms of monitoring, regulating and operating water-based delivery systems. For example, there are many opportunities for updating drought and flood contingency plans based on new information that could improve the overall resiliency, robustness and reliability of existing water management systems.

As an example, Figure 4 shows the Corps of Engineers' various operating procedures and manuals that govern reservoir management, that are part of a reservoir 'Master Manual' (U.S. Army Corps of Engineers, 2017). A water control manual is the guiding document that specifies how the Corps of Engineers operates their reservoirs. Each reservoir operates following congressionally mandated purposes, and the manuals specify how those multiple water management objectives are balanced. They also provide details on the reservoir's watershed characteristics, data collection networks, forecasting methods and stakeholder coordination.

It should be noted that the popularly advocated 'no regrets' climate adaptation practices inherently encompass current engineering 'best management practices', including contemporary risk-based design, safety factors and freeboard considerations that are routinely incorporated into the design of dams and flood works (Bhave *et al.*, 2016). Especially when the impacts of future changes remain highly uncertain, bottom-up 'no regrets' actions may offer the most realistic adaptation option. In this sense, undertaking infrastructure planning and design under upgraded risk-based methods within a traditional stationarity-based framework may be the most sensible approach to 'no regrets' climate adaptation.

CONCLUSIONS

Adaptation to climate change is currently the most practical and rational approach to dealing with the uncertainties of future climate, since international mitigation policies cannot effectively change the trajectory of global warming over the next few decades (Pielke *et al.*, 2007).

Current practices of risk-based planning and design standards have evolved incrementally, responding to each catastrophic natural disaster, some of which may be the precursors of a changing climate regime. Design standards and building codes encompass an accumulation of changes that reflect changing climate conditions over the past five decades, most notably because they are largely based on addressing climate extremes. These extremes may encompass the anticipated core of changes of a non-stationary climate regime, thereby indirectly addressing a large part of future unknown climate uncertainties. Risk analysis methods that form the basis of evolving design standards, progress in step with perceived climate changes, highlighting extreme events.

The combination of standards, codes and built-in engineering safety factors that aim to deal with a wide range of uncertainties, can be viewed as a collective foundation of a 'no regrets', precautionary approach to building more resilient infrastructure (OECD, 2018a, 2018b).

Planning and design are essentially separate functions with different objectives and decision criteria. The 'planning' function is centered on an evaluation framework that optimizes economic performance metrics for providing reliable water-based services, whether they be flood damage reduction, municipal water supply or hydropower. Once a suitable project is selected, project 'design' criteria focus on avoiding project and system failure and incorporate design safety factors and redundancies to account for a wide range of uncertainties.

A practical approach to climate adaptation and associated climate uncertainties would be to apply existing best management practices. These include adapting contemporary design procedures; risk management approaches with provisions for robustness, resiliency and adaptive management – i.e., nominal, 'no regrets' adjustments to changes in desired performance and service levels and technological adaptations with updates based on new

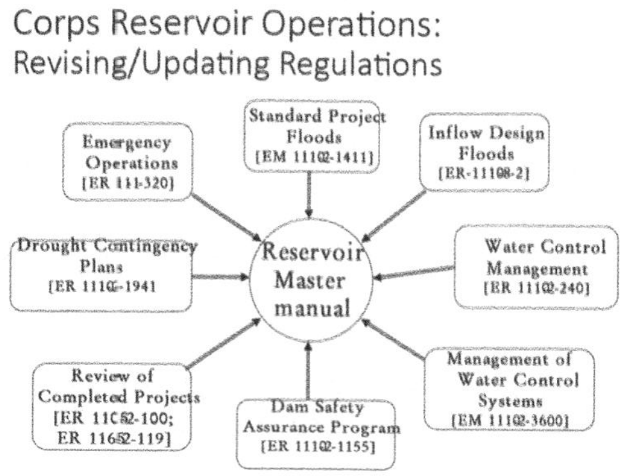

Fig. 4 | Updating reservoir master manuals (Stakhiv, 2011).

data and monitoring. Part of the ensemble of 'best management practices' includes climate adjustment factors, such as 'flood allowances', and upgrading design safety factors for critical components of infrastructure.

In particular, U.S. federal agencies have taken this route, and agencies such as the U.S. Army Corps of Engineers, the Bureau of Reclamation and the Federal Highway Administration have all updated both their planning and design procedures, employing a broader array of risk analytic tools and procedures that explicitly account for climate-related uncertainties. In addition, they have worked on better coordinating the risk-analytic methods used to assess and quantify uncertainties associated with climate change.

The core advantages of engineering codes and standards are central to 'best management practices' – i.e., uniformity of approach, replicability of analyses and consistency in outcomes of engineering design and throughout the globe, wherever hydraulic structures may be constructed. Such codified standards and practices serve as the basis for enhanced transparency and public trust and accountability.

Many other nations and international lending institutions, such as the World Bank and Asian Development Bank have upgraded their climate analysis procedures, have international engineering organizations in the form of new guidance for risk analysis and infrastructure planning, as well as developing new codes and standards for structural design.

DATA AVAILABILITY STATEMENT

All relevant data are included in the paper or its Supplementary Information.

REFERENCES

Anagnostopoulos, A., Koutsoyiannis, D., Christofides, A., Efstratiadis, A. & Mamassis, N. (2010). A comparison of local and aggregated climate model outputs with observed data. *Hydrological Sciences Journal 55*(7), 1094–1110. doi:10.1080/02626667.2010.513518.

Arnbjerg-Nielsen, K. (2008). Anticipated changes in extreme rainfall due to climate change. Recommendation paper No. 29. Water Pollution Comm. of the Soc. Danish Engrs (in Danish). http://ida.dk/svk/

Auld, H. (2008). Adaptation by design: the impact of changing climate on infrastructure. *Journal of Public Works and Infrastructure 1*, 276–288.

Ayyub, B. (2018). *Climate-Resilient Infrastructure Adaptive Design and Risk Management*. ASCE Committee on Adaptation to a Changing Climate. ASCE Manuals and Reports on Engineering Practice No. 140. doi:10.1061/9780784415191.

Ayyub, B. (2019). *Practical Resilience Metrics for Coastal Infrastructure Features*. Navigation Systems Research Program ERDC/CHL CR-19-1. U.S. Army Corps of Engineers, Vicksburg, MS, USA. Available at: https://erdc-library.erdc.dren.mil/jspui/handle/11681/32889.

Beven, K. (2016). Facets of uncertainty: epistemic uncertainty, non-stationarity, likelihood, hypothesis testing, and communication. *Hydrological Sciences Journal 61*(9), 1652–1665. doi:10.1080/02626667.2015.1031761.

Bhave, A., Mittal, N., Mishra, A. & Raghuwanshi, N. (2016). Integrated assessment of no-regret climate change adaptation options for reservoir catchment and command areas. *Water Resour Manage 30*, 1001–1018. doi:10.1007/s11269-015-1207-4.

Boukhris, O. & Willems, P. (2008). Climate change impact on hydrological extremes along rivers in Belgium. Flood Risk 2008 Conference, 30 September–2 October 2008. Oxford, UK. In: *Flood Risk Management: Research and Practice*. Samuels, P., Huntington, S., Allsop, W. & Harrop, J. (eds). CRC Press, Taylor and Francis Group, London, UK, pp. 1083–1091.

Bureau of Reclamation. (2019). *Best Practices in dam and Levee Safety Risk Analysis*. Manual, Joint Publication of Bureau of Reclamation and Corps of Engineers. Available at: https://www.usbr.gov/ssle/damsafety/risk/methodology.html.

Castellarin, A., Kohnová, S., Gaál, L., Fleig, A., Salinas, J. L., Toumazis, A., Kjeldsen, T. & Macdonald, N. (2012). *Review of Applied Statistical Methods for Flood Frequency Analysis in Europe, Milestone Report for WG2 of COST Action ES0901*. Tech. Rep., Centre for Ecology & Hydrology, Wallingford, UK.

COST (European Cooperation in Science and Technology). (2013). *A Review of Applied Methods in Europe for Flood Frequency Analysis in A Changing Environment*. WG4: Flood Frequency Estimation Methods and Environmental Change. FloodFreqCOST Action ES0901, p. 170. Available at: http://nora.nerc.ac.uk/id/eprint/501751/7/N501751CR.pdf.

DEFRA. (2006). Flood and coastal defense appraisal guidance (FCDPAG3), Economic appraisal supplementary notes to operating authorities – climate change impacts. Department for Environment, Food and Rural Affairs (DEFRA), London. 9 pp.

DEFRA. (2009). *Regionalised Impacts of Climate Change on Flood Flows – FD2020*. R&D Technical Report FD2020/TR. Department for Environment, Food and Rural Affairs, p. 113. Available at: http://randd.defra.gov.uk/Default.aspx?Menu=Menu&Module=More&Location=None&ProjectID=13958.

Diamantidis, D., Holický, M. & Sýkora, M. (2016). Reliability and risk acceptance criteria for civil engineering structures transactions of the VŠB – technical university of ostrava. *Civil Engineering Series 16*(2), 1–10. doi:10.1515/tvsb-2016-0008.

Ehsani, N., Vörösmarty, C., Fekete, B. & Stakhiv, E. (2017). Reservoir operations under climate change: storage capacity options to mitigate risk. *Journal of Hydrology 555*, 435–446.

Engineers Canada. (2018). *Public Guideline: Principles of Climate Adaptation and Mitigation for Engineers*. Engineers Canada, Ottawa, Canada, p. 27.

England, J., Julien, P. & Velleux, M. (2014). Physically-based extreme flood frequency analysis using stochastic storm transposition and paleoflood data on large watersheds. *Journal of Hydrology 510*, 228–245. doi:10.1016/j.jhydrol.2013.12.021.

England, J., Cohn, T., Faber, B., Stedinger, J., Thomas, W., Veilleux, A., Kiang, J. & Mason, R. (2018). *Guidelines for Determining Flood Flow Frequency – Bulletin 17C*. U.S. Geological Survey Techniques and Methods, Book 4, Chap. B5, p. 148. doi:10.3133/tm4B5.

Federal Emergency Management Agency (FEMA). (2020). *Guidance for Flood Risk Analysis and Mapping: Levees*. Guidance Doc 95, p. 141. Available at: https://www.fema.gov/flood-maps/guidance-reports/guidelines-standards.

Fiering, M. (1982). A screening model to quantify resilience. *Water Resources Research 18*(1), 27–32.

François, B., Schlef, K., Wi, S. & Brown, C. (2019). Design considerations for riverine floods in a changing climate – A review. *Journal of Hydrology 574*, 557–573. doi:10.1016/j.jhydrol.2019.04.068.

Government Accountability Office (GAO). (2013). *Climate Change: Federal Efforts Under Way to Assess Water Infrastructure Vulnerabilities and Address Adaptation Challenges*. GAO-14-23. Available at: https://www.gao.gov/products/GAO-14-23.

Government Accountability Office (GAO). (2021). *Puerto Rico Recovery: FEMA Made Progress in Approving Projects, But Should Identify and Assess Risks to the Recovery*. GAO-21-264. Available at: https://www.gao.gov/assets/gao-21-264.pdf.

Haasnoot, M., Kwakkel, J., Walker, W. & ter Maat, J. (2013). Dynamic adaptive policy pathways: a method for crafting robust decisions for a deeply uncertain world. *Global Environmental Change 23*(2), 485–498.

Haimes, Y., Li, D., Tulsiani, V., Lambert, J. & Krzysztofowicz, R. (1995). *Risk-Based Evaluation of Flood Warning and Preparedness Systems*. IWR Report 95-R-6, U.S. Army Institute for water Resources, Fort Belvoir, VA, USA. Available at: https://apps.dtic.mil/sti/pdfs/ADA524938.pdf.

Hallegatte, S., Shah, A., Lempert, R., Brown, C. & Gill, S. (2012). *Investment Decision Making Under Deep Uncertainty – Application to Climate Change*. World Bank Policy Research Working Paper 6193, World Bank, Washington, DC, USA.

Hatem, D. (1989). Impact of professional practice standards on liability of engineers. *Journal of Management in Engineering 5*(3), 249–256.

Hecht, J. & Vogel, R. (2020). Updating urban design floods for changes in central tendency and variability using regression. *Advances in Water Resources 136*. doi:10.1016/j.advwatres.2019.103484.

Hennegriff, W., Kolokotronis, V., Weber, H. & Bartels, H. (2006). Climate Change and Floods – Findings and Adaptation Strategies for Flood Protection (in German). KA-Abwasser, Abfall, 53(8) 770–779.

ICOLD (2020). General Principles and Framework for Dam Safety. ICOLD Draft Bulletin, pp. 36. itcold.it/wpsysfiles/wp-content/uploads/2020.

Institute for Water Resources. (2010). *Exploration of Tolerable Risk Guidelines for the USACE Levee Safety Program*. IWR Report 10-R-8. Alexandria, VA, USA. Available at: https://www.lrh.usace.army.mil/Portals/38/docs/civil%20works/IWR%20Tolerable%20Risk.pdf.

Institute for Water Resources. (2017). *Principles of Risk Analysis for Water Resources*. IWR Report 2017-R-01. U.S. Army Corps of Engineers, Alexandria, VA, USA. Available at: https://planning.erdc.dren.mil/toolbox/library/iwrserver/2017_R_01_PrinciplesofRiskAnalysisforWaterResources.pdf.

Intergovernmental Panel on Climate Change (IPCC). (2021). *Climate Change 2021: The Physical Science Basis. Contribution of Working Group I to the Sixth Assessment Report of the Intergovernmental Panel on Climate Change*. Edited by Masson-Delmotte, V., Zhai, P., Pirani, A., Connors, S. L., Péan, C., Berger, S., Caud, N., Chen, Y., Goldfarb, L., Gomis, M. I., Huang, M., Leitzell, K., Lonnoy, E., Matthews, J. B. R., Maycock, T. K., Waterfield, T., Yelekçi, O., Yu, R. & Zhou, B.. Cambridge University Press, Cambridge, UK.

International Standards Organization (ISO). (2015). *General Principles on Reliability for Structures*. Available at: https://www.iso.org/obp/ui/#iso:std:iso:2394:ed-4:v1:en.

Ittmann, J. (2020). *Liability of the Structural Engineer: Establishing, Quantifying, and Managing Risks. LSU Doctoral Dissertations*, p. 5126. Available at: https://digitalcommons.lsu.edu/gradschool_dissertations/5126.

Kalyanapu, A., Judi, D., McPherson, T. & Burian, J. (2011). Monte Carlo-based flood modelling framework for estimating probability weighted flood risk. *Journal of Flood Risk Management 5*, 37–48. doi:10.1111/j.1753-318X.2011.01123.x.

Kazama, S., Sato, A. & Kawagoe, S. (2010). Evaluating the cost of flood damage based on changes in extreme rainfall in Japan. In: *Adaptation and Mitigation Strategies for Climate Change*. Sumi, A., Fukushi, K. & Hiramatsu, A. (eds.). Springer, Tokyo, Japan. doi:10.1007/978-4-431-99798-6_1.

Kettner, A., Cohen, S., Overeem, I., Fekete, B., Brakenridge, G. & Syvitski, J. (2018). Estimating change in flooding for the 21st century under a conservative RCP forcing: a global hydrological modeling assessment. In: *Global Flood Hazard: Applications in Modeling, Mapping, and Forecasting*. Schumann, G., Bates, P., Apel, H. & Aronica, G. (eds.). AGU Geophysical Monograph Series. doi:10.1002/9781119217886.ch9.

Koutsoyiannis, D. (2014). Reconciling hydrology with engineering. *Hydrology Research 45*(1), 2–22. doi:10.2166/nh.2013.092.

Kundzewicz, Z. & Stakhiv, E. (2010). Are climate models 'ready for prime time' in water resources applications, or is more research needed? *Hydrological Sciences Journal 55*(7), 1085–1089. doi:10.1080/02626667.2010.513211.

Kunkel, K., Karl, T., Easterling, D., Redmond, K., Young, J., Yin, X. & Hennon, P. (2013). Probable maximum precipitation and climate change. *Geophysical Research Letters 40*(7), 1402–1408.

Lalonde, C. & Boiral, O. (2012). Managing risks through ISO 31000: a critical analysis. *Risk Management 14*, 272–300. doi:10.1057/rm.2012.9.

Lawrence, D. & Hisdal, H. (2011). Hydrological projections for flooding in Norway under a future climate. NVE Report 5-2011, Norwegian Water Resources and Energy Directorate, Oslo. 47 pp.

Lempert, R. (2019). Robust decision making (RDM). In *Decision Making Under Deep Uncertainty*. Marchau, V., Walker, W., Bloemen, P. & Popper, S. (eds.). Springer, Cham, Switzerland. doi:10.1007/978-3-030-05252-2_2.

Link, L. & Harris, J. (2007). Hurricane Katrina Interagency Performance Evaluation Task Force [IPET]. In: *World Environmental and Water Resources Congress 2007*. American Society of Civil Engineers. doi:10.1061/40927(243)609.

Liu, W. & Son, Z. (2020). Review of studies on the resilience of urban critical infrastructure networks. *Reliability Engineering & System Safety 193*, 106617. doi:10.1016/j.ress.2019.106617.

Major, D. & Stakhiv, E. (2020). *The Development of Water Resources Planning in the United States from the Reclamation Act of 1902 to the Principles and Standards of 1973*. IWR Maass-White Series. U.S. Army Institute for Water Resources, Fort Belvoir, VA, USA.

Matalas, N. & Fiering, M. (1977). Water resources systems planning. In: *Climate, Climatic Change, and Water Supply (Chap 6)*. National Research Council, The National Academies Press, Washington, DC, USA. doi:10.17226/185.

Mendoza, G., Jeuken, A., Matthews, J., Stakhiv, E., Kucharski, J. & Gilroy, K. (2018). *Climate Risk Informed Decision Analysis (CRIDA), Collaborative Water Resources Planning for an Uncertain Future*. UNESCO and ICIWaRM, Paris, France, p. 162. Available at: https://www.agwaguide.org/docs/CRIDA_Nov_2018.pdf.

Milly, P., Betancourt, J., Falkenmark, M., Hirsch, R., Kundzewicz, Z. & Lettenmaier, D. P. (2008). Stationarity is dead: whither water management? *Science 319*(5863), 573–574. doi:10.1126/science.1151915.

Mondal, A. & Mujumdar, P. P. (2017). Hydrologic extremes under climate change: non-stationarity and uncertainty. In *Sustainable Water Resources Planning and Management Under Climate Change*. Kolokytha, E., Oishi, S. & Teegavarapu, R. (eds.). Springer, Singapore. doi:10.1007/978-981-10-2051-3_2.

National Research Council. (2000). *Risk Analysis and Uncertainty in Flood Damage Reduction Studies*. The National Academies Press, Washington, DC, USA. doi:10.17226/9971.

National Research Counci. (2009). *The New Orleans Hurricane Protection System: Assessing Pre-Katrina Vulnerability and Improving Mitigation and Preparedness*. The National Academies Press, Washington, DC, USA. doi:10.17226/12647.

Office of Management and Budget (OMB). (1998). *Circular A- 119, Revised. Federal Participation in the Development and Use of Voluntary Consensus Standards and in Conformity Assessment Activities*. Circular-119-1.pdf (whitehouse.gov).

Olsen, R. (2015). *Adapting Infrastructure and Civil Engineering Practice to a Changing Climate*. ASCE Committee on Adaptation to a Changing Climate. doi:10.1061/9780784479193.

Organisation for Economic Co-operation and Development (OECD). (2018). *Climate Resilient Infrastructure*. OECD Environment Policy paper No. 14, p. 44. Available at: https://www.oecd.org/environment/cc/policy-perspectives-climate-resilient-infrastructure.pdf.

Organisation for Economic Co-operation and Development (OECD). (2018). *Cost-Benefit Analysis and the Environment: Further Developments and Policy Use*. OECD Publ., Paris, p. 454. doi:10.1787/9789264085169-en.

Oxford Economics. (2017). *Global Infrastructure Outlook. Infrastructure Investment Needs 50 Countries, 7 Sectors to 2040*. Global Infrastructure Hub.

Pechlivanidis, I., Arheimer, B., Donnelly, C., Hundecha, V., Huang, S., Aich, V., Samaniego, L., Eisner, S. & Shi, P. (2017). Analysis of hydrological extremes at different hydro-climatic regimes under present and future conditions. *Climatic Change 141*(3), 467–481. doi:10.1007/s10584-016-1723-0.

Peck, J. & Hoch, W. (1985). Liability of engineers for structural design errors: state of the art considerations in defining the standard of care. *Villanova Law Review 30*, 403. Available at: https://digitalcommons.law.villanova.edu/vlr/vol30/iss2/2.

Perrow, C. (1984). *Normal Accidents: Living with High-Risk Technologies*. Princeton University Press, Princeton, NJ, USA, p. 441.

Petroski, H. (1985). *To Engineer is Human: The Role of Failure in Successful Design*. St. Martin's Press, New York, NY, p. 247.

Petroski, H. (2006). *Success Through Failure: The Paradox of Design*. Princeton University Press, Princeton, NJ, p. 256.

Pielke, R. & Wilby, R. (2012). Regional climate downscaling: what's the point? *Eos Transactions American Geophysical Union 93*(5), 52–53. doi:10.1029/2012EO050008.

Pielke, Jr., R., Prins, G., Rayner, S. & Sarewitz, D. (2007). Lifting the taboo on adaptation. *Nature 445*, 597–598. doi:10.1038/445597a.

Prabhakar, S., Srinivasan, A. & Shaw, R. 2009 Climate change and local level disaster risk reduction planning: need, opportunities and challeneges. *Mitig. Adapt. Starteg. Global Change 14*, 7–33. doi.10.1007/s11027-008-9147-4.

Purdy, G. (2010). ISO 31000:2009–Setting a New standard for risk management. *Risk Analysis 30*(6), 881–886. doi:10.1111/j.1539-6924.2010.01442.x.

Ray, P. A. & Brown, C. M. (2015). *Confronting Climate Uncertainty in Water Resources Planning and Project Design: The Decision Tree Framework*. World Bank, Washington, DC, USA. Available at: https://openknowledge.worldbank.org/handle/10986/22544.

Read, L. & Vogel, R. (2015). Reliability, return periods, and risk under nonstationarity. *Water Resources Research 51*. doi:10.1002/2015WR017089.

Read, L. & Vogel, R. (2016). Hazard function analysis for flood planning under nonstationarity. *Water Resources Research 52*. doi:10.1002/2015WR018370.

Retief, J., Bernardo-Vijoen, C., van der Klashorst, E. & Diamantides, D. (2013). Extreme actions and climate change: experience gained in South Africa and Germany. In: *Safety, Reliability, Risk and Life-Cycle Performance of Structures and Infrastructures*. Deotides, G., Ellingwood, B. R. & Frangopol, D. M. (eds.). Taylor and Francis Group, London, pp. 1543–1550.

Rogers, P. (1997). Engineering design and uncertainties related to climate change. *Climatic Change 37*, 229–242.

Salas, J., Obeysekera, J. & Vogel, R. (2018). Techniques for assessing water infrastructure for nonstationary extreme events: a review. *Hydrological Sciences Journal 63*, 325–352.

Seemann, C. (1991). Understanding potential legal liabilities of water resource professionals and how to avoid them. In: J. Anderson (ed.). *Water Resources Planning and Management and Urban Water Resources*. ASCE, New York, NY, pp. 101–105.

Serago, J. & Vogel, R. (2018). Parsimonious nonstationary flood frequency analysis. *Advances in Water Resources 112*, 1–16. doi:10.1016/j.advwatres.2017.11.026.

Serinaldi, F. & Kilsby, C. (2015). Stationarity is undead: uncertainty dominates the distribution of extremes. *Advances in Water Resources 77*, 17–36. doi:10.1016/j.advwatres.2014.12.013.

Šraj, M., Viglione, A., Parajka, J. & Blöschl, G. (2016). The influence of non-stationarity in extreme hydrological events on flood frequency estimation. *Journal of Hydrology and Hydromechanics 64*(4), 426–437.

Stakhiv, E. (2011). Pragmatic approaches for water management under climate change uncertainty. *Journal of American Water Resources Association (JAWRA) 47*(6), 1183–1196. doi:10.1111/j.1752-1688.2011.00589.x.

SWWA (Svenskt Vatten). (2011). Precipitation data for design and analysis of urban drainage systems (in Swedish). Publication P104. Stockholm, Sweden.

Sykora, M., Diamantidis, D., Retief, J., Viljoen, C. & Roszas, A. (2017). On risk-based design of structures exposed to climatic action. In: *Life-Cycle of Engineering Systems: Emphasis on Sustainable Civil Infrastructure*. Bakker, F. & van Breugel (eds.). Taylor & Francis Group, London, pp. 472.

Sykora, M., Holicky, M., Jung, K. & Diamantidis, D. (2018). Human safety criteria for risk-based structural design. *International Journal of Safety and Security Engineering 8*(2), 287–298. doi:10.2495/SAFE-V8-N2-287-298.

Transportation Research Board (TRB). (2019). *Applying Climate Change Information to Hydrologic and Coastal Design of Transportation Infrastructure*. NCHRP Project 15-61. Available at: http://onlinepubs.trb.org/onlinepubs/nchrp/docs/NCHRP1561FinalReport.pdf.

Tsakiris, G. (2015). The status of the european waters in 2015: a review. *Environmental Processes 2*, 543–557. doi:10.1007/s40710-015-0079-1.

Tung, Y. -K., Yen, B. & Melching, C. (2006). *Hydrosystems Engineering Reliability and Risk Assessment*. McGraw-Hill, New York, NY, p. 483.

U.S. Army Corps of Engineers. (2017). *Management of Water Control Systems*. EM 1110-2-3600, October 10, 2017. Available at: https://www.publications.usace.army.mil/Portals/76/Publications/EngineerManuals/EM_1110-2-3600.pdf.

U.S. Army Corps of Engineers. (2019a). *Interim Approach for Risk-Informed Designs for Dam and Levee Projects*. Engineering and Construction Bull., Washington, DC. Available at: https://www.wbdg.org/FFC/ARMYCOE/COEECB/ecb_2019_15.pdf.

U.S. Army Corps of Engineers. (2019b). *Risk Assessment for Flood Risk Management Studies*. Engineer Manual (ER), Washington, DC, 1105-2-101, 15 July, 2019. Available at: https://www.publications.usace.army.mil/Portals/76/Users/182/86/2486/ER%201105-2-101_Clean.pdf.

U.S. Army Corps of Engineers. (2021). *Certification of the Hydrologic Engineering Center's Flood Damage Analysis Software Version HEC-FDA v1.4.3*. Available at: https://www.hec.usace.army.mil/software/hec-fda/documentation/FRM-PCX_Certification_HEC-FDA_v1.4.3_Sep_2021.pdf.

U.S. Water Resources Council. (1983). *Economic and Environmental Principles and Guidelines for Water and Related Land Resources Planning Implementation Studies*. U.S. Government Printing Office, Washington, DC, USA.

Underwood, B. S., Mascaro, G., Chester, M., Fraser, A., Lopez-Cantu, T. & Samaras, C. (2020). Past and present design practices and uncertainty in climate projections are challenges for designing infrastructure to future conditions. *Journal of Infrastructure Systems 26*(3), 04020026. doi:10.1061/(ASCE)IS.1943-555X.0000567.

Vogel, R. & Castellarin, A. (2017). *Risk, Reliability and Return Periods and Hydrologic Design Handbook of Applied Hydrology*, Vol. 78. Mc-Graw Hill Book Company, New York, NY, pp. 78-1–78-9.

Vogel, R., Yaindl, C. & Walter, M. (2011). Nonstationarity: flood magnification and recurrence reduction factors in the United States. *Journal of American Water Resources Association 47*(3), 464–474. doi:10.1111/j.1752-1688.2011.00541.x.

Vrouwenwelder, T. (2008). Treatment of risk and reliability in the eurocodes. *Structures and Buildings 161*(4), 209–214. doi:10.1680/stbu.2008.161.4.209.

Walker, W. E., Haasnoot, M. & Kwakkel, J. H. (2013). Adapt or perish: a review of planning approaches for adaptation under deep uncertainty. *Sustainability 5*(3), 955–979. doi:10.3390/su5030955.

Watkiss, P., Wilby, R. & Rodgers, C. 2020 Principles of climate risk management for climate proofing projects. *ADB Sustainable Development Working Paper Series 69*. Manila, Phillipines, 34 pp. doi:10.22617/WPS200203-2.

Wilby, R. & Dessai, S. (2010). Robust adaptation to climate change. *Weather 65*, 180–185. doi:10.1002/wea.543.

Willems, P. (2011). Revision of urban drainage design rules based on extrapolation of design rainfall statistics. In: 12th International Conference on Urban Drainage, Porto Allegre, Brazil, 10-15 September, 2011. 8 pp.

World Federation of Engineering Organizations (WFEO). (2015). *Model Code of Practice: Principles of Climate Change Adaptation for Engineers*. Paris, France. Available at: https://www.wfeo.org/wp-content/uploads/code-ofpractice/WFEO_Model_Code_of_Practice_Principles_Climate_Change_Adaptation_Engineers.pdf.

World Meteorological Organization (WMO). (2009). *Manual on Estimation of Probable Maximum Precipitation (PMP). WMO-No. 1045*. Manual on Estimation of Probable Maximum Precipitation (PMP) (wmo.int), Geneva, Switzerland, p. 245.

Wright, D., Bosma, C. & Lopez-Cantu, T. (2019). U.S. Hydrologic Design Standards Insufficient Due to Large Increases in Frequency of Rainfall Events. *Geophysical Res. Ltrs. 46*(14), 8144–8153.

Yannopoulos, S., Eleftheriadou, E., Mpouri, S. & Giannopoulou, I. (2015). Implementing the requirements of the european flood directive: the case of ungauged and poorly gauged watersheds. *Environmental Processes 2*, 191–207. doi:10.1007/s40710-015-0094-2.

Yen, B. & Tung, Y.-K. (1993). *Reliability and Uncertainty Analyses in Hydraulic Design. Subcommittee on Uncertainty and Reliability Analyses in Design of Hydraulic Structures of the Technical Committee on Probabilistic Approaches to Hydraulics*, American Society of Civil Engineers, New York, NY, USA, p. 305.

Yoe, C. (2017). Principles of risk analysis for water resources. *U.S. Army Institute for Water Resources*. Technical Report TR-2017-R-01. Fort Belvoi, Alexandria, VA, USA.

Yunika, A., Kok, M. & de Gijt, J. (2017). Specifying risk management standard for flood risk assessment: a framework for resources allocation. *MATEC Web of Conferences 138*, 09002. doi:10.1051/matecconf/201713809002.

First received 8 November 2021; accepted in revised form 8 November 2021. Available online 19 November 2021

© 2022 The Authors Water Infrastructure Planning, MDCU doi: 10.2166/wp.2021.241

Anticipated maximum scale precipitation for calculating the worst-case floods

Kuniyoshi Takeuchi [a,*] and Shigenobu Tanaka [b]

[a] University of Yamanashi, Takeda 4, Kofu, Yamanashi 400-8510, Japan
[b] Disaster Prevention Research Institute, University of Kyoto, Gokasho, Uji, Kyoto 611-0011, Japan
*Corresponding author. E-mail: takeuchi@yamanashi.ac.jp

KT, 0000-0003-1954-7535

ABSTRACT

Against increasing number of unprecedented heavy rains and typhoons reflecting climate change, the Japanese Government decided saving life as the top priority considering a 'worst-case' scenario. Accordingly, the Flood Risk Management Act was amended in 2015 to use the anticipated maximum scale precipitation (AMSP) for flood inundation calculation. In order to estimate the AMSP, the Ministry of Land, Infrastructure, Transport and Tourism (MLIT) chose historical maximum areal precipitation in the form of duration–area–depth (DAD) curves rather than climate change projections' dataset d4PDF. In this paper, policy development and detailed estimation procedures for the AMSP were reviewed and discussed. It was concluded that the current climate change projections are still not accurate enough to be used as the basis for real local operations, while long accumulated ground observations and ground-based radars are available in good quality all over Japan. But at the same time, historical maximum should always be updated as past records are renewed. Also, regional partitioning should not be done at too coarse of scale for proper regionalization of DAD. Such strategy would serve as a useful reference for other nations.

Key words: Anticipated maximum scale precipitation (AMSP), d4PDF, Flood hazard maps, Radar-AMeDAS precipitation, Regionalization of DAD, Worst-case scenario

HIGHLIGHTS

- To produce the flood hazard maps under the worst-case scenario, the anticipated maximum scale precipitation (AMSP) is used.
- The regionalized worst-case DAD, that is, the DAD estimated by the areal maximum precipitation observed in the history in a climatologically uniform region, is used to estimate the AMSP in any basin in the region.
- The ground observations by JMA since 1957 and the Radar-AMeDAS analyzed datasets since 1988 are considered more reliable than the d4PDF dataset assembled by climatological projections using GCMs and RCMs to be used for calculating inundation for flood hazard maps for local practice all over Japan.
- The areal maximum precipitation prescribed for a climatologically uniform region is subject to renewal in the future. Therefore, periodical updating of the AMSP is necessary. At the same time, reflection of climatological projections for the areal maximum and geographical partitioning of climatologically uniform regions should be considered.
- Ground observations and model-based climatological projections need to go hand-in-hand.

INTRODUCTION

Climate change influences in Japan have been appearing most significantly in the form of intensification of hydro-met hazards such as heavy rains and typhoons causing serious floods, landslides and debris flows every year especially since around the turn of the century. Responding to such serious hazards, the Japanese Government has been taking a new strategy to overcome the situation. It is to anticipate the worst-case scenario and to put a highest priority on saving lives, the same strategy already introduced for earthquakes and tsunamis after the Great East Japan Earthquake and Tsunami (GEJET) on 3 November 2011. As the situation is urgent, early warning and evacuation are emphasized as the necessary action for society to take to save life while construction of durable infrastructure progresses over a longer time span.

For issuing early warning, hydro-meteorological advancement in observation and forecasting technologies has been making great strides. Also, advancement in dissemination of information through TV, radio, smart phone, etc., is highly effective such as introducing new warning messages: 'Heavy rains that have never been experienced before are coming. Take an immediate action to protect your life'.

In order to support evacuation and to guide proper land use management, reliable hazard maps are indispensable. To facilitate this, the Ministry of Land, Infrastructure, Transport and Tourism (MLIT) decided to revise hazard maps to reflect the worst-case scenario of flood-related hazards including landslides, debris flows, tsunamis, storm surges, etc. They indicate spatial extent and magnitude of floods and associated evacuation centers in detail local maps that show individual houses for all areas of Japan.

As the basis of such hazard maps, hydrological analyses for flood inundation maps were made using worst-case heavy rains called the anticipated maximum scale precipitation (AMSP). This precipitation was estimated using the historical records rather than climate change projections using advanced GCMs and RCMs. This is because the model simulation results are not yet detailed enough to calculate local flood inundation maps. This is quite an important decision and needs continued scientific assessment.

This paper reviews and discusses the soundness of the concept of the AMSP and the engineering procedure to use the AMSP for calculating anticipated inundation for the worst-case flood hazard maps. The review provides some useful implication for other nations to produce their flood hazard maps. Before coming into detail policy and technology review and discussion, the background situation of recent flood-related disasters, the hydrological models used and the current situation of hazard maps in Japan are reviewed.

Recent flood-related disasters in Japan

In the past few decades, Japan has been experiencing increasingly extreme hydro-meteorological hazards and devastating flood-related disasters every year. Just one example from each year in the current decade illustrates the extraordinary development of flood-related disasters in Japan.

In September 2011, the Kii Peninsula Water Disaster triggered by Typhoon 12 brought heavy rain, over 2,000 mm in part of the peninsula. The number of fatalities and missing people reached 98 (MLIT, 2014a). It was only 6 months after the GEJET on 11 March 2011. In July 2012 and July 2017, twice, Northern Kyushu Torrential Rain Disasters occurred both during Baiu season, resulting in dead and missing of 32 and 42, respectively (JSCE, 2013; Cabinet Office, 2018). In October 2013, the Izu-Oshima Island was hit by Typhoon 16 and the dead and missing reached 43 people (Yamamoto *et al.*, 2014). In August 2014, Hiroshima Prefecture was hit by stationary linear rain bands formed by back-building phenomena which induced large-scale landslides and killed 77 people, the worst sedimentary disaster in Japan since 1983 July Heavy Rain hit West Shimane (JSCE, 2015).

In September 2015, Kanto and Northeast Japan Torrential Rain occurred. The stationary linear rain band was formed in Northern Kanto and caused the Kinu River flood at Josoh City, Ibaraki Prefecture. The total death toll was 8 (Cabinet Office, 2016). On 10 August 2016, Typhoon 10, the first typhoon ever to land on Northeastern Japan (first in Iwate Prefecture and later Hokkaido) directly from the Pacific Ocean that caused rare floods in Hokkaido and killed 29 people including the missing (FDMA, 2017). In July 2018, the West Japan Heavy Rain caused serious flood inundation, landslides and debris flows resulting in the dead and missing of 232 centered in Hiroshima and Okayama Prefectures (MLIT, 2018). This was the largest death toll since 2004, the year when the record breaking ten typhoons and two torrential rains hit Japan from June to October resulting in 232 deaths in West Japan, Tohoku and Hokkaido (MLIT, 2004) or, by a single event, since 1982 July torrential rain followed by Typhoon 10 hit West Japan resulting 439 death and missing (JMA website on 1982 July torrential rain and Typhoon 10).

In October 2019 again, Super Typhoon 19 (Hagibis) (JMA named East Japan Typhoon) hit Central to Eastern Japan causing serious floods and landslides by record-breaking rainfalls in large areas resulting dike breaches at 142 sites in 71 rivers and the dead and missing 107 (JSCE, 2020). In July 2020, the Baiu front stagnated over Japan and brought much rain and caused floods and landslides, first in Kyushu, especially Kumamoto Prefecture, then Chugoku and Tohoku regions and as a total 84 people died and 2 missing (Cabinet Office, 2021).

Thus, it is obvious that some fundamental actions should be taken for flood-related disaster management.

Hydrological models to simulate discharge and flood inundation

The precipitation-runoff model used for the design and operation of public infrastructure is the 'Storage Function Model (SFM)' originally proposed by Toshiaki Kimura in the 1950s (Kimura, 1961) at the Public Works Research Institute (PWRI). It is a physically based lumped model and has been nearly exclusively used in PWRI and consulting firms for designing public infrastructure for many years and accumulated a large variety of application cases in different scales, seasons, landcover, topography, geology, landscapes, storm types, etc. (Kimura, 1978). But it has never been formally published in an academic journal. The model was never officially designated as the only model to be used in infrastructure design, either. But it was referred to as one of the recommended models in the Ministry of Construction (former MLIT) publication 'River and Sabo Works Engineering Standard', first edition, in 1984 (Ministry of Construction, 1984) that is still in use with some updates every decade.

For inundation simulation, there is no such special model. Following the determination of AMSP, a calculation manual of inundation simulation to identify the anticipated flood inundation areas was issued in 2015 (MLIT, 2015d) which was the 4th revision of the original manual issued in 2001. Newly added were methods to calculate inundation duration, houses collapse areas, etc. In 2016, its simplified version was also issued for small- to medium-scale rivers (MLIT, 2016).

To get an inundation map for current flood hazard maps, the worst-case dike breach in some selected points are produced with 25 m resolution (suggested) and the resultant maps are overlaid to get the one that envelops all the worst cases (MLIT, 2015d).

Flood hazard maps

The first flood hazard map was produced in Japan by Masahiko Oya in the form of his work 'Water-disaster morphology classification map of Nobi Plain, Kiso River Basin' in 1956 (Oya, 1956). This map was basically a geomorphological map showing the classified topography and soil types of the area affected by repetitive floods in the past. This Nobi Plain geomorphological map was proven useful to indicate the area affected by the flood of the Ise Bay Typhoon in 1959 which brought the severest storm surge and floods in modern Japanese history. Since then, the water-disaster morphology classification maps have been produced in many river basins in Japan and eventually in various other countries (NIED website).

It was on the occasion of flooding in the Abukuma River in August 1998 when the first investigation was made to identify the effect of availability of flood hazard maps on the citizens' evacuation behavior in Koriyama City, Fukushima Prefecture by Katada (1999). He found that the people who had seen the flood hazard map before the flood had evacuated, on average, about an hour earlier than the people who had not.

With this evidence, MLIT moved to seriously consider the use of flood hazard maps in disaster management operations and the government made the amendment to the Flood Risk Management Act in June 2001 to make use of flood hazard maps compulsory in so-called flood forecasting rivers (298 minister designated rivers and 128 prefectural governor designated rivers that have large basins and potential to cause serious losses to national economy). In 2005, this was extended to so-called water level warning rivers (150 minister designated rivers and 1,491 prefectural governor designated rivers that have, in addition to flood forecasting rivers, potential to cause serious losses to national economy and assigned to issue the water level warning when river has reached to the inundation risk level) (MLIT website on flood forecasting rivers).

The specific floods to be considered for calculating inundation areas for hazard maps in each river basin were specified in the 'Basic Plan for Implementation of Construction Works' of each basin before 1997 according to the River Law of 1964. After 1997, the River Law of 1997 changed the 'Basic Plan' above into the 'Basic River Management Policy'. Incidentally, the difference between the 1964 and the 1997 river laws was in their decision-making process with or without inviting local stakeholders' requests before deciding the basic plan. Also, in the latter, it was clearly stated that the plan should be completed within 20–30 years and publicized in more detail in each river basin.

Issuance of hazard maps

Using the AMSP, discharge simulation is conducted followed by inundation simulation along rivers and lakes. In order to make hazard maps from inundation maps that have inundation area and depth information, evacuation centers should be added by cities, towns and villages in consultation with local communities. Then finally, the hazard maps are issued by those local governments.

By September 2018, the hazard maps under the 2001 amendment of Flood Risk Management Act were completed in 98% (1,316/1,340) of cities, towns and villages in Japan. Now, following the amendment of flood risk management act of May 2015, the revision of the maps using the AMSP is taking place. As of September 2018, only 20% (268/1,340) of cities, towns and villages were completed (MLIT website on flood hazard maps). Note that this completion includes identification of evacuation centers for each hazard for each local community.

REVIEW

Strategies to address a new stage of flood-related disasters

Preparation for the worst-case scenario

Intensification of flood-related hazards as stated in the previous section is believed in general to be associated with progression of climate change. The Ministry of Land, Infrastructure, Transport and Tourism (MLIT), Japanese Government considered that the nation came into a new stage of socio-geohazards relationships and established an informal committee on 'Disaster prevention and mitigation responding to the new stage' in October 2014 which submitted a report in January 2015 (MLIT, 2015a).

The basic strategy of the report 'Disaster prevention and mitigation responding to the new stage' was to specify that the nation had to prepare against the worst-case hazards, similar to earthquakes and tsunamis management already decided after the GEJET. The policy adopted for managing tsunami hazards was the L1 and L2 approach, where L1 stands for the level 1 tsunami as frequent as about once every 100 years for which all life, property and livelihood should be protected and L2, the level 2 tsunami as rare as about once every 1,000 years for which at least life should be saved (Central Disaster Management Council, 2011; MLIT, 2012).

The report recommendation was, similar to L2, to anticipate the maximum scale floods and prepare life-saving actions. It pointed out that since structural means cannot protect all life and property against the worst-case hazards, it is necessary to put highest priority to protect human lives and avoid the catastrophic losses even in the case where great property damages are inevitable. In practice, a standard was proposed to consider the maximum scale precipitation and avoid facing unanticipated events.

In order to implement such a policy against hydro-met hazards, the Government decided to revise the hazard maps throughout Japan, employing a new design precipitation standard called the 'Anticipated Maximum Scale Precipitation (AMSP)' and, in accordance, the 'Partial amendments to Flood Risk Management Act and others' was promulgated on 20 May 2015, where others include the Sewerage Act (MLIT website on Flood Risk Management Act).

Base data for estimating the AMSP

In the amendment, it was stated that the flood inundation maps should be drawn using the AMSP that conform with the criteria given by the Ministry of Land, Infrastructure, Transport and Tourism (MLIT). These criteria were presented by MLIT prior to enforcement of the act in the November 2015 guideline 'Design method for anticipated maximum scale precipitation to produce anticipated inundation (floods and inland floods), etc.' in July 2015 (MLIT, 2015b).

In this guideline, the precipitation data to be used was specified in the form of the depth–area–duration (DAD) curves estimated by the historically observed maximum areal precipitation records instead of simulated climate change projections. This selection was an important decision for estimating the AMSP, instead of using outcomes of great efforts that had been put to climate change projections, especially to impacts on heavy rains and typhoons using GCMs and RCMs.

The historically observed data used in the guideline were the Radar-AMeDAS dataset available since 1988 and other selected extreme records observed by MLIT and other ministries since 1957. The Radar-AMeDAS data are the radar observed precipitation calibrated by the ground observations measured by the Automated Meteorological Data Acquisition System (AMeDAS) and MLIT and other ministries observed precipitation data.

Note that this strategy does not at all claim that the future worst-case hazard can be represented by the historical maximum. But rather, as will be explained in the next section, if the historical maximum precipitation was less than 1/1,000 probability of occurrence, including 1/10,000 or even less, then it would be socially acceptable to consider it as the worst-case precipitation. Furthermore, if such an outrageous precipitation fell some part of the region, the same magnitude of precipitation is anticipated in any part of the region. It is also only a temporary measure to be replaced by the reliable climate projections in the future. This philosophy should not be interpreted simply as historical maximum can be considered as the worst-case especially where historical observations are limited.

Methodology for estimating the AMSP

Regional partitioning for regionalization unit

The method of identifying the AMSP was specified by the guideline (MLIT, 2015b) that uses the DAD estimated by the largest ever observed areal precipitation in the past that is available in the network of observation stations of Japan Meteorological Agency (JMA), instead of extreme values that are derived from rainfall projections under a future greenhouse gas emission scenario.

The detailed methodology utilizes the concept of regionalization of the worst-case DAD of hydrometeorological extremes. That is, instead of using the local historical maximum areal precipitation observed in the concerned basin, the regional maximum in a larger region where the concerned basin is situated is used. The larger region for each basin was determined by dividing Japan into 15 climatologically uniform regions based on historical statistics of rainfall including the dendrogram analyses of basin climate. There is no way to have really 'climatologically uniform' region but this term is used for convenience. The parameters used for dendrogram analyses are the ratio of slopes of depth–duration (DD) curve, average and coefficient of variation (CV) of annual maximum precipitation per 3 hours, 24 hours and 48 hours precipitation (MLIT, 2015c). Figure 1 depicts the resultant regional partition. Here, it is interesting to see Okinawa belonging to the North-west Kyushu region which is heavily affected by the climate of both Mainland China and Siberia. There are many other climatological partitions possible for different uses.

Regionalization of DAD diagram

The AMSP for any basin **B** is determined by the rainfall DAD map identified by the data collected from the larger region **K** introduced above that includes the concerned basin **B**. The methodology of identifying the DAD diagram is as follows: (refer **A**, **D** and **P** in Figure 2).

1. Select a partitioned region K (1, …, 15).
2. Select an area A in region K.
3. Select a duration D (1, 2, 3, 6, 12, 24, 48 and 72 h).
4. Within partitioned region K, move area A around as one area, not separated, but changing freely its shape and find the maximum ever observed areal average precipitation depth P for duration D in the whole data length.
5. Plot P against A with parameter D in the DAD diagram of region K.
6. Repeat the process 3–5 to cover all duration D.
7. Repeat the process 2–6 for all area A.
8. Repeat the process 1–7 for all region K.

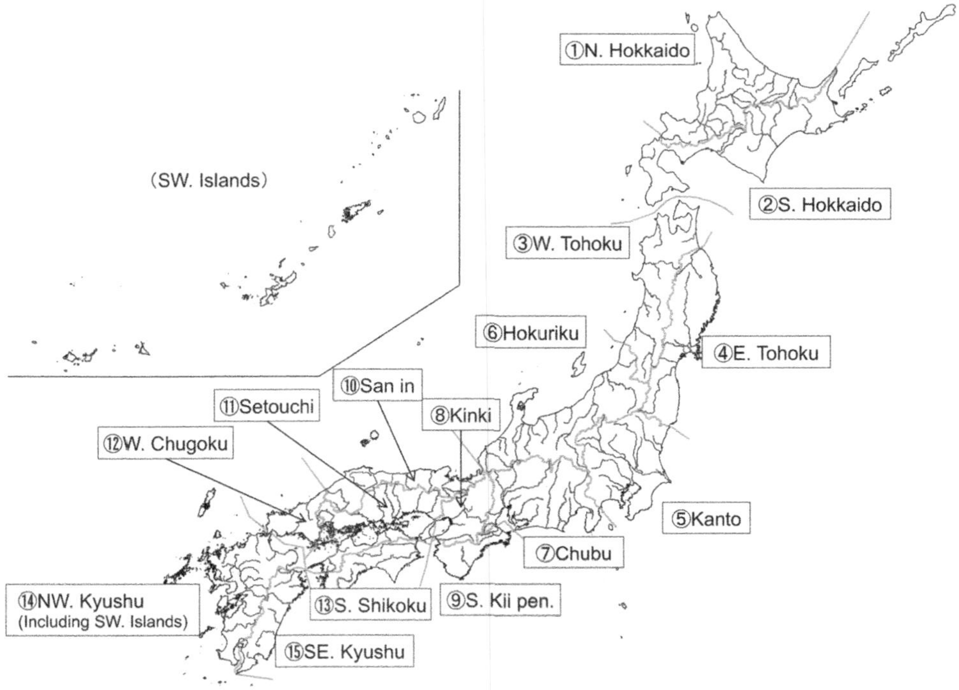

Fig. 1 | Fifteen climatologically uniform regions for AMSP classification (after MLIT (2015b)).

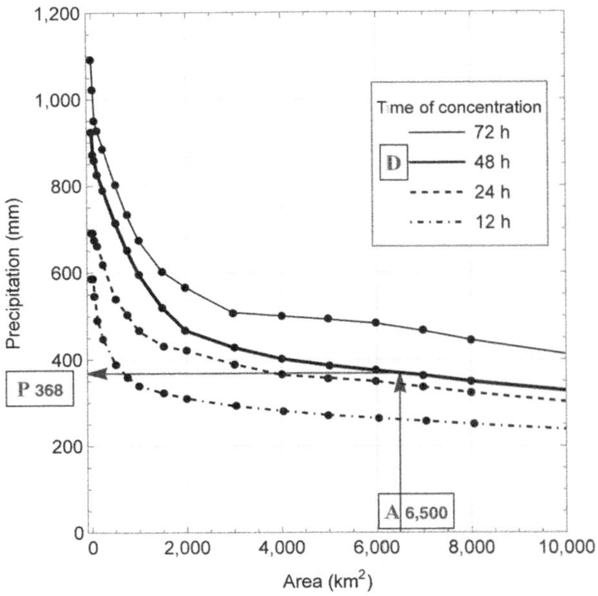

48 hours	
Area (km²)	Prec. (mm)
1	925
31	872
62	857
123	826
247	789
494	715
741	651
988	596
1,483	520
1,978	467
3,007	427
4,009	401
4,998	386
5,991	374
6,993	363
7,990	349

Interpolation method is used for obtaining precipitation on 6,500 km^2 because 6,500 km^2 is not shown. Using 374 mm and 363 mm on 5,991 km^2 and 6,993 km^2, respectively, precipitation 368 mm on 6,500 km^2 is obtained as follows.

$$374 + (363 - 374) \times \frac{6{,}500 - 5{,}991}{6{,}993 - 5{,}991} = 368$$

Fig. 2 | Specific calculation of AMSP from a DAD curve of a concerned region (**A**rea: 6,500 km^2, **D**uration: 48 h, to determine **P**recipitation: 368 mm, in Region: **K**) (after MLIT (2015b)).

In step 4, a 'grid point method' (referred to as the flexible elements method (FEM) in the guideline) was suggested as a concrete computer calculation method by WMO (1969). The DAD diagram of region K as depicted in Figure 2 is produced smoothly connecting the plotted points with interpolation between points and minor extrapolation at the end of lines for all 15 regions in Japan.

The AMSP for calculating flood inundation

In order to produce a hazard map of any target area, its anticipated inundation area should first be identified. Using the identified DAD diagram, the AMSP for the basin where the target area is located is selected simply reading value P at basin area A and its concentration time D. The procedure of identifying the AMSP for a basin of area 6,500 km^2 and concentration time 48 h resulting in a value of 368 mm is depicted in Figure 2. If the exceedance probability of the AMSP thus estimated for a basin is much more than 1/1,000 per year probability of occurrence, it is advised to adjust it to the one equivalent to 1/1,000 per year probability of occurrence (MLIT, 2015b).

The methodology of identifying the probability value of precipitation in Japan is unique and has not been formally published anywhere in an academic journal, but has been exercised in practice as a national standard for many years, at least since the Basic Plan for Implementation of Construction Works started in exercise in 1997. The latest official recommendation to use this methodology was given in the MLIT administrative directive 'The past examples and others to be referred to develop the basic river management policy' issued in 2015 by Water and Disaster Management Bureau, MLIT (MLIT Administrative Directive, 2015). It lists 'The Yabe River basic river management policy' (MLIT, 2014b) as a recommended example case. The Yabe River example suggests to utilize a set of probability distribution functions selected from 5 categories of 7 different probability distribution functions (12 types all together if distributions with different parameter estimation methods are counted) listed in Table 1 that clear the goodness-of-fit criteria Standard Least Squares Criterion SLSC < 0.040 (Tanaka & Takara, 1999). Among the acceptable distributions listed in Table 1, the distribution of the smallest jackknife estimator (Quenouille, 1949) is selected that gives the most stable estimate for resampled data, from which the precipitation return value is decided against a given non-exceedance probability. The parameter estimation method for each distribution is selected from the product moment maximum-likelihood method, PWM or L-moment as also shown in Table 1. The plotting position to be used is Cunnane's (Stedinger *et al.*, 1993).

Hyetograph identification by enlarging method

Given the design precipitation amount for the given duration (concentration time) in the target basin, the hyetograph for drawing the anticipated inundation map is selected again from the historical cases, the hyetographs of the worst-case that was experienced in the past and assumed to create the maximum losses in the future in the target basin. The worst-case usually corresponds to the precipitation event resulting in either the largest total volume of flood discharge or the highest peak discharge depending on the impacts on socio-economic damages in the basin.

As the total precipitation of the selected historical case in the concerned basin has less than or equal to the areal precipitation identified by the AMSP, the selected historical hyetographs are to be enlarged to meet the AMSP. Such modification is made by so-called enlarging method which has long been exercised in practice by MLIT. The procedure is depicted in Figures 3 and 4.

Figure 3 indicates the cases for modifying the shape of hyetograph, (a) the case where the duration of the selected historical hyetograph is less than the concentration time D of the sub-basin and (b) the case the duration of the selected is longer than D. If less than D, the historical hyetograph is simply enlarged at all time interval points of the observed values in equal proportion to meet the AMSP during D and if longer than D, only the central (largest) part within D is enlarged in proportion to meet the AMSP.

Figure 4 depicts the rare case that needs some additional modification. If the historical hyetograph that corresponds to the worst disastrous case has an exceptional shape that the normal enlarging method leads to an excessively high precipitation per hour or per 10 minutes, an arrangement is made to limit the maxima less than 220 mm/h and 60 mm/10 min, the highest values over which physical reality is unthinkable under any extreme situation (Ninomiya, 2001). This is to avoid the enlarging method to come into any exceptional case ending up in totally unrealistic values.

Table 1 | Probability distribution functions and their parameter estimation methods to be used to estimate probability precipitation (Tanaka & Takara, 1999; MLIT, 2014b).

Distribution	Parameter estimation method
GEV	LM
Gumbel distribution	LM
SQET	MLE
LN2	LM, PM
LN3	LM, PM Ishihara-Takase, QM, QM Iwai
LN4	PM
LPIII	PM with logarithmic sample, PM with original sample

GEV, Generalized extreme value distribution; SQET, Square root exponential-type distribution of maximum; LNn, *n*-parameter type log-normal distribution; LPIII, Log-Pearson type III distribution; LM, L-moment method; MLE, maximum-likelihood estimation; PM, product moment method; QM, quantile method.

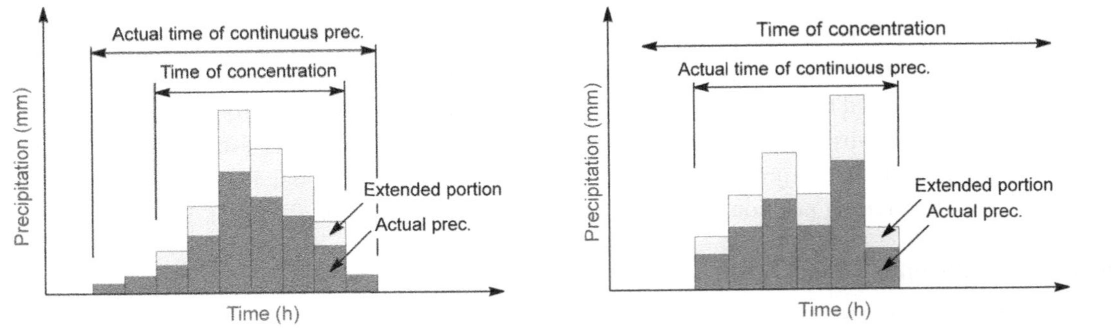

Case of actual time of a continuous prec. > time of concentration Case of actual time of a continuous prec. < time of concentration

Fig. 3 | Precipitation enlarging methods (MLIT, 2015b).

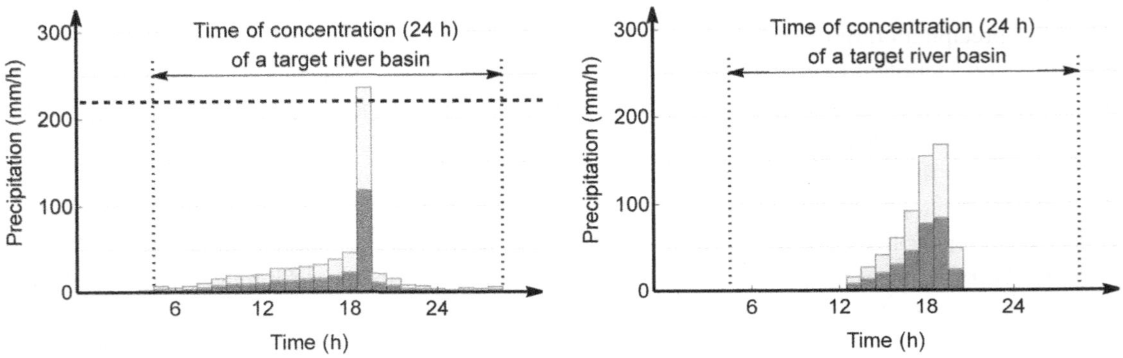

Fig. 4 | Examples of excessively irrational enlargement of temporal precipitation pattern (MLIT, 2015b).

Discharge and inundation simulation

The precipitation-runoff model to be used is the SFM. As it is a lumped model, if a target basin is larger than several hundred km^2 or so, the basin should be divided into sub-basins in the order of tens to a hundred km^2 area or so and the model should be applied to sub-basins and discharges from each sub-basin are routed to the concerned point. In case of the Tone River at Yattajima, for example, the upstream catchment (referred to as the Yattajima basin) is 5,150 km^2 which is divided into 39 sub-basins (MLIT website on the Tone River design flood verification).

The AMSP of the Yattajima basin is decided by the DAD of the Kanto region at area 5,150 km^2 and concentration time 3 days, which gives 491 mm/3 days. The next step is estimating the hyetographs necessary for SFM calculation for discharge from each sub-basin. The distribution of hyetographs over all sub-basins is selected as the one that gives most unfavorable discharge at Yattajima station by comparing several historical flood cases that caused large disaster damages in the Tone River basin in the past. The historical worst-case flood was by Typhoon Kathleen in September 1947 which should be included in comparison. The areal average precipitation of any case selected must be less than or equal to the AMSP at Yattajima so that the enlargement ratio is calculated and the same enlargement ratio is applied to all hyetographs of 39 sub-basins so as to make the areal precipitation of the Yattajima basin the same as the AMSP.

Various other parameters of the SFM of each sub-basin such as initial precipitation loss R_0, initial delay of discharge T_l, saturation precipitation Rsa, storage function constant K, storage function power p, etc., are decided as a robust set at each sub-basin under its representative flood cases (MLIT website on the Tone River design flood verification).

DISCUSSION

In the previous section, a series of developments in Japan for climate change adaptation were reviewed, namely, policy priority to save life anticipating the worst-case scenario, hazard maps based on the AMSP, use of observed precipitation data rather than climatological projections to estimate the AMSP, regionalization of DAD curves for the AMSP estimated by historical maximum areal precipitation, distribution of hyetographs in sub-basins selected from the past disaster cases, enlargement method for hyetograph adjustment, and finally discharge and inundation simulation for hazard maps. Among others, the following questions need further discussion:

1. Is it reasonable to choose observed data for estimating the AMSP while climatological projection technologies have been advancing remarkably?
2. Is it reasonable to choose the historical maximum areal precipitation as the AMSP of the worst-case scenario?
3. Is it reasonable to use the worst-case DAD estimated in a region to use for the AMSP at any basin in the region? In other words, is regionalization of the worst-case DAD reasonable?
4. Is it proper to select a distribution of sub-basin hyetographs from historical events? No need to consider any other distribution?

To follow are some critical discussions on those questions.

Observation records vs simulated climate projections

The basic conceptual issue that poses a dilemma for all planners, practitioners, engineers and scientists is how to develop new flood protection standards under a non-stationary climate. Should the new engineering standards be based on historical observations, using procedures developed for a stationary climate? Or, should the criteria and standards for flood frequency analysis and typhoons be based on climate change projections of GCMs and RCMs?

The methodology adopted in Japan is twofold: One is to protect life by early warning and evacuation by anticipating the worst-case hazards that should be estimated by the historical maxima, and the other is to protect life, property and livelihood as a whole as much as possible by strengthening the physical infrastructure considering the anticipated increase in hazards estimated by GCMs and RCMs.

This policy is considered reasonable as saving life is an urgent matter to be addressed without delay while strengthening physical infrastructure needs much time and money so that the investments should be adaptive to the progress of climate change as well as changing economic conditions. In order to invest a large amount of social capital for infrastructure development, such an adaptive approach is indispensable.

Now, are the historical ground observations trustable while model simulated projections are not? Indeed, in the research field, there have been significant research conducted for climate change projections using various GCMs and RCMs with the funding provided by the Ministry of Education, Culture, Sports, Science and Technology (MEXT), the Ministry of Environment and others in recent decades. The resultant projections were compiled into datasets called Database for Policy Decision-Making for Future Climate Change (d4PDF) which has been put to public use since 2016 (MIROC website on d4PDF). The datasets are available for the globe and Japan area. The ones for Japan area are 20 km mesh 60 years hourly projections with the global-mean surface temperature being current (1951–2010), plus 2 K (2030–2091) and plus 4 K (2050–2111) degrees for 50, 54 and 90 ensemble series, respectively, under RCP8.5 scenario. They are well prepared for use.

However, according to the reproducibility experiments of the annual maxima statistics of the present climate by Tanaka *et al.* (2019), the reproducibility was found poor in Kyushu, Shikoku and Kii Peninsula areas where the heavy rains matter most. In other regions, it was better. This finding seems quite reasonable because those areas have strong orographic effects with frequent visits of typhoons and frontal activities for which spatial resolution 20 km would not be fine enough. Thus, the use of d4PDF to estimate the AMSP is not recommended for producing hazard maps for local community use, which requires more detailed local information.

On the other hand, the reliable ground observations are available for long time periods since the start of meteorological observations by Japan Meteorological Agency in 1875. The number of observation stations were about 50 (average about 90 km grids) in 1900 and about 150 (average about 50 km grids) in 1970 covering all Japan (Ishii, 2019). From 1974, the Automated Meteorological Data Acquisition System (AMeDAS) started and, in 1979, it was completed with 1,316 stations (average about 17 km grids) every 10 min observation intervals for all of Japan.

Besides, JMA started the use of radars for precipitation measurement in 1954 and covered the entirety of Japan with 20 microwave radars (5 GHz band) by 1971 (Sato, 2007). Their measurements are calibrated by AMeDAS and other MLIT and prefectural observation data, and the 'Radar-AMeDAS Analyzed Precipitation' dataset have been produced since 1988. The resolution of these data was 5 km grids till 2001, 2.5 km till 2005, and has been 1 km since 2006 (JMA website on analyzed precipitation; Kuroyoshi *et al.*, 2005).

Since the Radar-AMeDAS Analyzed Precipitation dataset provides the radar-based interpolation where no ground observations available such as high mountainous areas, it is considered the most accurate observation data. It was therefore natural to choose this observed precipitation data as the basis of estimating historical maxima to calculate the AMSP.

Historical maximum precipitation for the AMSP in the worst-case scenario

The worst-case flood may mean different things in different nations. In the Netherlands, the probability of non-exceedance levels 1/4,000 or 1/10,000 per year are used to protect sea dikes against storm surges, and for rivers, 1/1,250 or 1/2,000 probability (Jonkman *et al.*, 2008). But, in Japan, such high safety levels would not be reasonable for infrastructure design considering the current level of safety achieved and the necessary investment to construct and maintain it. Also, the incremental height of the dike for increasing safety probability would be in general much higher in Japanese rivers than continental rivers. For example, the necessary increase of dike height from the safety level 1/15 to 1/50 per year probability, in the Tone River at Yattajima is 2 m and the same in the Saine River at Paris is 1 m. In Japan, the cost of safety increase is relatively more expensive than continental countries mainly because the steep slopes and narrow width of rivers relative to discharge.

But for hazard maps for emergency preparation and evacuation to save lives, the maps must be uniform in anticipating the worst-case scenario, avoiding no unanticipated events to happen. The key question is what is the real worst-case? The ultimate worst-case would be heavy rains based on the maximum vapor that the atmosphere can theoretically hold, falling at once everywhere in a basin. But such a case is clearly unrealistic and nobody would disagree. The historical maximum during some generations of experiences must be the most popular and psychologically acceptable worst-case in practice. Following this notion, the historical maxima were used to identify the DAD for the AMSP to be decided in the MLIT guideline (MLIT, 2015b). The data used were the partial maxima series of areal precipitation observed from 1957 to 1987 by JMA or MLIT precipitation observation stations and from 1988 to 2014 by Radar-AMeDAS datasets. However, 50 years of records are considered to be the minimum acceptable level of rainfall data for uniformly available data all over Japan. In order to comply with this constraint acceptable in using the historical maximum as the worst-case scenario, the following regionalization concept was added.

Regionalization of the worst-case DAD for any basins in the region

About the use of historical areal maximum precipitation, the following question would arise. That is, 'We were lucky that the heaviest rain occurred in a distant basin rather than here this time. But in the future, wouldn't it be possible for similar areal precipitation to occur in our basin?' The answer must be a big 'yes', at least within the same climatologically uniform region. This is the reason why the worst-case DAD identified in a large region is used for the AMSP of any sub-basin in the region. This is called here 'regionalization of the worst-case DAD' or simply 'regionalization of DAD'. That is, instead of using the local historical areal maximum observed in the concerned basin, the regional areal maximum in a larger region where the concerned basin is situated is used. This simply means that the heaviest rain area in a basin may come to any other basin if it is in the same region shown in Figure 1, where the entire Japan is divided into 15 climatologically uniform regions.

Such an assumption may be quite difficult to formally justify in a scientific sense. But there are some pieces of supporting evidence. One is that recent heavy rains observed are quite often the ones that 'have never been experienced before' in a particular basin, but the regional maxima are not so frequently broken even by recent new climate conditions such as stagnated linear rain bands or super typhoons. In fact, according to MLIT (2020), the Abukuma River basin received unprecedented areal precipitation by Super Typhoon 19 of 2019, and dike breach occurred at as many as 53 sites. Nevertheless, there were no cases of areal rainfall exceeding the DAD of East Tohoku Region. Also, it says that that same typhoon in the whole nation, impacted station records of 6-h rainfall were renewed at 89 stations, 12-h rainfall at 120 stations, 24-h rainfall at 103 stations and 48-h rainfall at 72 stations. However, the national records were broken only in Hakone's 12-, 24- and 48-h point rainfall and 24-h areal rainfall at the Shiroyama-dam catchment (1,201 km^2). It means that the national records are not often broken even under the current climate change developments (JMA website on nationwide ranking of historical records). The other evidence to support regionalization would be that a regional areal maximum rainfall often exceeds once over many thousands of years if evaluated by probability distributions listed in Table 1. It means that regionalized DAD gives extremes that are appropriate for the AMSP.

Thus, the regionalization of DAD for extreme areal precipitation would be a reasonable strategy for drawing the worst-case local inundation maps.

But the Hakone and Shiroyama-dam catchment cases have very important messages, too. The DAD for the AMSP in the Kanto region was issued in 2015. But only 4 years later, during the Super Typhoon 19 in 2019, the areal rainfall of the Shiroyama-dam catchment in Kanagawa and the point rainfall at Hakone, Kanagawa, were well above the past records, making the worst-case DAD no longer the worst (JSCE, 2020). As long as using the historical maxima, such cases are inevitable and unanticipated events will keep happening. It means that continuous periodic updating is necessary, although it is not a simple matter. The revision of the regional DAD means the renewal of hazard maps and community responses. To avoid frequent revision needs, climate change impact on defining of DAD should necessarily be reflected.

Another more serious question in relation to defining DAD would be the appropriateness of partitioning of the country into 15 regions shown in Figure 1. There are many other cases where similar geographical classifications are made. For instance, JMA seasonal weather forecasts use classification of 7 regions for 1–3 months precipitation forecasts and the warm season (June to August) forecast (JMA website on seasonal forecasts). If the number of partitions is smaller, more severe DAD would result for larger regions and heavier areal rainfall would be anticipated. In fact, if such an option is selected the cases that the AMSP being much less than 1/1,000 per year would appear much less frequently. In other words, selection of the number of regional partitioning is critical to this method and scientific analyses on the soundness of partitioning is necessary.

In both cases above, identification of areal maxima and regional partitioning, climate change impact should be considered. In other words, the criteria of identification should not be only the historical similarity but potential meteorological projections under climate change, including the occurrence of stagnated linear rain bands and super typhoons. This would enhance the rationale of regionalization of DAD.

Selection of the distribution of sub-basin hyetographs from historical cases

The distribution of hyetographs over sub-basins is the one which brought the worst disaster in the concerned basin in the past. In practice, several disaster cases are examined and the most unfavorable distribution pattern to the concerned basin outlet (Yattajima in the example above) is selected. In order to adjust the total amount into the AMSP, the enlargement rules depicted in Figures 3 and 4 are used. This method is quite heuristic, but it is reasonable as an engineering sense and there would not be any better way.

But again, as long as this process is based on historical events, more serious cases should be anticipated in the future. Especially, climate change will quite likely bring unexperienced areal rainfall patterns and hyetographs distributions over various sub-basins. This also implies the need of examining climate change impacts beyond the hazard.

CONCLUSION

Under the potential for climate change rapidly increasing the magnitude and frequency of hydro-meteorological extremes, it is necessary to take immediate action to prepare for saving lives and avoiding catastrophic economic losses prior to developing proper physical infrastructure.

For such a critical goal, it is quite reasonable to anticipate the worst-case scenarios for floods, landslides, debris flows, similar to earthquakes, tsunamis, volcanic eruptions, etc., so that people can prepare for such emergencies and evacuate early enough to protect their lives.

In Japan, the historical precipitation observations have been accumulated since the establishment of JMA in 1875 and the radar-AMeDAS database since 1988 in high density and accuracy all over Japan. Under such circumstances, the use of observed data would be more reliable to identify local scale extreme statistics as compared with the current level of climate projections by GCMs and RCMs assembled in 20 km mesh d4PDF datasets for Japan area.

The use of historical areal maxima as the AMSP through regionalization of DAD in the worst-case scenario is a unique heuristic approach. It considers that the heaviest areal precipitation that has occurred in a climatically uniform region can occur anywhere in the region. It is a reasonable assumption to use historical records for the worst-case scenario appropriate for local communities.

Nevertheless, climate change is occurring. The stationary assumption does not hold and hydrologists cannot challenge ‘Stationarity is dead’ although this does not ‘wither water management’ (Milly *et al.*, 2008). The state of the art in climate change projections should be assessed not just by scientific achievements but from the local practice point of view. Unfortunately, the current state is not satisfactory enough for local practical use all over Japan. Under such circumstances, the current Japanese engineering solution is an art, a beautiful art but it is an intermediate state to be more closely connected with meteorological sciences. Connection would still be difficult and meanwhile the accumulated ground observations and engineering experiences can well serve society.

In developing countries, however, there is often a lack of observation data and it is impossible to use historical data for drawing local anticipated inundation maps. In such cases, the interpolation and extrapolation of observed data in the surrounding area and the use of reanalysis data is an inevitable alternative. The direct use of GCM projection data would also be an option using the CMIP6 (ESGF Website) or d4PDF datasets. But even in such cases, the ground observation is the indispensable basis for real actions. Even a few years of observations would provide a precious means for verification to those estimated or projected data.

In any region where no ground observation network is installed, it is an urgent matter to develop it in a decent density. The ground observation data are necessary to formulate development policy, to use scientific models as well as to use remote-sensing data. But, in order to estimate the worst-case meteorological hazards through regionalization, such as for once in several hundred years to a thousand years, the long observation data of at least 50 years extent would be necessary as in the case of Japan (WMO, 2009).

It cannot be overstated that the importance of ground observations, in developed or developing countries, should not be overshadowed by advanced sciences, models and remote sensing. They rather increase their effectiveness.

ACKNOWLEDGEMENT

The authors are indebted to River Planning Division, Water and Disaster Management Bureau, MLIT for their kind cooperation by providing various information and documents.

REFERENCES

Cabinet Office (2016). Damages by September 2015 Kanoto-Tohoku torrential rains. 19 February 2016. Available at: http://www.bousai.go.jp/updates/h27typhoon18/pdf/h27typhoon18_28.pdf (accessed May 31, 2019).

Cabinet Office (2018). *Special Report of 2017 Northern Kyushu Torrential Rain Disaster*. Disaster Prevention White Paper. Available at: http://www.bousai.go.jp/kaigirep/hakusho/h30/ (accessed June 29, 2019).

Cabinet Office (2021). Damages by July 2020 Torrential Rains. 7 January 2021. Available at: http://www.bousai.go.jp/updates/r2_07ooame/pdf/r20703_ooame_40.pdf (accessed April 13, 2021).

Central Disaster Management Council (2011). Report of the committee for technical investigation on countermeasures for earthquakes and tsunamis based on the lessons learned from the '2011 off the Pacific coast of Tohoku Earthquake'. 28 September 2011. Available at: http://www.bousai.go.jp/kaigirep/chousakai/tohokukyokun/index.html (accessed May 31, 2019).

ESGF Website. Available at: https://esgf-node.llnl.gov/projects/esgf-llnl/ (accessed April 29, 2021).

FDMA (2017). Damages by 2016 Typhoon 10. Report No. 43. Fire and Disaster Management Agency. 8 November 2017. Available at: https://www.fdma.go.jp/disaster/info/assets/post816.pdf (accessed June 29, 2019).

Ishii, M. (2019). Meiji 150 years, historical observation records and climatic analyses. *Tenki 66*(8), 91–98.

JMA website on Analyzed Precipitation. Available at: https://www.jma.go.jp/jma/kishou/know/kurashi/kaiseki.html and https://www.data.jma.go.jp/add/suishin/cgi-bin/catalogue/make_product_page.cgi?id=KaisekiU (accessed April 14, 2021).

JMA website on 1982 July Torrential Rain and Typhoon 10. Available at: https://www.data.jma.go.jp/obd/stats/data/bosai/report/1982/19820701/19820701.html (accessed November 28, 2019).

JMA website on nationwide ranking of historical records. Available at: https://www.data.jma.go.jp/obd/stats/etrn/view/rankall.php?prec_no=&block_no=&year=&month=&day=&view= (accessed 14 April 2019).

JMA website on seasonal forecasts. https://www.jma.go.jp/bosai/season/ (accessed April 29, 2021).

Jonkman, S. N., Kok, M. & Vrijling, J. K. (2008). Available at: Flood risk assessment in the Netherlands: a case study for Dike Ring South Holland. *Risk Analysis 28*(5), 1357–1373. doi:10.1111/j.1539-6924.2008.01103.x.

JSCE (2013). Report of investigation of 2012 Northern Kyushu torrential rain disaster. JSCE investigation team of 2012 Northern Kyushu torrential rain disasters, reported in February 2013. Available at: http://committees.jsce.or.jp/report/node/56 (accessed May 31, 2019).

JSCE (2015). *Report of Investigations of 2014 August Hiroshima Torrential Rain Disaster*. Japan Society of Civil Engineers.

JSCE (2020). Investigation report on heavy rain disasters by 2019 Typhoon 19. Committee on Hydroscience and Hydraulic Engineering, Japan Society of Civil Engineers. Available at: https://committees.jsce.or.jp/report/system/files/taifu19.pdf (accessed April 29, 2021).

Katada, T. (1999). Evacuation behavior and the use of hazard maps at the time of Abukuma River flood of the end of August 1998. *Textbook of disaster management seminar* '99. Organized by Tohoku Regional Bureau, Ministry of Construction, pp. 17–28.

Kimura, T. (1961). *Flood Discharge Routing by Storage Function Method.* Public Works Research Institute.

Kimura, T. (1978). Recent progress in the storage function method. *Proceedings of the Japanese Conference on Hydraulics 22,* 191–196.

Kuroyoshi, R., Kunitsugu, M., Tsujimura, Y. & Makihara, Y. (2005). Outline of radar AMeDAS analyzed precipitation. *JMS 2005 Annual Convention Pre-proceeding,* 207. Available at: https://dl.ndl.go.jp/info:ndljp/pid/10590665 (accessed April 15, 2021).

Milly, P. C. D., Betancourt, J., Falkenmark, M., Hirsch, R. M., Kundzewicz, Z. W., Lettenmaier, D. P. & Stouffer, R. J. (2008). Stationarity is dead: whither water management? *Science 317,* 573–574.

Ministry of Construction (1984). *River and Sabo Works Engineering Standard.*

MIROC website on d4PDF. Available at: http://www.miroc-gcm.jp/~pub/d4PDF/design_en.html (accessed April 14, 2021).

MLIT (2004). *Water Disaster Report 2004* (in Japanese). Available at: https://www.mlit.go.jp/river/pamphlet_jirei/bousai/saigai/kiroku/suigai2004/index.html (accessed May 31, 2019).

MLIT (2012). *MLIT White Paper, Section 3 Policy Change after the Earthquake Disaster.* Available at: http://www.mlit.go.jp/hakusyo/mlit/h23/hakusho/h24/html/n1131000.html.

MLIT (2014a). 2011 great water disaster of the Kii Peninsula. Report of disaster response by Kinki Regional Development Bureau, Ministry of Land, Infrastructure, Transport and Tourism (MLIT), first edition 2013, revised in 2014. Available at: https://www.kkr.mlit.go.jp/bousai/qgl8vl0000008ajd-att/kiihantou-kirokushi.pdf (accessed May 31, 2019).

MLIT (2014b). The basic river management policy of the Yabe river. Material on the basic high water and others (reference material), Water and Disaster Management Bureau, MLIT. 31 August 2014. Available at: https://www.mlit.go.jp/river/basic_info/jigyo_keikaku/gaiyou/seibi/pdf/yabegawa93-6.pdf (accessed December 11, 2019).

MLIT (2015a). Report of the informal committee on disaster prevention and mitigation responding to the new stage. January 2015.

MLIT (2015b). Design method for anticipated maximum scale precipitation to produce anticipated inundation (floods and inland floods) etc, Water and Disaster Management Bureau, MLIT, July 2015. https://www.mlit.go.jp/river/shishin_guideline/pdf/shinsuisoutei_honnbun_1507.pdf (accessed 14 April 2019).

MLIT (2015c). Reference material: Region dividing method considering anticipated maximum scale precipitation, Water and Disaster Management Bureau, MLIT, July 2015. https://www.mlit.go.jp/river/shishin_guideline/pdf/kouu_chiiki_1507.pdf (accessed 14 April 2019).

MLIT (2015d). Manual for producing the anticipated flood inundation area, Version 4. Water and Disaster Management Bureau and National Institute for Land and Infrastructure Management, MLIT. July 2015.

MLIT (2016). Manual for producing the anticipated inundation area in small to medium scale rivers. Water and Disaster Management Bureau, MLIT. March 2016.

MLIT (2018). Outline of losses by July 2018 torrential rain disaster. *Distribution material* no. 2, 1st sub-committee on counteraction for water disasters by large scale wide area torrential rain. River Committee, Social Infrastructure Management Council, MLIT. 28 September 2018. Available at: http://www.mlit.go.jp/river/shinngikai_blog/shaseishin/kasenbunkakai/shouiinkai/daikibokouikigouu/1/index.html (accessed June 29, 2019) (in Japanese).

MLIT (2020). Disaster damages by Typhoon 19. *Committee on nation's long perspective,* 2nd, material 1-5. Available at: https://www.mlit.go.jp/policy/shingikai/content/001317859.pdf (accessed April 29, 2021).

MLIT Administrative Directive (2015). The previous examples and others to be referred to develop the basic river management policy, Water and Disaster Management Bureau, MLIT, March 12, 2015. Available at: https://www.cao.go.jp/bunken-suishin/doc/tb_h26fu_14_mlit02.pdf (accessed December 13, 2019).

MLIT website on flood forecasting rivers. Available at: http://www.mlit.go.jp/river/bousai/main/saigai/tisiki/syozaiti/pdf/yohousyuutikasen_1603.pdf (accessed May 31, 2019).

MLIT website on flood hazard maps (in Japanese). Available at: http://www.mlit.go.jp/river/bousai/main/saigai/tisiki/syozaiti/pdf/shinsui-hm_h3009.pdf (accessed May 31, 2019).

MLIT website on flood risk management act: Partial amendment of flood risk management act and others. 20 May 2015. Available at: http://www.mlit.go.jp/river/suibou/suibouhou.html (accessed May 31, 2019).

MLIT website on the Tone River design flood verification. Available at: https://www.mlit.go.jp/common/000165843.pdf (accessed April 23, 2021).

NIED website. Available at: http://ecom-plat.jp/suigai-chikei/index.php?gid=10018. National Research Institute for Earth Science and Disaster Resilience (accessed May 31, 2019).
Ninomiya, K. (2001). Torrential rain and precipitation system (Tokyo-do Publishing Co., 2011).
Oya, M. (1956). Water-disaster morphology classification map of Nobi Plain, Kiso River Basin (in Japanese). Available at: http://ecom-plat.jp/suigai-chikei/index.php?gid=10018 (accessed May 31, 2019).
Quenouille, M. H. (1949). Approximate tests of correlation in time-series. *Journal of the Royal Statistical Society: Series B 11*, 68–84.
Sato, S. (2007). Development of meteorological radar in Japan. *Tenki 54*(9), 17–20.
Stedinger, J. R., Vogel, R. M. & Foufoula-Georgiou, E. (1993). Frequency analysis of extreme events. In: *Handbook of Hydrology, Chap. 18*. Maidment, D. R. (ed.). McGraw-Hill, New York, pp. 18.1–18.66.
Tanaka, S. & Takara, K. (1999). Goodness-of-fit and stability assessment in flood frequency analysis. *Annual Journal of Hydraulic Engineering 43*, 127–132.
Tanaka, T., Kawai, Y. & Tachikawa, Y. (2019). Evaluating reproducibility of annual maximum basin-averaged rainfall of d4PDF in all class A rivers in Japan. *Proc. JSCE B1 75*(2), I_1135–I_1140. https://doi.org/10.2208/jscejhe.75.2_I_1135.
WMO (1969). Manual for depth-area-duration analysis of storm precipitation, WMO – no.237. TP. 129, p. 49.
WMO (2009). Guidelines on analysis of extremes in a changing climate in support of informed decisions for adaptation. Climate data and monitoring WCDMP-No. 72. Available at: https://www.ecad.eu/documents/WCDMP_72_TD_1500_en_1.pdf (accessed April 29, 2021).
Yamamoto, H., Kobayashi, H. & Yamamoto, M. (2014). Characteristics of heavy rainfall and landslides in Izu Oshima by Typhoon No. 26, 2013. *Journal of Natural Disaster Science 32-4*, 337–351. doi: J.JSNDS32-4337-351.

First received 27 October 2021; accepted in revised form 27 October 2021. Available online 12 November 2021

© 2022 The Author Water Infrastructure Planning, MDCU doi: 10.2166/wp.2021.266

Screening for nonstationary analysis

Paul H. Kirshen
School for the Environment, University of Massachusetts Boston, 100 Morrissey Blvd, Boston, MA 02125, USA
E-mail: paul.kirshen@umb.edu

ABSTRACT

Adjustments in the designs of water resources systems due to climate change and other nonstationarities are warranted because the benefits of effective adaptation are well recognized. Therefore, the time and resources invested in these analyses are well worth the effort. Before a major investment in an effort is made, however, it is reasonable to determine if the problem is of sufficient complexity or the value of additional information is high enough to warrant the inclusion of complex, sophisticated methods that explicitly include nonstationarity and associated decision-making under deep uncertainty. There exist several planning level conditions such as the lifetime of the project, its criticality, and its reversibility that may indicate detailed analysis is not needed. There are also sequential analysis and screening steps that can be applied to determine the complexity of the methodology needed. Finally, the use of decision analysis can also help determine if additional, detailed analysis, or data collection are necessary. The use of one or several of these methods should be considered as initial steps before undertaking a vulnerability assessment and developing an adaptation strategy for a water resources system.

Key words: Climate change, Decision-making, Nonstationarity, Preliminary analysis, Screening analysis, Water resources planning

HIGHLIGHTS

- Including nonstationarity in the design of water resources systems may require considerable resources and finances to apply but given the high cost of failures of most water-based systems and the generally high benefit–cost ratios of adaptation, the expense and effort of appropriate planning are certainly justified. There are circumstances, however, when even though nonstationarity is acknowledged, a full detailed assessment of it is not necessary.
- The decision on the complexity of the analysis can be based upon a project's lifetime, criticality, reversibility, flexibility/adaptability, robustness, and the use of forward design standards.
- Top-Down and Bottom-Up screening methodologies also exist to assist in the analysis decision. Some rely upon determining if a system is operating below capacity and/or its historical performance, assessing the climate sensitivity of the needs of stakeholders, and changes in the design conditions.
- Several recent methods such as the World Bank Decision Tree Process and Climate Risk Informed Decision Analysis (CRIDA) include screening as part of their complex processes.
- The use of decision analysis can help determine if additional, detailed analysis, or data collection are necessary.

INTRODUCTION

It is widely recognized that all human and natural systems are sensitive to climate; thus as the climate changes, the services provided by these systems also change. The adjustment to the impacts of these services changes is known as adaptation. Water resources and coastal systems are some of the most sensitive systems to climate change because of their strong linkages to climatic driving forces and because of the many important services and also threats they provide to human and natural systems. Here, we focus upon the analysis of their services to human systems, but similar analysis can be applied to natural systems. Similar analysis can also be applied to respond to other nonstationary drivers such as demographic and land use changes; again not discussed here.

The impacts of climate change upon the services of water resources and coastal systems and their adaptation to the impacts have been the subject of intense research and analysis at least since the late 1980s and are now considered for most new projects and existing projects. Both top-down (e.g., scenario analysis) and bottom-up (e.g., context first) approaches can be applied. In all cases, however, the major challenge is incorporating the uncertainties of the nonstationary climate changes into the analysis. It is not possible to confidently describe them with probability functions (Mendoza *et al.*, 2018). If this was possible, then adaptation planning to threats and opportunities would become a less complex challenge and the well-known and established systems analysis approaches described in Maass *et al.* (1962), Major (1969), Loucks & van Beek (2017), and others could be applied – some with proven success (Brown *et al.*, 2015). Of course, as recently pointed out by Loucks (2021), there remain significant challenges in inserting the results of sophisticated water resources analyses into management policies.

Since climate change unfortunately will result in more threats to water resources and coastal systems than opportunities, here we only focus upon managing the threats though, of course, similar analysis could be applied to capturing any limited opportunities associated with climate change. Opportunities might include increased annual rainfall in presently humid areas, longer growing seasons, and increased offshore winds for renewable energy. Of course, unfortunately many of these opportunities may be accompanied by more droughts, more plant disease, and more intense tropical storms.

As noted above, there exist many approaches to analyze system vulnerabilities and adaptation options with some of them included in this volume. These approaches continue to evolve as the field of water resources systems analysis expands to meet these new challenges (Kasprzyk, *et al.*, 2018) but some have already had some successful and impactful applications (Mauroner *et al.*, 2021). Many of them require considerable resources and finances to apply but given the high cost of failures of most water-based systems and the generally high benefit–cost ratios of adaptation, the expense and effort of appropriate planning are certainly justified. The challenge is applying the correct method to the problem; guidance to this is also presented in this volume.

There are circumstances, however, when even though nonstationarity is acknowledged, a full detailed assessment of it is not necessary. This chapter starts with a discussion of these cases when a detailed assessment of nonstationarity may not be warranted based upon planning conditions, and then moves to sequential screening methods to assist in determining the level of nonstationarity analysis that should be considered, and concludes with the use of decision analysis to aid in this decision.

PLANNING CONDITIONS

There are situations where even though nonstationarity is recognized, it may not be necessary to carry out a detailed assessment of its consequences. These situations can be characterized by a project's lifetime, criticality, reversibility, flexibility/adaptability, robustness, and the use of forward design standards. There will be, of course, cases where multiple conditions exist. In any case, however, if there is any doubt, a more complex method of nonstationarity analysis must be used given the large vulnerabilities of some systems and the significant positive benefits of reasonable adaptation strategies.

Project lifetime

If an asset is not threatened now and, based upon a preliminary analysis of the more extreme range of climate change threats, will never be impacted over the planning horizon, then a detailed nonstationarity analysis is not needed. It may be needed, however, if the inputs necessary for that system to function or its outputs are threatened by climate change. For example, an electrical generation station far from the flood zone may never be threatened and thus no analysis is needed unless its cooling water supply is threatened or the demands on it might increase if other power plants do not function and more demands are placed on it.

Similarly, if a project has a short lifetime, is not exposed now to a threat, and will be upgraded before the estimated earliest time when it may be threatened, then nonstationarity analysis can be implemented later as part of the new design. The caveats about its inputs and outputs being threatened as above also apply.

If the project lifetime or the planning period is relatively short – perhaps 20–30 years – then the project can be designed with trend analysis applied to recent data. For example, Serago & Vogel (2018) suggest exploiting historical trends in discharge data to update flood design discharges to reflect current conditions.

Criticality

If an asset is not critical to system performance or there are many alternatives to it or adaptive capacity is high, then detailed analysis is also not needed. A simple case is a road segment which, if flooded, only results in minor consequences because there are many alternative routes to destinations. This is in contrast to a road that is the sole connection of the neighborhood to medical facilities.

Reversibility

Hobbs *et al.* (1997) report upon the results of a workshop that identified project characteristics that may warrant the consideration of complex nonstationary analyses because they are sensitive to climate change. These include:

- Options involving commitments to resources that have long-lived effects and are costly to alter or are irreversible.
- The benefits and costs of options are large and could be altered by climate change.
- The resource commitment is a unique part of the system for which there are no ready substitutes so adjustments to changed conditions are difficult to make later on.

Flexibility/adaptability

The American Society of Civil Engineers Manual of Practice (MOP) on Climate-Resilient Infrastructure (American Society of Civil Engineers, ASCE, 2018) examines how to consider nonstationarity in infrastructure design. While recognizing that there are many situations when complex design approaches are necessary, there are also situations where less complex methods may be applicable.

The most relevant may be the Observational Method (OM) historically used in the geotechnical field where uncertainty conditions are high. The approach is to identify system changes during the design phase that can be implemented over time as the uncertain conditions change over the project lifetime or as stated in the MOP:

> 'The Observational Method [in ground engineering] is a continuous, managed, integrated, process of design, construction control, monitoring and review that enables previously defined modifications to be incorporated during or after construction as appropriate. All these aspects have to be demonstrably robust. The objective is to achieve greater overall economy without compromising safety' (page 52).

Implementing such a process requires an accompanying observation system. The article also points out that the data collected from the system can also resolve some of the uncertainties of the performance of future systems, i.e. some of the uncertainties can be reduced. Projects most suitable for OM include those with multiple stages or construction processes and in situations where the changes are gradual such as sea level rise. In situations with possibly rapid changes, it may not be possible to have obtained the observational data in time to act.

The steps involved in the OM include:

1. 'Project design is based on the most-probable weather or climate condition(s) rather than the most unfavorable. The most-credible unfavorable deviations from the most-probable conditions are identified.
2. A course of action or design modification is devised (in advance) for every foreseeable unfavorable weather or climate deviation from the most-probable condition(s).
3. The performance of the project is observed over time (using preselected quantities) and the response of the project to observed changes is assessed.
4. Design and construction modifications (previously identified) can be implemented in response to observed changes' (page 53).

The modification to a system at any one time can be the construction of an element of a type different from previous modifications. For example, a levee may initially be modified by adding more material to it, but then modified by building a flood wall on top of it or implementing means to accommodate any flooding that overtops the levee.

An example of this is the recent decision of the City of Boston USA not to plan to install a massive harbor-wide barrier system to protect it from present and future coastal flooding (Kirshen *et al.*, 2018, 2020; Climate Ready Boston, 2021). It was found that the system was not cost-effective, and after a few decades of operation, the gates would have to close perhaps as frequently as once per week because of sea level rise – resulting in increased chance of operational failure of the gates, and ecological, recreational, and shipping impacts. It was not necessary to analyze this system in more detail for its performance under nonstationarity conditions because one alternative to the barrier was shore-based systems such as a shoreline elevated with parks and open space (nature-based solutions) or protected by floodwall or levees. These engineered systems can be implemented relatively quickly when needed, can be expanded over time as climate changes, are more cost-effective than a harbor-wide barrier system under a range of climate change scenarios, and provide many social and ecological co-benefits. Detailed nonstationarity analysis is not needed for selecting their initial elevations because of the existence of reasonable engineering principles for their selection based upon possible future flood elevations, economies of scale, and site design constraints. The major concern is ensuring that space is reserved for the shore-based systems so they can be implemented when needed.

As noted, one of the disadvantages of the approach is that many of the costs of modifying a project in an urban area such as access, disruption, interference with other infrastructure systems can be high. In addition, it is not possible to modify some systems so their expansion can be staged. Still, some of these can be avoided if they are recognized in the planning phase and options preserved to act on future possible conditions. For example, development could be limited in an area where a future adaptation plan may have to be implemented.

Robustness

Detailed nonstationarity analysis also may not be necessary if it is shown that the reasonable solutions over time to the climate stress are relatively insensitive to different climate change scenarios. For example, in an analysis of water supply and conservation alternatives to meet the present and future demands in Amman Jordan, Ray *et al.* (2012) showed that the set of surface water and groundwater resources, imported water, reclaimed water, water conservation, and other sources to meet future demands did not change significantly over time given a range of climate change scenarios. However, the relative quantities of the sources varied over time.

Forward design standards

Complex adaptation analysis also may not be needed if previous reasonable analysis has been codified and can be followed in the design process. An example of this is the requirements of Federal Executive Order (EO) 13690. As reported in ASCE (2018), the EO set minimum flood protection requirements for federally funded buildings and infrastructure. It gave flexibility to select one of three approaches for establishing a Design Flood Elevation:

1. Climate-Informed Science Approach (CISA) – use best available, actionable hydrologic and hydraulic data, and methods that integrate current and future changes in flooding based on climate science and other factors or changes affecting flood risk to determine the vertical flood elevation and corresponding horizontal floodplain in a manner appropriate to policies, practices, criticality, and consequences.
2. Freeboard Value Approach (FVA) – use the present Base Flood Elevation (or 1 percent annual flood determined using best available data) and an additional height to calculate the freeboard value. The additional height will depend on whether or not the action is a critical action.
3. The 0.2-percent-annual-chance Flood Approach (0.2 PFA) – use the 0.2-percent-annual-chance flood elevation (also known as the 500-year flood elevation).

In situations where previous analysis has been carried out, its application could be straightforward. For example, in Boston, USA, the City's planning and permitting authority, the Boston Planning and Development agency, requires that 2 feet of freeboard be added to the elevation of the highest sea level rise projection and associated storm flooding determined by a scientific organization for critical facilities and buildings with ground floor residential units and one foot for all other building types (Boston Planning & Development Agency, 2017).

Related to this is use of traditional design approaches using products that have been enhanced to include future climates. ASCE (2018) contains several methods for enhancing precipitation Intensity–Duration–Frequency relationships and flood frequencies (and some methods are also reviewed in this volume). If the designer is confident that one of the scenarios of the enhancement is reasonable or at least adjustments to the design can be made if it turns out not to be sufficiently accurate, then the enhanced relationships can be used in the design process.

SCREENING METHODOLOGIES

Asian Development Bank (ADB)

ADB (2013) offers guidance to investigate the complexity of climate changes stresses. ADB suggests a stakeholder workshop to address what might be the changes caused by climate change and how will these affect utilities and services in the context of the Pressure:State:Reponse framework. They present some principles for ranking stresses. For example, if a system is currently operating below capacity, then vulnerability is likely to be low. Conversely, if already operating at peak capacity, and occasionally fails under severe weather events, then vulnerability is likely to be high. An example of the outcome of such a process could be as in Figure 1.

US Agency for International Development (USAID)

USAID (2014) recommends a two-step Scoping and Assessment process to determine the vulnerability of a system to climate and other stresses. There are several intermediate steps in the process to decide if a more detailed analysis is necessary. The Scoping step consists of the following:

- 'Frame the planning process by identifying the development goals important to the country, community, or sector you are working with; and identify inputs and enabling conditions necessary to achieve those development goals.
- Consider the impacts of climate and non-climate stressors on those inputs, and consider the needs and opportunities for addressing the various stresses affecting inputs and undermining development' (page 9).

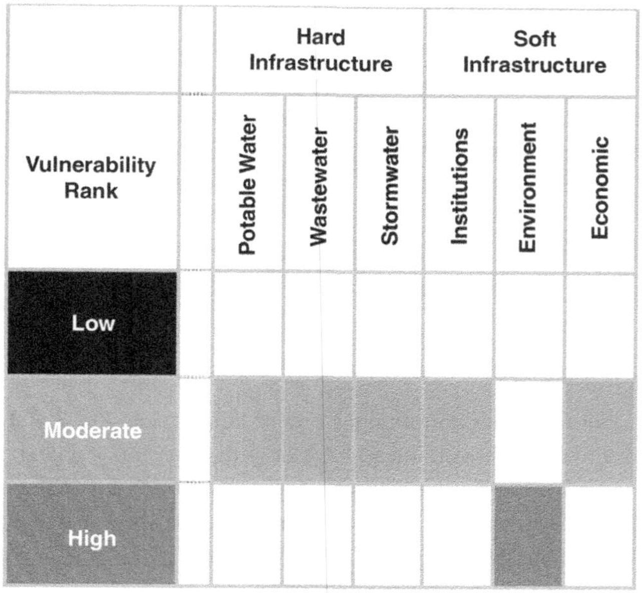

Fig. 1 | Example of the outcome of vulnerability ranking (ADB, 2013).

This is procedure may result in either complex or less complex solutions or perhaps the need to immediately move a more detailed vulnerability assessment – but if there is not a need right now, the next step is the Assessment step.

The Assessment step includes the analyses of exposure to the threat, sensitivity to the threat, and adaptive capacity to manage the threat. These analyses result in the vulnerability of the system. The level of detail and hence the complexity of the method depends on factors such as the information needed to make the decision, the time and resources available, and the importance of the decision.

US Climate Resilience toolkit

The above screening approaches can be characterized as bottom-up because they start with a set of questions about existing capacity and performance. In contrast, the US Climate Resilience Toolkit (2021) presents a top-down, staged approach based upon the exposure-sensitivity-adaptive capacity approach to investigate a system's vulnerability to climate change and subsequent adaptation strategies. Initially stakeholders, using scenarios of climate change projections, determine if weather and climate threatens valued assets. If so, the next step is to determine the vulnerability – they recommend setting priority for action based upon how close in time and/or condition an asset is to a tipping point – 'a point when incremental change in a system results in a new, irreversible response'.

Federal Highway Administration

Once an asset has been shown to be possibly exposed to a threat and is deemed critical, then an examination can be made of its sensitivity to the threat. Federal Highway Administration (2017) describes several methods of screening for sensitivity. These include:

Historical performance and agency knowledge

If performance of assets has been lowered in past weather events, particularly extreme weather events, then it indicates climate sensitivity. Maintenance records correlated with weather events are also useful. Discussions with design and maintenance staff can also yield insights.

Infrastructure design standards and guidelines

Many smaller scale assets are designed for particular events and frequencies. If possible changes in the events are known, then a comparison can indicate sensitivity. For example, in Boston, USA, most of the storm drainage network was designed approximately 100 years ago using around 4 inches of rainfall in 24 h. Recent analysis indicates that this value is now over 5 inches (Douglas *et al.*, 2016). If a more complex system has been designed using maximum expected present value net benefits that analysis can also be recalculated with current information.

The sensitivity can also be weighed by a qualitative assessment of the probability of the damage to the asset from an event – that is, the risk. Shown in Figure 2 is an example of this applied to a particular transportation asset.

World Bank decision tree framework

The entire four phase methodology of the bottom-up, Context First process of The Decision Tree Framework of Ray & Brown (2015) is described in this volume so here only the screening phases (1 and 2) are summarized. Phase 1 is known as Project Screening. In this phase, a thought-exercise is carried out to determine the initial possibility that the project may be sensitive to nonstationary climate and other drivers. The process is guided by the Climate Screening Worksheet in Appendix B of Ray & Brown (2015). This leads the project analyst to

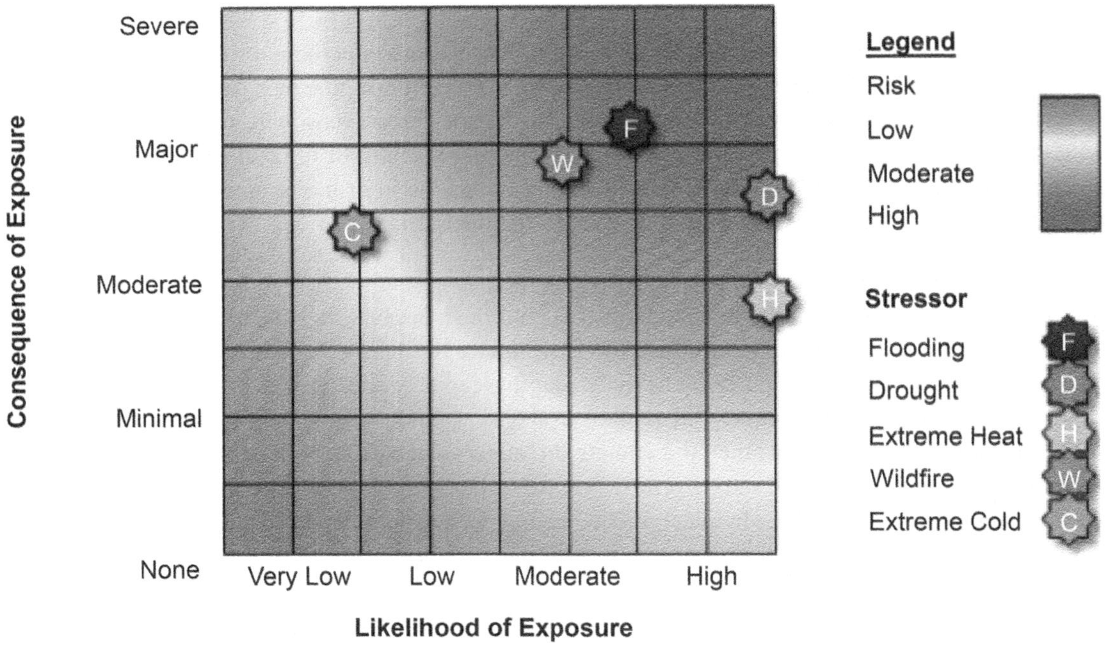

CAMPO developed risk summary fact sheets for each asset studied that explain the extreme weather risks it faces. This is an example Risk Summary Matrix for SH 71E at SH21. Flooding and drought present the greatest risks to this asset, while wildfire and extreme heat present moderate risks. Though flooding is projected to occur less often that drought, it would have greater consequences for the asset. (Source: CAMPO)

Fig. 2 | Example of consequence of exposure (sensitivity) weighted by likelihood of exposure to estimate risk (FHWA, 2017).

consider the 'Four C's' of the project: 'available project design choices; the consequences or performance metrics that will be used to evaluate the project's success; the connections through which choices and consequences are linked; and finally, the uncertainties that affect the conclusions that can be drawn about the profitability of the project'. If the result of this guided process is that the project may be significantly affected by future climates, then the project undergoes more extensive and complex analysis in Phase 2 to determine if a full detailed analysis of the project's vulnerability and adaptation options under deep uncertainty is carried out. If not significantly impacted by future climates, but perhaps instead by other uncertain drivers, then project analysis can proceed without consideration of a nonstationary climate.

Phase 2, known as Initial Analysis, recommends 'Rapid Project Scoping' to assess the systems sensitivity to climate change. It requires the use of a computer-based water resources model of the major elements of the system – this could be a complex watershed model or a less complex tool based upon, e.g., Excel. While this may require 1–2 months of time to complete, it is inexpensive compared to the actual design, construction, and operation of a large water resources project – even more valuable when the benefits lost or harm due to construction are also included. Details are given in Ray & Brown (2015).

The Phase 2 process is carried out in seven steps that result in the elasticities of stakeholder-defined Performance Indicators (PI) to the nonstationary climate inputs, then probability density functions (PDFs) of changes on PIs due to hydrologic changes, and eventually the risks and probabilities of changes of the PIs as a function of possible climate changes. The PI values can then be compared to acceptable thresholds of them. If the PI values are not acceptable, then the project is climate sensitive and must be analyzed in more detail with a DMDU method. This methodology can also be carried out for other nonstationary drivers such as population change if desired.

Climate risk informed decision analysis (CRIDA)

This process (Mendoza *et al.*, 2018) for inclusion of nonstationary analysis into decision-making starts with a bottom-up assessment of the system to determine if it may fail under certain climate and other conditions – the stress test of Brown *et al.* (2012) performed with a water resources system model. This is then compared to the uncertainty of this occurring in a figure such as Figure 3. If the outcome of this preliminary analysis (known as the 'Level of Concern') falls in Quadrants I or II of Figure 3, then existing analytical planning methods can be used that do not rely heavily upon nonstationary inclusion.

The previous analysis of Kirshen *et al.* (2018, 2020) used in the comparison of harbor-wide barriers versus shore-based Natural-Based Solutions to protect metro Boston from present and increased coastal flooding falls into Quadrant II of Figure 3.

Hobbs *et al.* (1997)

Based upon the results of using decision analysis to determine adaptation strategies (see the below section on 'Decision Analysis'), Hobbs *et al.* (1997) developed a five step procedure on when to apply decision analysis. The procedure can also be used to determine when detailed nonstationarity analysis can be used.

1. Determine if the project has characteristics that may make it sensitive to climate change (already mentioned in the above section on 'Planning Conditions'): 'irreversibility; playing a unique role in the system; long lived benefits and costs that would be significantly affected by climate change; and the ability to delay a decision to obtain more information on climate change'.
2. Assuming a climate change scenario, perhaps more extreme than average, determine if the projects net benefits would change significantly.
3. Determine the loss of net benefits if a decision was made assuming no climate change but climate change occurred (i.e. regret).

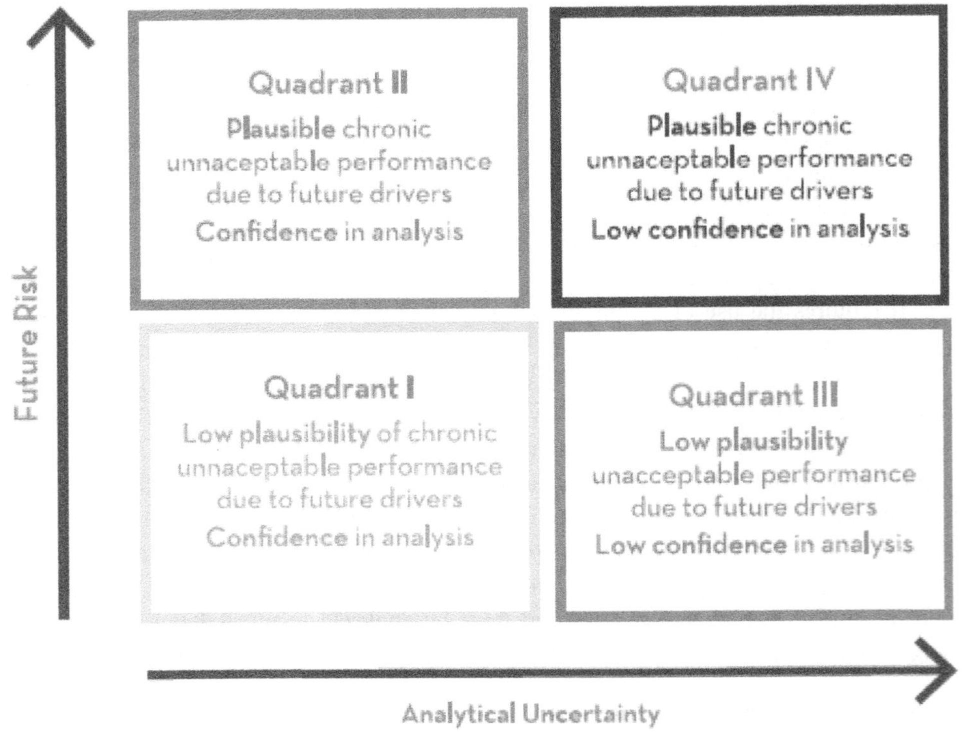

Figure 2.1. Establishing a "level of concern" in the planning process. Future risk in this document refers to the plausibility that a future driver is realized that surpasses a performance threshold. Analytical uncertainty results in a reduced confidence for the decision maker in making a decision given the available information.

Fig. 3 | The CRIDA level of concern matrix (Mendoza *et al.*, 2018).

4 and 5. If results of (3) are significant, then construct a decision tree with multiple climate change scenarios and with the option of delaying the project to wait for additional information. The decision tree approach follows below.

DECISION ANALYSIS

The Decision Analysis process breaks a problem out into a systematic display of all relevant choices. Oftentimes it is combined with the use of Bayes' Law so the impact and value of additional information can be added to the decision process; in some cases, the most reasonable solution is to pay for and/or delay to obtain more information. Decision trees are also commonly used to illustrate all the various decision pathways, the probabilities and economic consequences of various outcomes occurring, and provide a structured approach to making a decision (de Neufville, 1990). These processes can be used to help decide on the complexity of the method warranted for vulnerability and adaption planning.

In a now classic paper, Hobbs *et al.* (1997) uses decision analysis to determine the value of including climate change uncertainty in a water resources management strategy with an example of the Great Lakes in the USA. With the goal or objective of minimizing cost, the expected cost of ignoring uncertainty is the difference between the expected value of making the decision assuming no climate change and the expected value of making the decision assuming the possibility of climate change occurring and this is factored into the decision process. In the case study, using Bayes' Law to include information on the possibility of climate change given observations 20 years in the future, it is shown that it is better to wait 20 years to review the monitoring than take a construction action now. This method results in not only the most reasonable decision to make given what is known now, but also the value of information. Based upon the results of research on the case studies, Hobbs *et al.* (1997) recommended the five step approach described in the 'Screening Methodologies' section above.

In another application of decision analysis, Rosner *et al.* (2014) developed an adaptation planning process that considers the possibilities and consequences of both stationary and nonstationary future conditions. This is done by explicitly analyzing the consequences of analyzing flood trends and making either Type I or Type II errors. Type I errors occur when it is concluded a trend exists when it does not and results in over-preparedness and thus more expensive than necessary. Type II error is rejecting a trend when it actually exists and results in under-preparedness, and consequently additional damages. Such an approach could be used to help decide if more detailed analysis of trends or similar climate changes is necessary in water resources adaption planning. As shown in Figure 4, the decision tree used in this process includes branches with the possibilities of these types of errors occurring with their probabilities related to the significance level of the null hypothesis and the power of the alternative hypothesis. It is shown in an example that consideration of both types of errors occurring results in a better decision than ignoring the possibilities. Bayesian analysis could also be added to this process to investigate the value of additional information. While not directly including decision-making regarding climate change, Hecht *et al.* (2020) show how to include ecosystem impacts as well into this type of analysis.

SUMMARY

The impacts of climate and other changes on the services provided by water resources systems must be considered in examining the operation of existing systems and those in the planning phase. Adjustments in their designs may be warranted because the benefits of effective adaptation are well recognized. Therefore, the time and resources invested in these analyses are well worth the effort. Before a major investment in an effort is made, however, it is reasonable to determine if the problem is of sufficient complexity or the value of additional information is high enough to warrant the inclusion of complex, sophisticated methods that explicitly include

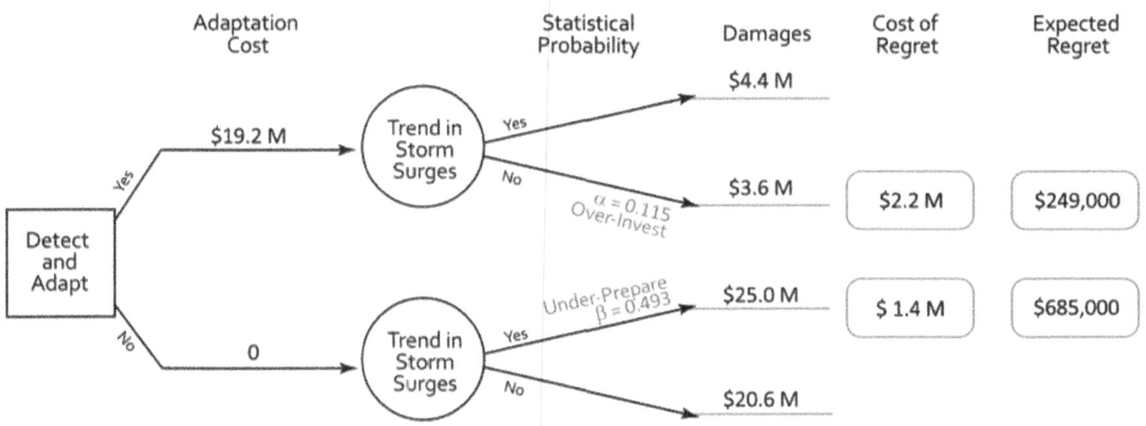

Fig. 4 | Decision tree for risk-based coastal flood adaptation in a nonstationary world case study. A decision point is denoted by a square, a chance outcome by a circle. Alpha, probability of Type I Error, Beta, probability of Type II error.

nonstationarity and associated decision-making under deep uncertainty. There exist several planning level conditions such as the lifetime of the project, its criticality, and its reversibility that may indicate detailed analysis is not needed. There are also sequential analysis and screening steps that can be applied to determine the complexity of the methodology needed. Finally, the use of decision analysis can also help determine if additional, detailed analysis, or data collection are necessary. The use of one or several of these methods should be considered as initial steps before undertaking a vulnerability assessment and developing an adaptation strategy for a water resources system.

ACKNOWLEDGEMENTS

The author appreciated the helpful comments and suggestions of Eugene Stakhiv. He also benefitted from conversations with Richard Vogel and Jory Hecht.

REFERENCES

American Society of Civil Engineers (2018). In: *Climate-Resilient Infrastructure, Adaptive Design and Risk Management, Manual of Practice 140*. Ayyub, B. (ed.). ASCE Committee on Adaptation to a Changing Climate, Reston, VA.

Asian Development Bank (2013). *Increasing Climate Change Resilience of Urban Water Infrastructure*. Asian Development Bank, Mandaluyong City.

Boston Planning and Development Agency (2017). *Climate Resiliency Guidance*. Available at: http://www.bostonplans.org/getattachment/5d668310-ffd1-4104-98fa-eef30424a9b3 (accessed April 30, 2021).

Brown, C., Ghile, Y., Laverty, M. & Li, K. (2012). Decision scaling: linking bottom-up vulnerability analysis with climate projections in the water sector. *Water Resources Research 48*, 9. doi:10.1029/2011WR011212.

Brown, C. M., Lund, J. R., Cai, X., Reed, P. M., Zagona, E. A., Ostfeld, A., Hall, J., Characklis, G. W., Yu, W. & Brekke, L. (2015). The future of water resources systems analysis: toward a scientific framework for sustainable water management. *Water Resources Research 51*, 6110–6124. doi:10.1002/2015WR017114.

Climate Ready Boston (2021). *Resilient Boston Harbor*. Available at: https://www.boston.gov/environment-and-energy/resilient-boston-harbor (accessed April 30, 2021).

De Neufville, R. (1990). *Applied Systems Analysis*. McGraw-Hill, New York.

Douglas, E., Kirshen, P., Hannigan, R., Herst, R., Palardy, A., DeConto, R., Fitzgerald, D., Hay, C., Hughes, Z., Kemp, A., Knopp, R., Anderson, B., Kuang, Z., Ravela, S., Woodruff, J., Barlow, M., Collins, M., DeGaetano, A., Schlosser, C., Ganguly, A., Kodra, E. & Ruth, M. (2016). *Climate Change and Sea Level Rise Projections for Boston*. The Boston Research Advisory Group, for Climate Ready Boston, City of Boston.

Federal Highway Administration (2017). *Vulnerability Assessment and Adaptation Framework*, 3rd edn. Federal Highway Administration, Cambridge, MA.

Hecht, J., Vogel, R., McManamay, R., Kroll, C. & Reed, M. (2020). Decision trees for incorporating hypothesis tests of hydrologic alteration into hydropower-ecosystem tradeoffs. *J. Water Resources Planning and Management 146*(5). https://ascelibrary.org/doi/abs/10.1061/%28ASCE%29WR.1943-5452.0001184.

Hobbs, B., Chao, P. & Venkatesh, B. (1997). Using decision analysis to include climate change in water resources decision making. *Climatic Change 37*(1), 177–202.

Kasprzyk, J., Smith, R., Stillwell, A., Madani, K., Ford, D., McKinney, D. & Sorooshian, S. (2018). Defining the role of water resources systems analysis in a changing future. *Journal of Water Resources Planning and Management 144*(12). https://ascelibrary.org/doi/10.1061/%28ASCE%29WR.1943-5452.0001010.

Kirshen, P., Borrelli, M., Byrnes, J., Chen, R., Lockwood, L., Watson, C., Starbuck, K., Wiggin, J., Novelly, A., Uiterwyk, K., Thurson, K., McMann, B., Foster, C., Sprague, H., Roberts, H., Jin, D., Bosma, K., Holmes, E., Stromer, Z., Famely, J., Shaw, A., Hoffnagle, B. & Herst, R. (2018). *Feasibility of Harbor-Wide Barrier Systems: Preliminary Analysis for Boston Harbor*. Sustainable Solutions Lab, University of Massachusetts, Boston.

Kirshen, P., Borrelli, M., Byrnes, J., Chen, R., Lockwood, L., Watson, C., Starbuck, K., Wiggin, J., Novelly, A., Uiterwyk, K., Thurson, K., McMann, B., Foster, C., Sprague, H., Roberts, H., Jin, D., Bosma, K., Holmes, E., Strummer, Z., Famely, J., Shaw, A., Hoffnagle, B. & Herst, R. (2020). Integrated assessment of storm surge barrier systems under present and future climates and comparison to alternatives; a case study of Boston USA. *Climatic Change 162*(2), 445–464. doi:10.1007/s10584-020-02781-8.

Loucks, D. P. & van Beek, E. (2017). *Water Resources Systems Planning and Management: An Introduction to Methods, Models and Applications*. Springer, Switzerland.

Loucks, D. (2021). Science informed policies for managing water. *Hydrology 8*, 66.

Maass, A., Hufschmidt, M., Dorfman, R., Thomas, H., Marglin, S. & Fair, G. (1962). *Design of Water-Resource Systems: New Techniques for Relating Economic Objectives, Engineering Analysis, and Governmental Planning*. Harvard Univ. Press, Cambridge, Mass.

Major, D. C. (1969). Benefit-cost ratios for projects in multiple objective investment programs. *Water Resources Research 5*(6), 1174–1178.

Mauroner, A., Timboe, I., Matthews, J., Taganova, J. & Mishra, A. (2021). *Planning Water Resilience from the Bottom-Up to Meet Climate and Development Goals*. UNESCO and AGWA, Paris, France and Corvallis, USA.

Mendoza, G., Jeuken, A., Matthews, J., Stakhiv, E., Kucharski, J. & Gilroy, K. (2018). *Climate Risk Informed Decision Analysis (CRIDA)*. UNESCO, Paris, and International Center for Integrated Water Resources Management, Alexandria, VA.

Ray, P. & Brown, C. (2015). *Confronting Climate Uncertainty in Water Resources Planning and Project Design: The Decision Tree Framework*. World Bank, Washington, DC.

Ray, P., Kirshen, P. & Watkins, D. (2012). Staged climate change adaptation planning for water supply in Amman, Jordan. *Journal of Water Resources Planning and Management. 138*, 403–411.

Rosner, A., Vogel, R. & Kirshen, P. (2014). A risk-based approach to flood management decisions in a nonstationary world. *Water Resources Research*, published online March 7, 2014.

Serago, J. & Vogel, R. (2018). Parsimonious nonstationary flood frequency analysis. *Advances in Water Resources 112*, 1–16.

US Agency for International Development (2014). *Climate-Resilient Development: A Framework for Understanding and Addressing Climate Change*. US Agency for International Development, Washington, DC.

US Climate Resilient Toolkit (2021). Available at: https://toolkit.climate.gov/ (accessed April 30, 2021).

First received 18 November 2021; accepted in revised form 18 November 2021. Available online 30 November 2021

© 2022 The Author

Water Infrastructure Planning, MDCU doi: 10.2166/wp.2021.246

How Hurricane Katrina influenced the design of hurricane protection and risk reduction systems and national approaches to risk and resilience: Part 1. Hurricane Katrina and New Orleans: A forensic assessment, risk and reliability analysis, and key lessons learned

Lewis E. Link[a,b]

[a] Department of Civil and Environmental Engineering, University of Maryland, College Park, MD, USA
[b] Interagency Performance Evaluation Task Force, Washington DC, USA
E-mail: elink@umd.edu

ABSTRACT

A systems perspective is presented of what happened during and after Hurricane Katrina (2005) and the potential for reducing the likelihood of large losses in the future. This work was the basis for the rapid repair of the damage resulting from Katrina and ultimately the development and construction of a new risk reduction system for the region and a major shift in engineering guidance and practice related to public water infrastructure. The work was primarily accomplished through the Interagency Performance Evaluation Task Force (IPET) established by the Chief of Engineers, US Army Corps of Engineers, to conduct a comprehensive forensic analysis of what happened and why, and to an engineering risk and reliability assessment of the hurricane protection system in place when Katrina struck.

Key words: Flood risk, Forensic analysis, Lessons learned, Hurricane Katrina, Risk and reliability analysis

HIGHLIGHTS

- Hurricane Katrina caused extensive and unprecedented economic damages and loss of life in Southeast Louisiana. An interagency task force was formed to find out what happened and why.
- The resulting forensic analysis of the storm and resulting forces, protective structures and their performance, the resultant flooding, and associated economic and life losses portrays a vivid picture of the consequences of being unprepared.
- The associated risk and reliability analysis conducted for New Orleans and vicinity provided a strategic baseline for understanding why New Orleans was so vulnerable and a foundation for configuring both short-term repairs and a longer-term initiative for a dramatic reduction of flood risk for the future.
- The lessons learned from this effort were the roots of major changes in the approach taken for flood mitigation through the application of risk and reliability and a better understanding of both the hurricane threat and the failure modes of coastal structures.

PART 1: INTRODUCTION

The paper is segmented into three components: the first describing the Katrina event, the impact on the existing infrastructure in New Orleans and Southeast Louisiana, and the losses that resulted in terms of both life and property. This provided the basis for understanding the failure modes and vulnerabilities of the existing infrastructure and led to the design of immediate repairs and upgrades to those structures prior to the next hurricane season.

The second component describes the risk and reliability assessment made to understand the sources of the structural failures and the losses in both geographical and socioeconomic terms. This analysis became a key input for developing a strategy for the design and construction of a new system to dramatically reduce flood risk in the future, discussed in a separate paper in this volume. The third component is a summary of lessons learned from these analyses that are relevant to other critical weather infrastructures and systems that have exposure to flooding.

The IPET was established (October 2005) to examine the performance of the New Orleans and Southeast Louisiana hurricane protection system (HPS) and provide real-time input to the immediate repairs and long-term rebuilding of the system. The results of the IPET effort were initially published as a draft in June 2006 and in final form in June 2009 following extensive review and augmentation. The IPET Final Report consisted of nine volumes (Performance Evaluation of the New Orleans and Southeast Louisiana Hurricane Protection System, Vols I–IX and a supplemental report on risk) available on the Web at https://biotech.law.lsu.edu/katrina/ipet/ipet.html. In addition, a separate study of somewhat reduced scope was conducted by a team led by the University of California at Berkeley. Their report, 'Investigation of the Performance of the New Orleans Flood Protection Systems in Hurricane Katrina on August 29, 2005, Independent Levee Investigation Team Final Report', July 31, 2006, is also available on the Web at http://www.ce.berkeley.edu/projects/neworleans/.

PART 2: HURRICANE KATRINA FORENSIC ANALYSIS

The following section describes the fundamental character of the infrastructure in place when Hurricane Katrina struck, the character of the hurricane itself and the forces the infrastructure experienced, the performance of that infrastructure, and the resulting flooding and related losses.

How the New Orleans HPS was planned, designed, and constructed

Planning and construction

Although federally authorized, the Lake Pontchartrain and Vicinity Project was to be a joint federal, state, and local effort, with the federal government paying 70% of the costs and the state and local interests paying 30%. The local interests included the State of Louisiana Department of Transportation and Development, the New Orleans Sewerage and Water Board, and the local levee boards. The Corps of Engineers was assigned responsibility for project design and construction, and the local interests were responsible for the operation and maintenance of the levees and flood control structures. This was one of the first major cost-sharing projects for the Corps of Engineers and the nation.

During the first 17 years of the project, it was focused on what has become known as the 'barrier plan'. The barrier plan included a series of levees along the lakefront, concrete floodwalls along the Inner Harbor Navigation Canal (IHNC), and a variety of control structures, including barriers and flood control gates located at the Rigolets and Chef Menteur Pass areas that connect Lake Pontchartrain to Lake Borgne. These structures were intended to prevent storm surges from entering Lake Pontchartrain and overflowing the levees along the lakefront. A number of project delays and cost increases occurred as a result of technical issues, environmental concerns, legal challenges, and local opposition to various aspects of the project.

A December 1977 court decision enjoined the Corps from constructing the barrier complexes and certain other parts of the project until a revised environmental impact statement was prepared and accepted. The Corps conducted a 'Re-Evaluation Study,' published in 1984, in response to the court order and examined the feasibility of providing protection mostly by means of raising and strengthening levees and floodwalls. The Energy and Water Development Act of 1992 mandated the use of parallel protection and set the stage for the construction of the levee and I-wall structures that were in place prior to Katrina. Note that the original authorization for protection occurred in the 1965 time frame and the final resolution of how to provide protection for a large portion of the metropolitan area of New Orleans was not determined until 1992, over a quarter of a century later.

The construction of the HPS, in place when Katrina struck, occurred over the subsequent 40 years. Figure 1 shows the general layout of the major components of the HPS in the immediate vicinity of New Orleans. Some components of the system were not scheduled to be completed until 2015, primarily the West Bank and Vicinity Project. At no time did the entire New Orleans and Vicinity area have a reasonably uniform level of protection around its perimeter. At no time did any individual parish or basin have the full authorized protection planned for in 1965. As of May 2005, the Lake Pontchartrain and Vicinity Project included about 125 miles of levees, major floodwalls, flood-proofed bridges, and a mitigation dike on the lake's west shore. Progress on the project varied by area: 90% complete in Orleans Parish, 70% complete in Jefferson Parish, 90% complete in the Chalmette area, and 60% complete in St. Charles Parish.

The history of this HPS has been one of continuous fragmented decision-making, funding, and deficiencies. This situation was a product of the overall incremental legislative water resources development process; the magnitude of the investments needed to accomplish such projects; the piecemeal allocation of resources; the time and complex processes required to resolve differences in local and federal priorities; and the traditional step-by-step construction process for structures such as levees in subsidence-prone areas such as New Orleans.

The hazard

By 2005, three separate hurricane protection projects had been designed and partially constructed in the New Orleans and the Southeast Louisiana region: Lake Pontchartrain and Vicinity Project, the West Bank and Vicinity Project, and the New Orleans to Venice Project. Each project had a specific hazard as a criterion for design.

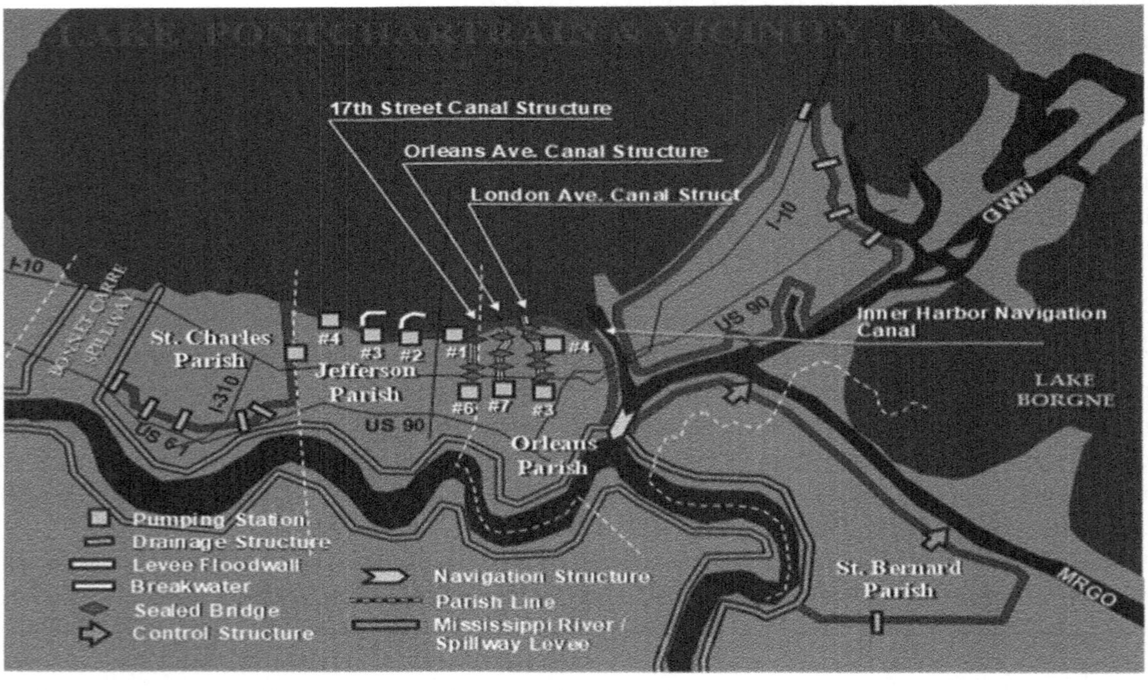

Fig. 1 | Map of structures and features of New Orleans and the pre-Katrina Hurricane Protection System (*Source*: IPET Report, Vols I and III).

The Lake Pontchartrain and Vicinity Hurricane Project was intended to protect areas around the lake (in Orleans, Jefferson, St. Bernard, and St. Charles Parishes) from flooding caused by a storm surge or rainfall associated with a hurricane that would be roughly the same as what is today classified by the Saffir–Simpson scale as a fast-moving Category 3 hurricane. The basis for this was the Standard Project Hurricane (SPH) developed for the Corps of Engineers by the US Weather Bureau (now the National Weather Service; see NOAA Technical Report NWS 23, 1979).

The SPH is a steady-state storm based on an analysis of meteorological parameters of past large hurricanes. The assumption of steady-state precludes the consideration of some of the dynamic behaviors we now know characterize hurricanes such as decreasing in intensity and increasing in diameter as they approach shore. For the initial definition of the SPH used in the design of the New Orleans hurricane protection structures, hurricanes were considered that occurred during the period 1900–1956. The Central Pressure Index (CPI) was the primary intensity criterion, and the 1% recurrence CPI (100 years) was chosen for the initial SPH definition. When the additional consideration of the likelihood of a storm of that description hitting the area near New Orleans is added, it was estimated to be equivalent to a 200- to 300-year recurrence event. Note that the character of the 1% recurrence CPI has changed dramatically between the initial SPH estimates and the time of Katrina occurring.

The SPH was intended to represent the most severe meteorological conditions considered 'reasonably characteristic' for the region. A maximum wind speed at landfall was also associated with the SPH; for Lake Pontchartrain and Vicinity, it was assumed to be 100 miles/hour. Following Hurricane Betsy in 1965, the wind speed criterion was revised but all other characteristics remained the same. The 1965 version of the SPH was used for the design of both the Lake Pontchartrain and Vicinity and New Orleans to Venice Projects. In 1979, NOAA issued a report that significantly revised the SPH criteria, resulting in a more intense storm event (1% recurrence CPI went from 934.6 mb in 1965 to 926.2 mb in 1979, see NOAA Technical Report NWS 23, 1979), and this became the basis for the design of the West Bank and Vicinity Project. All activities with respect to the Lake Pontchartrain and Vicinity project continued to use the original SPH criteria through the time of Hurricane Katrina. After Karina, which had a CPI of approximately 905 mb, a NOAA preliminary analysis revised the 1% recurrence CPI to approximately 901 mb. Hence, from a CPI perspective alone Katrina would be associated with a 1% recurrence event in 2005.

The SPH was a basis for estimating the surge and wave conditions that might exist for the New Orleans area. The storm was assumed to travel a number of paths, and the maximum surge and wave elevation generated from these different scenarios (estimated for each location of interest) was assumed to be the design water conditions for the project.

Design of the HPS structures

The HPS includes approximately 350 miles of protective structures, 56 miles of which are floodwalls. The majority of the floodwalls are I-walls with small sections of T-walls and a very small number of L-walls, basically similar to T-walls with a horizontal component on only one side at the base.

An aspect of the HPS is that the designs constructed, with the exception of the few sections constructed with T-walls, do not explicitly include protection against overtopping. The structures were, in effect, designed to perform at water elevations up to overtopping, but not beyond. This is not atypical in flood protection measures when cost/benefit justification is a primary decision factor.

Some sections of levees were constructed with higher quality (from an erosion perspective) materials than others, basically clays hauled in versus lower quality soils dredged (the method termed hydraulic fill) from adjacent areas. Along the outfall canals, I-walls were added to existing levees to gain higher protective elevations. Structures along the 17th Street and London Avenue Canals were designed with less conservative assumptions than those for the Orleans Canal. Soil strengths assumed for the clay layer underlying the 17th Street Canal were higher than warranted given the measurements available. For all of the canals, the design analysis used acceptable (at the time) but dated approaches for stability analysis and a factor of safety of 1.3. Failure modes examined were traditional for the time but did not explore other possibilities. The design for the Orleans Canal structures was more conservative than those for the 17th Street and London Avenue Canals. The levee section was wider and the freeboard of the exposed floodwall less. Orleans Canal had fewer real estate access restrictions, being adjacent to open public land. In the end, all of these differences played a large role in the ultimate performance of the structures during Katrina.

It is important to note the complexity of the HPS. Besides the structures considered part of the hurricane protection projects cited above, other features such as the main-line levees of the Mississippi River, levees and walls that form the outer periphery of the HPS, pump stations, bridges, navigation structures, and drainage and control structures are all factors in the ultimate ability to protect New Orleans from flooding. Some of these structures are not part of the HPS and fall under different jurisdictions, making their management during a flood more complicated.

The elevations of the hurricane protection structures in 2005 were significantly below the originally authorized heights, due in part from errors in initially constructed elevations, in part from rapid subsidence, and in part from sections where the authorized hurricane protection structures are not yet in place.

Hurricane Katrina event

The path followed by Hurricane Katrina, as shown in Figure 2, caused severe surge and wave conditions on the east side of the HPS, from Lake Pontchartrain to southern Plaquemines Parish. Katrina struck early on the morning of 29 August 2005, after building up water levels to the east of New Orleans for several days. Katrina was a Category 5 storm with up to 139-knot (160-mph) sustained surface winds until it was approximately 170 miles from landfall. When it reached landfall at Buras, LA, surface wind speeds were at about 100 knots (115 mph), but the long path through the Gulf, and its intensity and size, had built up record levels of surge and waves, larger than any previous storm to strike the New Orleans area, or the North American continent.

Katrina (a Category 3 storm at landfall) generated substantially higher surges than Camille in 1969, a Category 5 storm at landfall in the area where they both made a direct hit. Whereas the Saffir–Simpson scale (https://www.weather.gov/mfl/saffirsimpson) is a good surrogate for wind levels and damage from hurricane winds, it is not a particularly good metric to represent the surge and wave generation potential for hurricanes. Surge and wave levels are particularly sensitive to the width of the storm; the fetch, or distance it travels over water; the path the storm takes, the geometry of the coastline and the continental shelf, and the offshore character of the storm. Hurricane Katrina had far greater wave and storm surge generation potential than the SPH storms used to design the HPS in 1965. Note that both were classified as fast-moving Category 3 storms at landfall, according to the Saffir–Simpson scale for hurricanes.

Katrina swept through the New Orleans area rapidly, making a second landfall at Pearl River, MS with surface wind speeds around 100 knots (115 mph). Over a 24-h period, sections of New Orleans near the intersection of Lake Pontchartrain and the IHNC received over 14 inches of rainfall. This rainfall was approximately 20% of the total volume of water that flooded the New Orleans metropolitan area. The east and south-facing levees of New Orleans East, St. Bernard, and Plaquemines Parishes absorbed the brunt of the storm, experiencing surges and waves significantly beyond their design levels. Overtopping was common and, depending on location, persisted for hours.

Literally, all of the gauging instruments to measure water conditions were destroyed by Katrina. Other than high-water marks, and the devastation, there were few measurements to confirm the actual water-level time histories

Fig. 2 | Hurricane Katrina path and intensity history (intensity defined based on flight-level wind speeds, as opposed to surface winds) (*Source*: IPET Report, Vols I and IV, NOAA).

resulting from the storm. The IPET used the ADCIRC model (Luettich *et al.*, 1992) with a very high-resolution computational grid to model the storm and predict the time history of the surge levels that occurred at different locations around the region. Figure 3 shows the maximum surge levels predicted for Katrina. The high-water marks were used to confirm the accuracy of the model results, and in most cases, they agree to within a foot or two. Surge levels ranged from about 10–12 ft along the south shore of Lake Pontchartrain to 20 ft along the Plaquemines Levees. Even enclosed areas such as the IHNC experienced water levels above 14 ft, not including waves.

Winds from Katrina generated a record-wave environment. Again, the lack of measurements caused IPET to model the wind-generated waves to determine the conditions created by the storm. IPET applied a nested approach that used the Wave Analysis Model (https://journals.ametsoc.org/view/journals/phoc/18/12/1520-0485_1988_018_1775_twmtgo_2_0_co_2.xml) to generate wind-wave fields for the entire Gulf and the Steady State Spectral Wave model (https://apps.dtic.mil/sti/pdfs/ADA588527.pdf) to model nearshore waves in and around New Orleans. The most significant finding was that the waves along the east side of New Orleans levees were ocean-generated waves, with wave periods in the 15–16-s range, much more capable of overtopping structures than the design-assumed local wind waves with periods of 5–6 s.

Figure 4 presents a simple comparison of the surge generated by Katrina with that used for the HPS design. Note that the surge in Lake Pontchartrain was roughly the same as the design levels assumed for the HPS. On

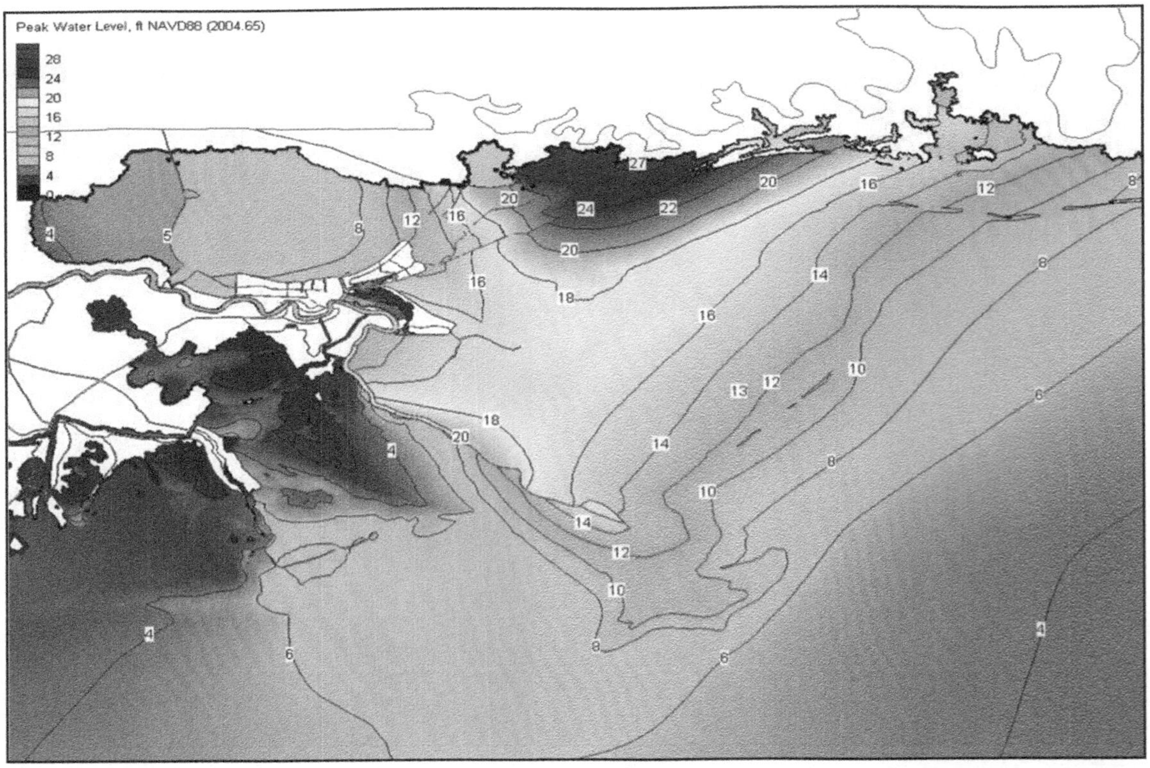

Fig. 3 | Maximum computed storm surge levels using the ADCIRC model, Mississippi to Louisiana region, water levels in feet NAVD88 (2004.65) (*Source*: IPET Report, Vols I and III).

the east side of New Orleans, Katrina-generated surges were significantly greater than the design criteria, ranging from 17 to 20 ft compared to the 12–14 ft assumed in the HPS design.

Wave conditions assumed for the SPH design and those estimated from Katrina have a similar relationship. With the exception of Plaquemines Parish, Katrina-generated wave heights were not much different from the design assumptions. In Plaquemines Parish, Katrina did generate significantly higher waves. Wave periods were a different story. In Lake Pontchartrain, the period of Katrina-generated waves was similar to the design assumptions, in the 6–7-s range. Along the St. Bernard and Plaquemines Levees, however, Katrina generated 14–16-s period waves, approximately three times that assumed for the design wave periods. This is very significant in that the longer period waves were largely generated in the Gulf and propagated to shore and have much greater potential for runup and overtopping of coastal structures.

Performance of the HPS in place when Katrina struck

Figure 5 shows the locations of the most severe damage to the HPS. Over 220 miles of the protective structures were damaged by Katrina-generated surges and waves, as well as 34 of 71 pumping stations. Approximately 41 miles of structures were judged to be severely damaged. Initially, there were a total of 50 major breaches identified, causing a dramatic reduction in protective elevation and losing the ability to prevent the inflow of external water. Of the 50 major breaches, four were caused by foundation-induced failures and the remainder from a combination of overtopping and scour. Three of the four foundation breaches occurred in the outfall canals and one in the IHNC. I-wall structures were particularly vulnerable as were levee sections created from hydraulic fill, and

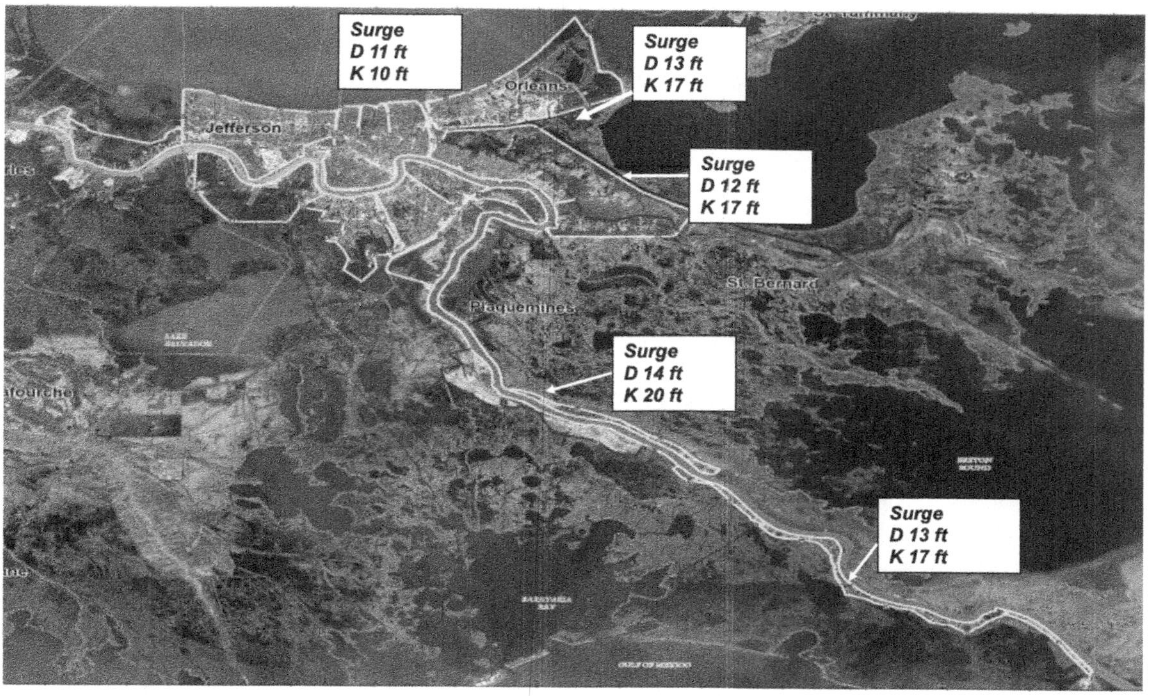

Fig. 4 | Comparison of general surge levels assumed in the SPH-based design (D) and estimated from Katrina (K) (*Source*: IPET Report, Vols I and III).

transitions were either elevation or strength differences occurred from changes in structure type or capability. Transitions between types of flood protection structures were also vulnerable, especially where the transition included a significant change in elevation between the structures.

The storm surge and waves first attacked the Plaquemines Levees well before Katrina's landfall, causing significant overtopping and erosion before dawn. The Lake Borgne Levees were soon hit with similar conditions, and eventually, both Plaquemines and St. Bernard Levees would be overtopped by both high surge and high, long-period waves. The persistent east-to-west winds had also built up a significant surge level at the convergence of the Gulf Inter-Coastal Water Way (GIWW) and the IHNC. Wind-generated waves reached at least 4 ft in the IHNC, contributing to very high water and dynamic loading on structures. The surge and waves had a devastating effect on the sections of the levees along the GIWW and MRGO that were constructed with materials dredged from the adjacent channels using hydraulic fill. Even though the levees were capped with clay, they were no match for the energetic environment they experienced. The overtopping waves created very high-flow velocities down the back sides of the levees, reaching 10–15 ft/s. These velocities were two to three times those experienced on the waterside of the levees (4–6 ft/s). The potential for erosion being related to the cube of velocity, it is no wonder that the back sides of the levees, especially where they were comprised of erodible materials, was scoured away leading to, in many cases, complete breaching.

Early in the morning of 30 August, a section of I-wall along the Lower Ninth Ward breached. Underlain by the same marsh deposits and clay as the 17th Street Canal, the rising water and waves caused the wall to deflect enough to open a crack that created a direct avenue for high water pressures to reach the foundation. The weak clays underneath, now only reacting with the mass of soil on the protected side of the levee, could not

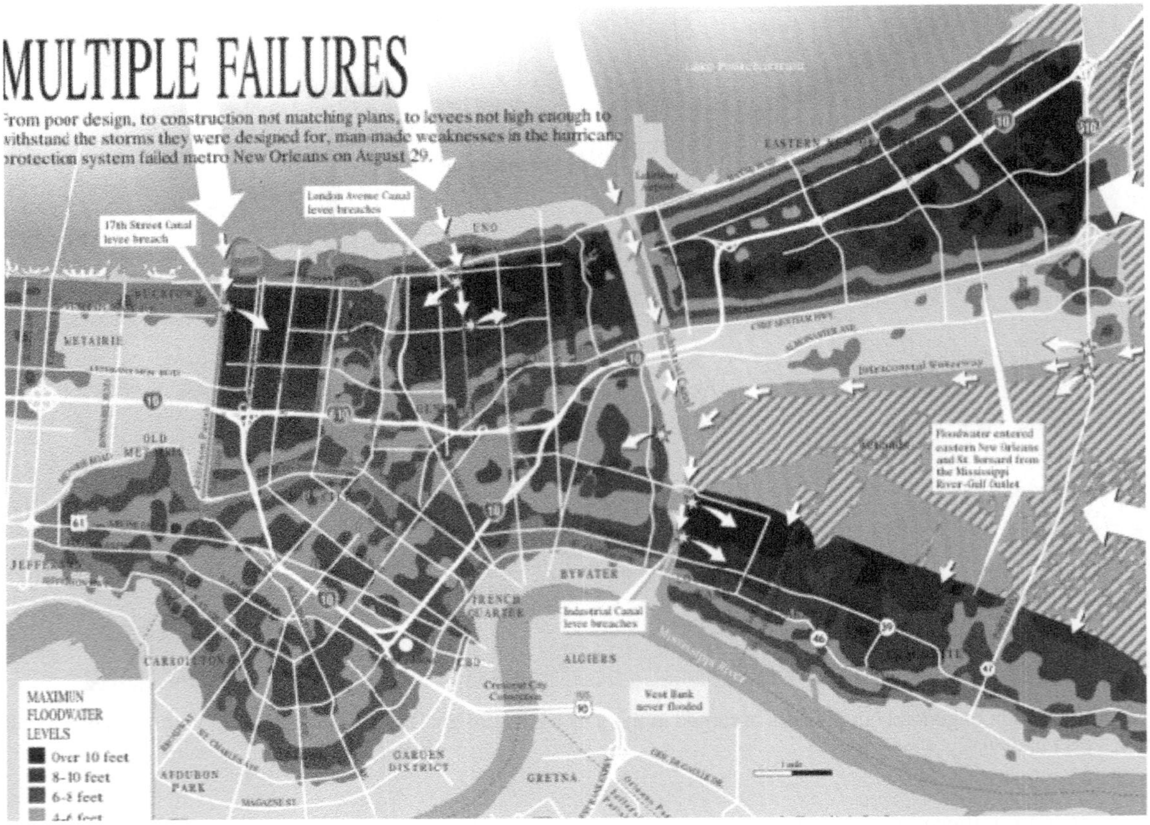

Fig. 5 | Locations of principal structural failures and resulting flood depths from hurricane Katrina in New Orleans (*Source*: Times-Picayune).

withstand the force and displaced backward, a process that would repeat itself on the 17th Street Canal. The water levels in the IHNC were approximately 9.5–10.5 ft when the foundation failure occurred. No overtopping had occurred and the design water elevations had not been reached at either breach location. Note that the factor of safety (1.3) for the canal structures was irrelevant because the failure mode that caused the breaching (deflection of the I-wall and hydrostatic pressure reaching the weak clay layer) had not been considered in the design process. See Figure 6 for a summary of major structural breaches on the outfall canals and IHNC and resulting in maximum flood depths.

Not long after the 17th Street Canal breach started, the south breach on the London Avenue Canal was initiated. As in the case of the 17th Street Canal and IHNC failures, water elevations below the design levels caused a crack to form on the waterside of the floodwall and allowed high pressures to be introduced directly into the foundation materials of the levee, this time relic beach sand. The porous sand quickly conveyed the pressure under the levee and caused significant uplift on the protected side. It also is likely that significant subsurface erosion occurred under the levee and caused a blowout on the protected side through which much sand and water flowed, decreasing the support for the levee and floodwall and causing a narrow failure. The north breach on the London Avenue Canal suffered a similar fate around the same time. This breach was much wider and involved less erosion, failure being caused by a loss of stability from the uplift. Water levels in the London Avenue Canal reached about 9 ft, below the design levels and well below the height of the I-walls.

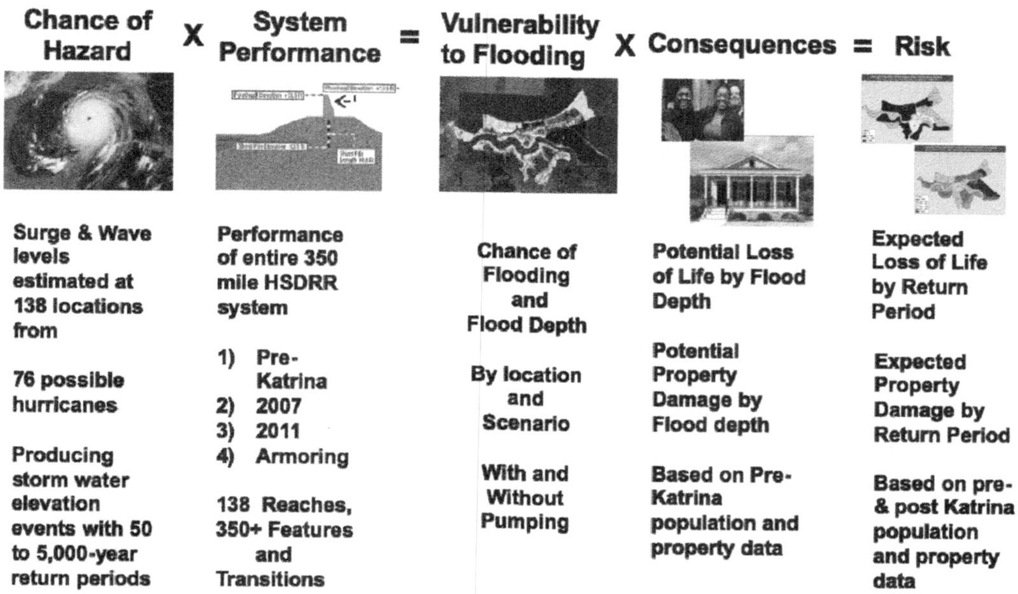

Fig. 6 | Risk assessment methodology used by IPET to evaluate both the HPS and planned HSDRRS (*Sources*: IPET Report, Vols I and VIII; IPET Supplemental Report 1992).

During mid-morning, the I-walls along the IHNC were overtopped and erosion behind the wall reduced their stability, causing three separate sections to fail. There was also a levee failure along the west side of the IHNC that caused additional flooding into the Upper Ninth Ward. There were no T-wall failures with the exception of a small section in southern Plaquemines Parish.

The flooding resulting from the overtopping and breaching was catastrophic, with almost 80% of New Orleans was inundated. Pumping stations were for the most part not operating due to prior evacuation of operators, loss of power, or loss of clean cooling water for the pumps. The pump stations in New Orleans were not designed to operate during major storms. A few stations, notably in Orleans Parish, may have continued to operate if the flooding had not been so extensive. Had the pumps been able to operate, the extent of flooding may not have been impacted greatly, but the duration of flooding could have been significantly reduced. Using temporary pumps and slowly bringing the permanent pumps on line after Katrina required 53 days to unwater the city.

The consequences of the flooding were enormous, dwarfing the losses from previous disasters. Approximately 78% of the direct property losses from flooding were experienced by residential property. This is in part due to their location in some of the lowest elevations of the city. When coupled with the approximately $4.5–$5.6 billion in public infrastructure damages, the total direct property losses for New Orleans alone reach nearly $25 billion.

Loss of life in the Gulf region was staggering, with almost 1,600 fatalities accounted for and hundreds more missing and presumed dead. The New Orleans metropolitan area suffered over 1,000 storm-related deaths. The majority of deaths in Orleans Parish were caused by drowning and preexisting medical conditions. The deaths in St. Bernard Parish were primarily attributed to drowning. Loss of life was highly and inversely correlated with evacuation. Of those who remained, the elderly were particularly vulnerable with three of every four persons who died being over 60 years old. In fact, the flooding in general was disproportionately cruel to the poor, the elderly, and the disabled, groups least likely to be able to care for themselves.

The flooding and resultant prolonged loss of services caused what became more of a migration than an evacuation, casting long shadows on the region's ability to recover. Only 8 of 73 neighborhoods did not flood, while 34 were completely inundated. Residential property losses were a staggering 78% of the total. Commercial property losses were approximately 11% of the total, while industrial losses were under 2%. Clearly, the people of New Orleans suffered the most direct losses, and these losses represent perhaps the greatest challenge to recovery, not just in terms of property damages. The extensive flooding caused a breakdown in the area's social and cultural structure, significantly complicating recovery and redevelopment. Critical social institutions such as schools and hospitals were very slow to reopen.

PART 3: KATRINA RISK AND RELIABILITY ASSESSMENT

To gain a true systems perspective of what happened during Katrina and the potential for reducing the likelihood of large losses in the future required a different type of analysis. One that allowed a broad look at the geographically distributed infrastructure system associated with New Orleans within the context of the complex threat represented by the large variety of hurricane events possible for the region. This required a comprehensive risk and reliability analysis, one of the first attempts for such a complex infrastructure system and equally complex hazard. Risk in this context is the potential for loss (expected losses in terms of economic and/or life) and includes the likelihood of the conditions and forces (by location) generated by the hazard, the performance of structures and measures in place to deal with the forces, the likelihood of level and character of flooding, and the expected losses from that flooding. A key aspect of this is fabricating a hazard definition relevant to the future, not just the past so that alternative approaches to dealing with the hazard can be evaluated effectively.

Risk and reliability analysis approach

The following discussion provides a general overview of the risk assessment performed to determine the vulnerability of New Orleans and vicinity to flooding from hurricanes and to estimate the difference in risk, by location, for the pre-Katrina HPS conditions, the post-Katrina (June 2007) HPS conditions, and ultimately the performance of the planned Hurricane and Storm Damage Risk Reduction System (HSDRRSS) that was completed in 2011. The risk assessment process, its application with respect to the pre-Katrina (2005) and post-Katrina HPS (2007), and a presentation and discussion of results are presented in detail in the IPET Report, Vol. VIII, Engineering and Operational Risk and Reliability Analysis. The analysis of the risk given the construction of the HSDRRS and a comparison of that scenario to the risk existing in 2005 and 2007 are provided in the IPET Supplemental Report, 'A General Description of Vulnerability to Flooding and Risk for New Orleans and Vicinity: Past, Present, and Future' (Supplemental Report of the Interagency Performance Evaluation Task Force, 2009).

The risk information provided is focused on hurricanes as the primary source. The surge and wave conditions generated by hurricanes are the principal hazard, although rainfall amounts are also factored into the flood analysis. The reliability or performance of the many components of the HPSs is examined through the individual and collective performance of levees, walls, gates and closures, transitions, and pumping stations. The chance of flooding is estimated based on these factors and the rainfall associated with hurricanes. Losses are estimated based on estimated depths of flooding at different frequencies of occurrence. This analysis did not consider flooding resulting from high flows on the Mississippi River not directly related to hurricane surge.

It is important to note that this effort involved developing and applying a prototype method to estimate risk for a large, complex, and geographically distributed system. In many respects, this was a first effort of its kind. The goal of assessing risk is to facilitate rational decision-making. The decisions that risk assessment supports include:

- Policy-level decisions on how best to expend resources to minimize the risk of flooding from hurricanes.
- Planning-level decisions concerning the relative vulnerability of different areas to focus efforts on areas of greatest risk.

- Understanding the sources of risk to include the least capable structures and the most exposed population or assets.
- Planning-level decisions concerning the value of different alternatives for reducing the chance of flooding and losses.
- Insights for design-level decisions on the location and character of structures where to put gates or raise walls.
- Communicating risk to the public, supporting personal decisions on how to prepare for and respond to the possibility of flooding from hurricanes.

The IPET assessment examined risk for the HPS as it existed before Katrina (2005) and for the short-term repairs and upgrades to the HPS as it existed in June 2007 (HSDRRS Design Guidelines 2007). A follow-on assessment was accomplished for the 'as planned' HSDRRS Design Guidelines (2012) to both estimate its ability to reduce risk and for the evaluation of additional structural resilience measures such as armoring the levee surfaces.

Risk assessment for New Orleans and Southeast Louisiana provides a broad picture of the relative chance of a flooding in different areas and the losses that could occur as a result of flooding. Risk assessment in the current context is intended to support planning decisions. It is not intended to support engineering decisions. A risk assessment over a large, complex geography like New Orleans requires many generalizations and assumptions, compared to the details of engineering design. Nonetheless, while risk assessment does not generate design information, it does inform design by defining hazards and suggesting alternative approaches to providing protection. It also provides a means to broadly examine the effectiveness of new design approaches and criteria by factoring in the improvements in system performance through the characterization of the fragility of individual system elements.

Risk assessment methodology

Figure 6 shows the general risk assessment methodology used by IPET. The risk assessment methodology consisted of four steps. First defining the hazard, the forces that can ultimately cause flooding and losses. Second, defining system performance, how the individual structures and mitigation measures would respond to the hazard forces. Third, estimating the amount of flooding that would occur in individual areas. Fourth, examining consequences of flooding in terms of potential loss of life and loss of property.

The hazard is the event or condition with the potential for causing an undesirable consequence. In the present context, the hazard is surge and wave conditions caused by hurricanes; it is not the hurricane themselves. To assess the hazard, it is first necessary to identify the range, character, and frequency of hurricanes that may strike the southern Louisiana coast. IPET used state-of-the-art methods, including supercomputer models, to compute the surge and wave conditions that a wide variety of hurricanes would produce around New Orleans. This analysis led to estimates of the frequency of extreme surges and waves everywhere around the HPS. These estimates are for current climate conditions, and they do not project climate variations into the future. Figure 7 shows the peak surge elevations in feet with respect to the geodetic datum for the 500-year recurrence interval. Of particular importance is the complexity of the surge levels, demonstrating that only sophisticated analytical modeling could reveal this heterogeneity. The analysis resulted in the ability to look at surge and wave conditions for a wide range of recurrence intervals.

System performance is the response of the HPS or HSDRRS to the hazard, that is, to surge and wave conditions generated by the hurricanes. The system performance is assessed by modeling the reliability of the HPS or HSDRRS under loads generated by surge and wave. This leads to an estimate of the likelihood that the HPS or HSDRRS can withstand those loads and, correspondingly, to an estimate of the chance of flooding at various places across the city and region. This chance of flooding is sometimes called the vulnerability.

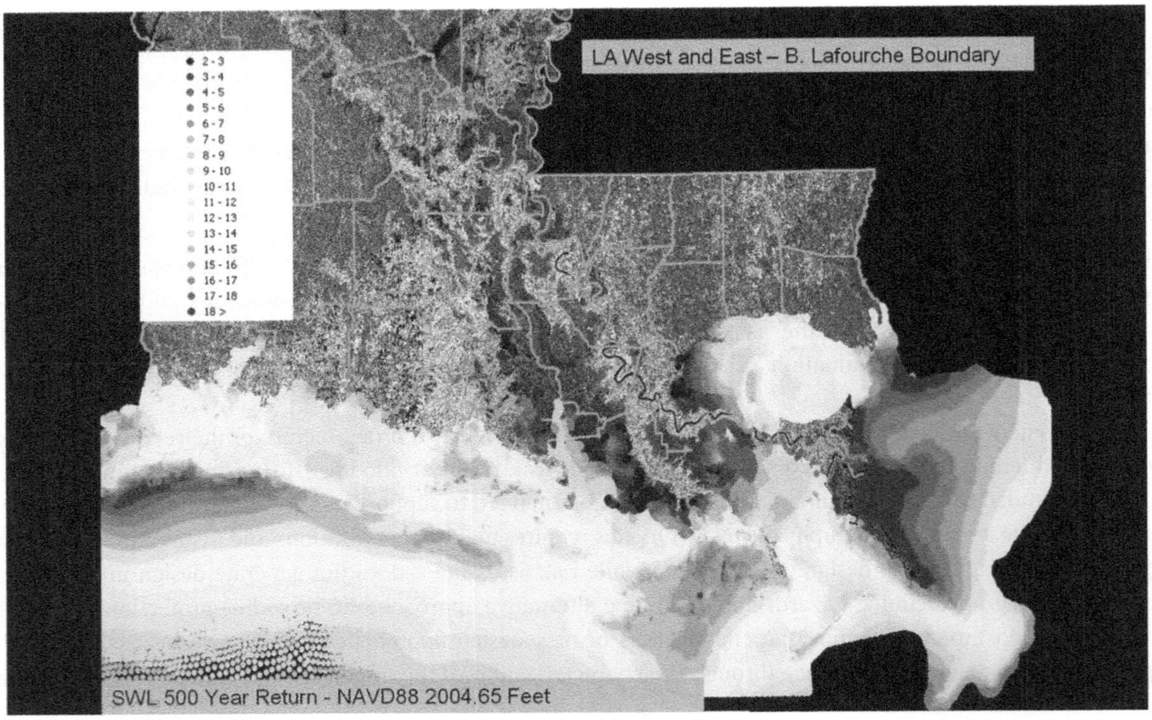

Fig. 7 | Spatial character of 0.2% (500-year) surge height from IPET modeling using the JPM-OS process with 76 hurricane data set (*Source*: IPET Final Report, Vol. VIII).

The reliability analysis starts with a detailed inventory of the engineering characteristics of every section and component of the HPS or HSDRRS. Then, the potential for overtopping and breaching is estimated for the spectrum of surge and wave conditions forecast in the hazard assessment. Combining the potential for overtopping and breaching with the frequencies of the corresponding hazards leads to an assessment of vulnerability to flooding from the spectrum of hurricanes that are possible for the region. The calculation also includes the chance of water entering through open gates and the amount of rainfall associated with hurricanes. Figure 8 shows an example of the resultant flood depth maps generated from these analyses.

The consequences of flooding, measured by the potential loss of life and property damage, are estimated by defining the distribution of people and structures within each sub-basin, the elevations of all structures and the surrounding land, and the value of the properties; and then by applying actuarial information and models to approximate losses.

Consequences were estimated for different depths of flooding to determine expected losses across the spectrum of hurricanes. The results are summarized for three chances of occurrence, specifically, the 1/50-, 1/100-, and 1/500-year floods. For loss of life estimation, a sophisticated simulation model was developed using geospatial census databases and evacuation plans. For property damage estimation, historical data from flood control and coastal protection projects across the nation were used to develop flood-depth versus damage relationships.

Risk was calculated by combining the chance of undesirable consequences occurring with the magnitude of those consequences should they occur. This allowed an estimate of risk by area, based on the character of the HPS/HSDRRS and other measures that may influence who and what is exposed to flooding. Risk was calculated by multiplying the chance of flooding to a certain depth by the losses expected by the flooding. Losses can be expressed as potential loss of life or potential loss of property.

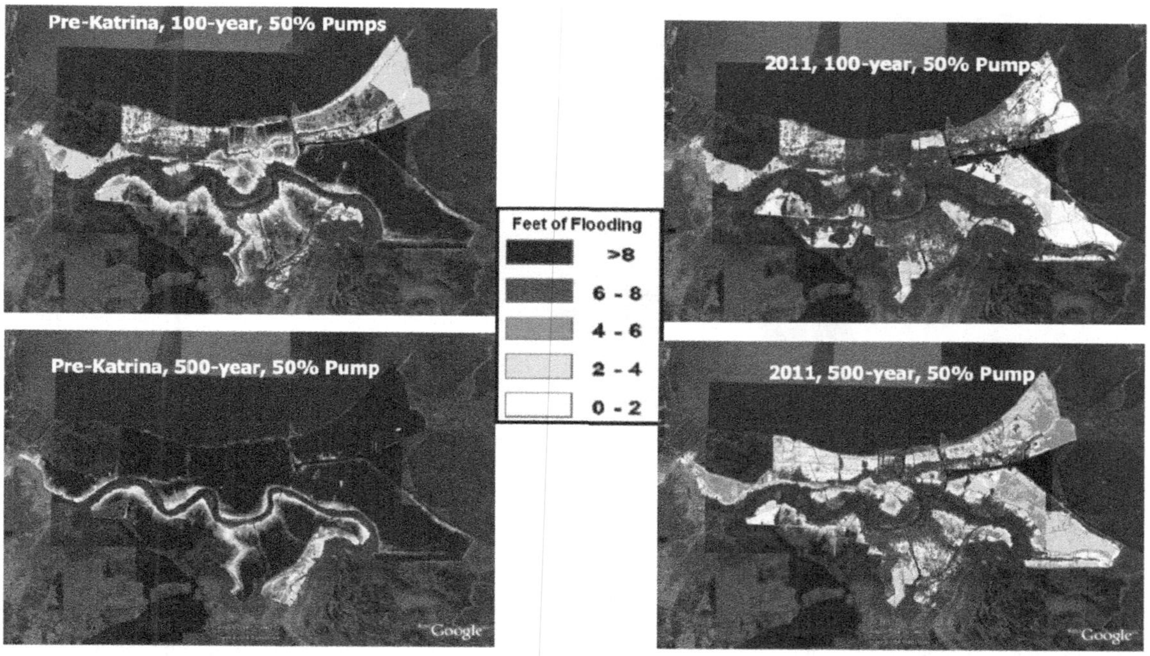

Fig. 8 | Flood depth maps for 1 and 0.2% exceedance floods, comparison between pre-Katrina HPS and the HSDRRS System (2011) (*Source*: IPET Supplemental Report, 2009). The darkest areas represent water depths greater than 8 ft.

Risk was estimated for the HPS as it existed (1) prior to Katrina and (2) after the repair and rebuilding of the HPS through June 2007. Risk was also assessed for a projected 100-year HSDRRS design scheduled for 2011. The combined risk assessments for the three scenarios are presented in the IPET Supplemental Report 'A General Description of Vulnerability to Flooding and Risk for New Orleans and Vicinity: Past, Present, and Future' (Supplemental Report of the Interagency Performance Evaluation Task Force, 2009).

Figures 9 and 10, respectively, show the expected loss of life and loss of property (in terms of percentage of value) for the 1% or 100-year recurrence level for the 2007 (Repaired HPS) and 2011 (planned HSDRRS) scenarios. Since a principal purpose of the risk assessment was to determine how risk is changing with respect to the capabilities of the HPS, both the pre-Katrina and post-Katrina risks were estimated using the pre-Katrina distribution of population and property. The significant value of the HSDRRS is evident in the dramatic reduction of potential losses for both loss of life and loss of property.

PART 4: KEY LESSONS LEARNED

There is a more extensive list of lessons learned published in the IPET Report, Vol. I. The lessons presented below represent those that are deemed big picture issues that all flood protection and risk reduction efforts should consider.

Correct elevations and reference datum are essential

All hurricane and flood control protection structures should be designed, constructed, and maintained relative to an up-to-date local sea level reference datum. Areas experiencing variable subsidence, such as New Orleans, are likely to have systematic datum and elevation accuracy issues that need frequent attention. It is important to have

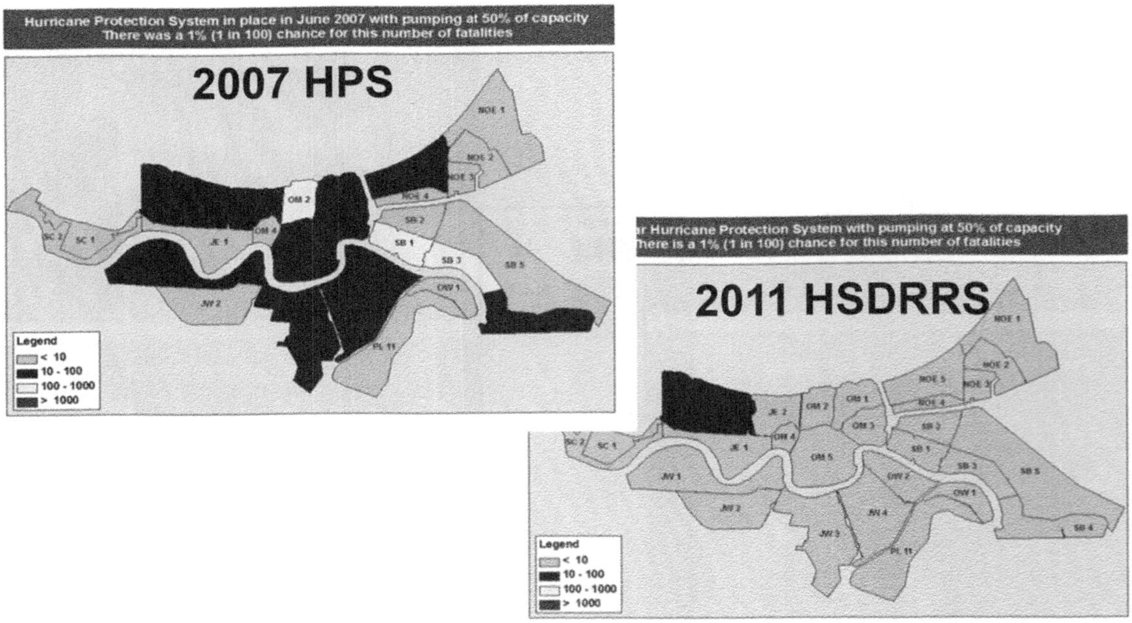

Fig. 9 | Expected loss of life for 1%, 100-year frequency flooding for pre-Katrina HPS and post-Katrina HSDRRS (*Source*: IPET Supplemental Report 1992).

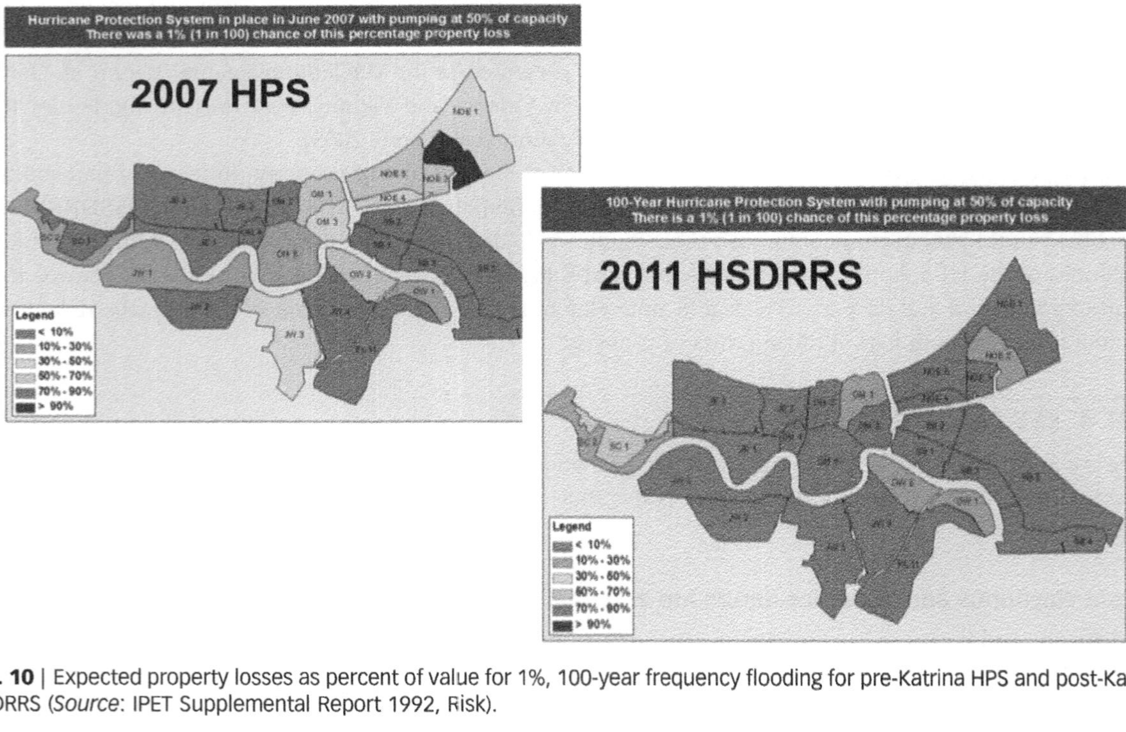

Fig. 10 | Expected property losses as percent of value for 1%, 100-year frequency flooding for pre-Katrina HPS and post-Katrina HSDRRS (*Source*: IPET Supplemental Report 1992, Risk).

appropriate monitoring stations (for tide and subsidence) in place and associated up-to-date guidelines for the application of this information to existing and new projects.

Systems planning and design methods are needed

Planning and design methodologies need to allow for the examination of system-wide performance. It is obvious from the IPET analysis that the piecemeal development of the New Orleans HPS provided a system in name only. This is especially true of the sections that have not been completed, transitions between types of protection that differ in capability (thereby representing weak points), and differences in the relative levels of reliability that created areas with a greater likelihood of failure. The system-based approach should have a time dimension to allow consideration of the potential changes in requirements or conditions over the life of the project and to examine approaches to build in adaptive features and capabilities. Subsidence, changing population demographics, and the changing patterns of hurricane intensity and frequency are obvious examples of the time-dependent challenges all projects face. All components that contribute to the performance of the overall system must be treated as an integral part of the system; pump stations are one example in New Orleans.

Frequent update to guidance and review of projects is critical

Design methods and designs need frequent review to determine whether they represent best practices and knowledge. Designs in coastal flood damage reduction projects need to include the concepts of resilience, adaptation, and redundancy to accommodate unanticipated conditions or structural behaviors.

Modern hazard characterization approaches are essential

More comprehensive probabilistic methods that consider a broader variety of storm characteristics and storm-generated conditions should be used as a basis for planning and design. The joint probability method–optimal sampling approach described in Volume VIII is recommended as a technically credible approach. Meteorological designations such as the Saffir–Simpson scale by themselves are not adequate to characterize the distributed surge and wave conditions that an HPS will face. The traditional methods of assessing the frequency of occurrence of hurricanes, dependent primarily on historical data, are too simplistic to capture important characteristics of the hurricane hazard such as time- and space-dependent storm intensity and track patterns. The wave and storm surge modeling (using the best available physics-based spatial prediction models) provided considerable insight into how water surrounding such a complex physical system responds to an equally complex hurricane wind system. It is the surge and wave environment that constitutes the hazard, not the meteorological event. Resolution of wave conditions and wave contributions to surge is a critical element in the estimation of design levels of structures in coastal areas.

The use of central pressure deficit as a metric is not adequate to characterize the storm surge potential of a hurricane. Other factors such as storm size (radius to maximum winds), coastal geometry, coastal bathymetry, and storm path are essential elements to characterize surge levels at a location. Given these additional factors, IPET estimated Katrina recurrence interval to be approximately 400 years.

Designs need to better consider unknowns

The design approaches taken for the outfall canals were not conservative enough to deal with the unknowns; in this case, the excessive floodwall deflection was not considered in the design. Floodwall design methods need to consider a broader spectrum of possible behaviors, and resilience to overtopping should be considered as a fundamental performance characteristic. Research is needed to understand the full performance limits of structures and to discover new approaches for creating adaptive designs. Design methods and assumptions need continuous review and update. Factors of safety are irrelevant if the appropriate failure mode is not included in developing that metric. This was the case for the failure of the outfall canal I-walls.

Planning methods should facilitate the examination of system-wide performance

In addition, HPS should be deliberately designed and built as integrated systems to enhance reliability and provide consistency in levels of protection. Components such as the interior drainage and pumping need to be an integral part of the system because of the important role they can play in limiting the amount and duration of flooding. Resilience in pumping capacity is especially important. IPET analyses showed that the pumping systems, if operable, could have significantly reduced the total flooding and its duration.

Resilience to catastrophic breaching can provide huge benefits in reduced loss of life and property

It is clear that a resilient HPS can provide enormous advantages. Resilience, in this case, refers to the ability to withstand, without catastrophic failure, forces and conditions beyond those intended or assumed in the design. For IPET purposes, resilience refers to the ability to withstand higher than designed water levels and overtopping without breaching. As demonstrated in this analysis of Katrina, approximately two-thirds of flooding and half of the losses were the result of breaching, i.e., the significant loss of protective elevation in structures. While overtopping alone from Katrina would have created dramatic flooding and losses, the difference is staggering in many regards. Reductions in losses of life, property, and infrastructure; associated reductions in the displacement of individuals, families, and the workforce, coupled with reduced disruption to businesses and social and cultural networks and institutions, would have a dramatic impact on the ability of a community and region to recover.

Even with the structures at lower than desired elevations, if they had been robust enough to not fail catastrophically, the loss of life and loss of property could have been half of what actually occurred.

System-based planning should include all aspects of hurricane response

A broad and system-based planning capability can increase the effectiveness of integrating evacuation, recovery, and reconstruction aspects into the HPS. In particular, a risk-based approach can provide an effective means to examine approaches to manage both the probability of an adverse event and the exposure to losses as well as the consequences. The spatial analysis of consequences and the ability to relate consequences to physical performance are powerful tools for making difficult decisions concerning hurricane protection.

Risk assessment provides a viable means to understand the relative vulnerability of protected areas

The combination of the likelihood of stormwater levels (surge and waves), the likelihood of structural failure at different water levels and dynamic loadings, the likelihood of flooding based on the expected performance of the system, and the consequences of that flooding provides a comprehensive information set on residual risk. This information defines the relative vulnerability of each area as well as the sources of that vulnerability.

It is critical to estimate the inherent uncertainty in the individual components of the risk assessment and in the final risk products

Risk assessment combines a variety of data types and incorporates numerous models. Each of these has an inherent degree of uncertainty, in their values, in the ability of the models to replicate the processes they represent, and in the end products themselves. The level of uncertainty must be estimated and incorporated in an overall uncertainty analysis to understand the variance associated with the risk assessment results. This provides some measure of the confidence one can have in using the risk data. The uncertainty quantification accomplished as a component of the risk and reliability analysis demonstrated that 71% of the overall epistemic uncertainty could be attributed to the hazard estimation process and 29% to structural performance estimation.

Vulnerability and risk are key to assessing the value of options

Having a quantitative estimate of vulnerability and risk is extremely important information for understanding both the current situation and the relative value of alternative risk reduction measures for the future. Both the public at large and public officials at all levels benefit by having a common situational awareness of residual risk. The common picture provides a valuable focus on communication for current risk reduction measures as well as those that should be examined for the future.

There is always risk

No amount of effort will eliminate risk. It is critical to monitor and communicate the residual risk so that individuals and public officials can make effective decisions to manage it. Communicating risk is perhaps the most difficult task of them all. Risk assessments for large and complex systems typically have large uncertainties associated with the results. It is important to understand that these uncertainties are not the result of the risk assessment, but result from the data and models used in the assessment process, which are largely the same fundamental tools and inputs used for non-probabilistic deterministic analyses. The difference is that the deterministic approach does not give the investigator knowledge of either the size or sources of uncertainty. The probabilistic approach provides the statistically most likely values for results, which is the best estimate given knowledge available. It also provides an understanding of the sources of greatest uncertainty that allows subsequent development to focus on reducing the largest sources.

DATA AVAILABILITY STATEMENT

All data referenced or discussed in this paper are available within public domain documents, principally the IPET reports, Performance Evaluation of the New Orleans and Southeast Louisiana Hurricane Protection System, Volumes I–IX and a Supplemental Report on the Vulnerability to Flooding and Risk for New Orleans and Vicinity, available on the at https://biotech.law.lsu.edu/katrina/ipet/ipet.html.

REFERENCES

HSDRRS Design Guidelines (2007) *Hurricane and Storm Damage Risk Reduction System Design Elevation Report*. Draft Report, US Army Corps of Engineers, New Orleans District.

HSDRRS Design Guidelines (2012) *Hurricane and Storm Damage Risk Reduction System Design Guidelines*. INTERIM, US Army Corps of Engineers, New Orleans District, Engineering Division (with Revisions).

IPET Supplemental Report (2009) *A General Description of Vulnerability to Flooding and Risk for New Orleans and Vicinity: Past, Present, and Future*. Supplemental Report of the Interagency Performance Evaluation Task Force, US Army Corps of Engineers.

IPET Supplemental Report (1992) *Guidelines for Risk and Uncertainty Analysis in Water Resources Planning*, Vol. I. Principles, with Technical Appendices, IWR REPORT 92-R-1.

Luettich, R., Westerink, J. & Scheffler, N. (1992) ADCIRC Report 1, Theory and Methodology. Coastal Engineering Research Center, US Army Corps of Engineers, Vicksburg, MS.

NOAA Technical Report NWS 23 (1979) Meteorological Criteria for Standard Project Hurricane and Probable Maximum Hurricane Wind Fields, Gulf and East Coasts of the United States, Washington, D.C.

Performance Evaluation of the New Orleans and Southeast Louisiana Hurricane Protection System. Final Report of the Interagency Performance Evaluation Task Force, US Army Corps of Engineers. Volume I – Executive Summary and Overview, June 2009; Volume II – Geodetic Vertical and Water Level Datums, March 2007; Volume III – The Hurricane Protection System, June 2009; Volume IV – The Storm, March 2007; Volume V – The Performance – Levees and Floodwalls, June 2007; Volume VI – The Performance – Interior Drainage and Pumping, June 2007; Volume VII – The Consequences, March 2007; Volume VIII – Engineering and Operational Risk and Reliability Analysis, June 2009; Volume IX – General Appendices, June 2009.

First received 15 November 2021; accepted in revised form 15 November 2021. Available online 29 November 2021

Water Infrastructure Planning, MDCU doi: 10.2166/wp.2021.347

How Hurricane Katrina influenced the design of hurricane protection and risk reduction systems and national approaches to risk and resilience. Part 2: Designing the Hurricane and Storm Damage Risk Reduction System and resulting long-term engineering guidance and practice changes

Lewis E. Link[a,b]

[a] Department of Civil and Environmental Engineering, University of Maryland, College Park, MD, USA
[b] Interagency Performance Evaluation Task Force, Washington DC, USA
E-mail: elink@umd.edu

ABSTRACT

Following Hurricane Katrina, the US Army Corps of Engineers, supported in part by the risk and reliability analysis conducted by the Interagency Performance Evaluation Task Force (IPET), made a major shift from 'protection' to 'risk reduction' as the principal goal in flood mitigation. The mitigation of the flood risk in Southeast Louisiana was embodied in the design and construction of the 'Hurricane and Storm Damage Risk Reduction System', the post-Katrina initiative for New Orleans flood mitigation. It also spawned a major overhaul of many of the Corps of Engineers' technical guidance and engineering practice documents, incorporating risk as a key measure in the planning and design processes. The criteria applied for the design of the HSDRRS are discussed, with summaries of the associated major changes in Corps engineering guidance and practice relevant to flood mitigation.

Key words: Design criteria, Engineering practice, Flood risk, Hurricane Katrina, New Orleans, Risk mitigation

HIGHLIGHTS

- The losses from Hurricane Katrina and subsequent forensic and risk analyses for New Orleans resulted in substantial changes to the knowledge base and guidance for planning and designing coastal risk reduction systems.
- The new knowledge was a substantial input to the design of the Hurricane and Storm Damage Risk Reduction System that is currently in place in New Orleans.
- The processes used were considerably different from past practice and resulted in a more robust and resilient system with some capacity to deal with uncertainty and change. This paper highlights the new processes.
- The same knowledge base was the stimulus for significant changes to the US Army Corps of Engineers engineering guidance and practice documents. A major change was the incorporation of risk-based decision making in many aspects of project development and execution.

INTRODUCTION

The losses from Hurricane Katrina and subsequent forensic and risk analyses for New Orleans resulted in substantial changes to the knowledge base and guidance for planning and designing coastal flood risk reduction systems.

The new knowledge, generated by an exhaustive and comprehensive forensic study of the hurricane protection system around New Orleans subjected to Hurricane Katrina (IPET Final Report, 2009; IPET Supplemental Report,

2009), represented a landmark shift in design approaches for the Corps of Engineers. It substantially changed the way the inputs to the design of the Hurricane and Storm Damage Risk Reduction System that is currently in place in New Orleans were formulated and analyzed. The processes used were considerably different from past practices and resulted in a more robust and resilient system with some capacity to deal with uncertainty and change.

The same knowledge base was the stimulus for significant changes to the US Army Corps of Engineers (USACE) engineering guidance and practice documents. A major change was the incorporation of risk-based decision-making in many aspects of project development and execution. The documents updated with this information are highlighted. An example of these documents is Engineer Regulation ER 1105-2-101 Risk Analysis for Flood Risk Management Studies, USACE (7/17/2017).

One of the major shifts in philosophy, post-Katrina, was the change from 'Protection' to 'Risk Reduction' relevant to mitigation measures that deal with storm surges and waves. Going forward the public needed to understand that the term 'protection' was not appropriate because it implied a 'fail-safe' system and associated level of protection. Under the conditions of climate uncertainty, it is impossible to guarantee that an area is protected against all flooding – only that it is protected against an expected 'design' event of relatively uncertain frequency. For example, the US standard for floodplains is a 100-year event. Hence, an area not prescribed to be within the 100-year floodplain was not 'protected' but has a risk of flood losses that are less than those areas within the floodplain. In the case of the New Orleans Hurricane and Storm Damage Risk Reduction System, the risk reduction objective was defined by a new combination of stochastically generated hurricane storm tracks and storm surge levels in order to define a new 100-year event. Risk reduction was the more appropriate objective, focusing on reducing the likelihood of losses, with the realization that there will always be some chance of flooding and losses from potentially larger and statistically likely storm events regardless of what mitigation measures are deployed. That remaining potential loss is defined as the residual risk.

The lessons learned through the IPET forensic and risk analyses and the Corps of Engineers applications of that knowledge in the immediate repair and short-term upgrades to the Pre-Katrina era Hurricane Protection System were leveraged into the design and construction of a new risk mitigation infrastructure system termed the Hurricane and Storm Damage Risk Reduction System (HSDRRS). The HSDRRS was a first in many aspects for US flood mitigation efforts, using probabilistic quantification of the flood hazard using joint probability analysis methods, incorporating new failure mode and fragility analyses for structures, explicitly including uncertainty in consideration of structure design and adding 'structural resilience' to preclude overtopping failure of structures to name a few. Figure 1 provides a map of the HSDRRS depicting the major structural components that were constructed in addition to broad increases in flood wall and levee elevation and strength.

The reports, Hurricane Storm Damage Risk Reduction System Elevation Report (USACE 2012) and Hurricane Storm Damage Risk Reduction system Design Guidelines, Interim, (New Orleans District 2012), provide a detailed documentation of the coastal and hydraulic engineering analysis performed to determine the project design elevations for three legacy projects within the spatial domains of the HSDRRS: Lake Pontchartrain & Vicinity, West Bank & Vicinity, and New Orleans to Venice projects, including the portions of the Mississippi River Levees coincident with these projects.

Congress authorized the Secretary of the Army to:

- repair and restore these projects;
- accelerate the completion of unconstructed portions;
- armor critical elements; and
- in the case of the existing Lake Pontchartrain & Vicinity project and the existing West Bank & Vicinity project, raise levee and floodwall heights where necessary and otherwise enhance the existing Lake Pontchartrain &

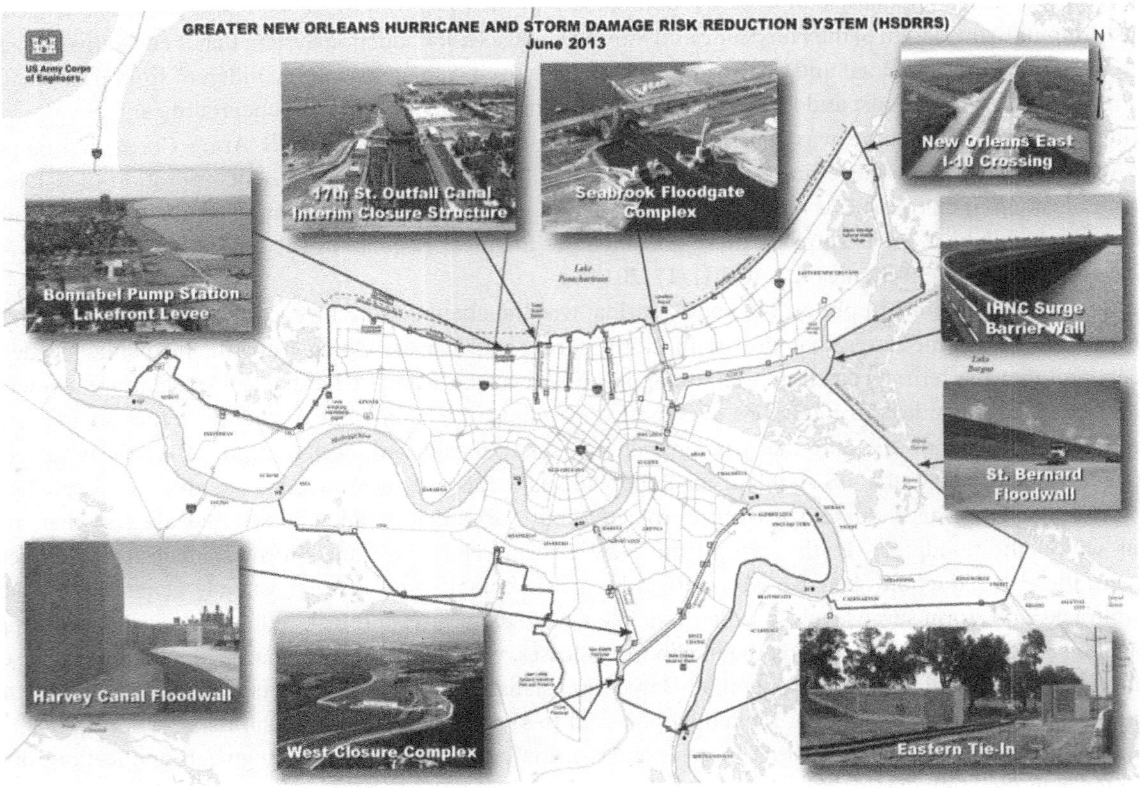

Fig. 1 | Map of the HSDRRSS depicting the major structural components that were constructed in addition to broad increases in flood wall and levee elevation and strength, June 2013 (*Source*: Corps of Engineers).

Vicinity project and the existing West Bank & Vicinity project to provide the levels of protection necessary to achieve the certification required for participation in the National Flood Insurance Program.

For the HSDRRS coastal and hydraulic engineering analyses, new processes and procedures were supported in part based on the IPET analyses and results. A team consisting of the USACE, the Federal Emergency Management Agency (FEMA), the National Oceanographic and Atmospheric Administration (NOAA), the private sector, and academia developed a new process for estimating hurricane inundation probabilities, the Joint Probability with Optimal Sampling (JPM-OS) method. This process is documented in the IPET Final Report, Vol. VIII and in Journal Papers by Resio & Westerink (2008), Resio *et al.* (2008, 2013), Resio & Irish (2015). The results were applied to USACE work by the Interagency Performance Evaluation Team (IPET) risk analysis, the Louisiana Coastal Restoration and Restoration project, and FEMA Base Flood Elevations for the production of Digital Flood Insurance Rate Maps (DFIRM) for coastal Louisiana (LA) and Texas.

As recommended in the IPET analysis (IPET Final Report, Vol. I) three major flaws in the pre-Katrina Hurricane Protection System had to be addressed in the formulation of the HSDRRS: (1) the hurricane protection system in New Orleans did not perform as a system; (2) redundancy should be a component of the system; and (3) consideration should be given to the performance of the system if the design event or system requirements are exceeded. These became fundamental aspects of the planning for the HSDRRS.

A systems approach was used in the coastal and hydraulic engineering analyses. Surge and wave models were inclusive of the entire flood mitigation area and all related components of the mitigation effort. Analyses included the evaluation of the effects of subsidence and sea-level rise on surge elevations and waves. The construction of the hurricane protection system to the design elevations and cross-sections in this report ensures that the components (walls, levees, pumping stations, drainage systems, etc.) have a common capacity and capability based on the hazard. Redundancy was also included in the system planning and design through the presence of secondary structural measures that were located within the perimeter of the main HSDRRS. The existing levee/floodwall system in the Inner Harbor Navigation Canal/GIWW (IHNC/GIWW) and along the three outfall canals was designed to provide a useful measure of redundancy to the flood risk reduction system behind the primary line of protection at the perimeter (such as with overtopping). On the West Bank, the Harvey and Algiers Canal areas on the West Bank have levee/floodwalls along the interior drainage outlets that provide redundancy with respect to the external perimeter structures and associated pumping and water control structures. Consideration was given in the analyses to structural resiliency, the performance of the system if the design event or system requirements are exceeded. This was primarily through the consideration of armoring of the back side of levees where overtopping may occur in rare events.

Levee/structure design height

The HSDRRS design elevations are sufficient to provide protection from a hurricane event that would produce a 1% exceedance surge elevation and associated waves. The design elevations presented in this report are determined for each location of interest, using the 1% annual exceedance still water elevation, 1% annual exceedance wave height, and 1% annual exceedance wave period, assuming simultaneous occurrence of maxima of surge level and wave characteristics. These assumptions are conservative and are in line with a resilient design approach recommended by the IPET.

The design criteria for the levee and structure elevations also consider wave overtopping limits. Guidelines for establishing the overtopping rate threshold (i.e., the threshold associated with the onset of levee erosion and damage) in different types of embankments can be found in Engineering Manual (EM) 1110-2-1100 (Part VI), Table VI-5-6. These threshold values are consistent with those that are adopted by the Technical Advisory Committee on Flood Defence in the Netherlands (Technische Adviescommissie voor de Waterkeringen).

The following wave overtopping rates were established for the New Orleans HSDRRS systems:

- For the design still water, wave height, and wave period, the maximum allowable average wave overtopping of 0.1 cubic feet per second per foot (cfs/ft) at 90% level of assurance and 0.01 cfs/ft at 50% level of assurance for grass-covered levees.
- For the design still water, wave height, and wave period, the maximum allowable average wave overtopping of 0.1 cfs/ft at 90% level of assurance and 0.03 cfs/ft at 50% level of assurance for floodwalls with appropriate protection on the backside.

Uncertainty of knowledge of variables and functions

Uncertainty may be represented by a specific probability distribution with associated parameters, or sometimes expressed simply as standard deviation. Present guidance, USACE (2012), supplements freeboard assumptions by providing upper and lower bounds of required levee performance based on specified levels of assurance of protection against the design flood using the design criteria stated above. A probabilistic approach (Monte Carlo) was used in applying these criteria by using a spectrum of potential wave overtopping situations that incorporates uncertainty in the still water elevation and wave characteristics. This resulted in the final structural elevation determinations.

The IPET identified resilience (in this case it could also be termed robustness for structures) as one of the 'Overarching Lessons Learned' from Hurricane Katrina. Engineers are working to develop guidance to define resiliency and the level of resilience needed for levees and structures. For the HSDRRS design, resiliency was defined as the ability of the levee or structure to provide protection during events greater than the design event without total failure. The minimum criterion for resilience was that levees and structures do not catastrophically breach when design criteria are exceeded. Resilience also includes designing for possible changes in conditions, with the flexibility to adapt to future design conditions. Guidance considered for the HSDRRS structures included, ensuring that the height of all barriers is sufficient to prevent free flow at the 0.2% annual exceedance event. The 0.2% annual exceedance event was selected because it represents the approximate recurrence of Hurricane Katrina and is approximately equivalent to a 1 in 500-year event. The 0.2% water (surge and wave) elevations were determined for each location of interest using the JPM-OS hazard analysis for a large number of potential hurricane events in the region.

Structures/pump stations

Pump stations throughout the New Orleans area have been constructed and are operated and maintained by local government agencies. There are no existing federal pump stations in the HSDRRS of greater New Orleans. Prior and present hurricane protection projects do not rely significantly on the ability to pump out water from rainfall and overtopping of levees and walls. In urban and urbanizing areas, the provision of a basic drainage system to collect and convey local runoff from rainfall is usually considered a non-federal responsibility. Within the New Orleans area, however, Congress authorized a federal project to improve interior drainage, the Southeast Louisiana Urban Flood Control Project.

Recognizing the damage that may result from a weakened or inoperable storm drainage system, the Corps' New Orleans District also examined alternatives to reduce the consequences of interior flooding. They include:

- Completion of the Southeast Louisiana Urban Flood Control Project, a federal project to improve interior drainage in New Orleans and surrounding communities.
- Design and construction of positive shut-off gates at pump stations to block backflow.
- Providing fronting protection at pump stations to improve resilience and survivability of pump stations through storm surge events.
- Storm proofing selected pump stations to improve discharge capabilities during storm events.

Subsidence

Planning for anticipated subsidence, both short-term and long-term, was included in the design of the HSDRRS. During the design of individual reaches, geologists and geotechnical engineers examined site-specific soil conditions and estimate long-term settlement and subsidence in the barriers. For levees over soft foundations, engineers typically recommend construction in several lifts. This allows the foundation soils to consolidate and gain in shear strength. When future lifts are constructed to higher elevations, the footprint of the levee system does not need to increase. Final construction lifts are typically constructed with added height in the anticipation of long-term settlement. This added height assures that the levee crown elevation will be at or above the design elevation.

Future conditions

Design elevations have been calculated for both existing conditions and future conditions (year 2057). Existing conditions represent conditions that exist at the completion of the HSDRRS. Future conditions include changes in still water levels (SWLs) and wave characteristics due to subsidence and sea-level rise. Historical subsidence, projections of sea-level rise, and previous studies have been used to estimate future changes in SWLs. Natural subsidence rates, including sea-level rise, have been mapped by the New Orleans District for the Louisiana Coastal Area Plan. A relative sea-level rise of 1.0 foot (ft) over 50 years has been used for the purposes of the HSDRRSS design efforts. The effect of increasing sea-level rise on surge levels has been further investigated and resulted in the 1.5–2.8 ft increase in surge level, applied as future conditions. Moreover, the wave characteristics have also been corrected for the increasing water depth.

The New Orleans District, USACE, has also planned regular reassessment of design parameters in order to assure the effectiveness of the system in future years. Changes in sea level and land loss are some of the factors that need to be periodically revisited. As the inventory of storms increases, periodic assessment of the JPM-OS method should also be undertaken. The system should also undergo a reassessment after major events or significant changes in design and analysis methodologies. The intent is to conduct such reviews no less than once every 10 years.

Once initial construction was completed, the responsibility to operate, maintain, repair, replace, and rehabilitate barriers was turned over to the local sponsor in most cases. Periodic inspections and annual reviews submitted to the USACE will assure proper performance. To ensure that requirements are well understood, an operation, maintenance, repair, replacement, and rehabilitation manual is developed for each project and serves as the basis for future monitoring, inspection, and reporting. In addition, the Corps of Engineers will conduct periodic surveys of levees as part of an improved quality assurance program.

LONG-TERM CHANGES IN ENGINEERING GUIDANCE AND PRACTICE RESULTING FROM KATRINA

The USACE is the owner and regulator of 708 large dams and regulator of over 2,500 levee systems (14,000 miles) in the USA. The Corps' mission and infrastructure responsibilities also include hydropower, water supply, navigation, and environment. The characteristics of the infrastructure under USACE governance have been described by Halpin & Escuder (2015). They state that 'from a risk perspective, the infrastructure is diverse, spanning over five orders of magnitude in probability of failure, consequences of failure, and annualized risk. Approximately one-third of the portfolio is actionable for undesirable risks.' Also, 'Historically, management solely via compliance via a traditional "one-size-fits-all" engineering standards was viewed as not improving understanding or decisions, nor was it cost effective.'

The same paper described the USACE strategy of adaptive learning to implement risk governance. After a decade of effort and gradual growth and improvement, the Corps' expertise and experience in risk assessments and risk-informed decision-making is substantial. The Corps has worked strategically with other agencies, industry, and international partners to further build expertise and overall governance to include considerable development of new policies and guidance. The paper cites that 'risk-informed decisions have reshaped the very culture of the agency'.

Hurricane Katrina served as one broad stimulus for these actions and the evolution of risk management and analysis within the Corps and other federal agencies. The Corps of Engineers had accepted risk as an important measure in the 1990s, as evidenced by the development of policy and guidance documents addressing risk use for Corps analyses such as 'Guidelines for Risk and Uncertainty Analysis in Water Resources Planning, Volume I, Principles', (IWR Report 92-R-1, March 1992), Engineer Regulation ER 1110-2-1150, Engineering and Design for Civil Works Projects, August 1999, and EM 1110-2-1619, Risk-Based Analysis for Flood Damage Reduction Studies, 1996. While these became part of the Corps governance policy documents, risk-based analysis and management did not rise to prominence until after the Katrina event.

Organizational changes

A primary organizational change within the Corps was the establishment of new national technical centers for Risk Management, Modeling, Mapping, Consequence Estimation, and Infrastructure Modification. These centers have assisted in shaping and implementing risks informed decision-making and continuing to adapt through training, policy, and methodology development. They provided the capability to begin conducting routine risk assessments and providing support to decision-making. For example, USACE moved from a solely standards-based approach for its dam safety program to a dam safety portfolio risk management approach. The standards-based approach is included in the risk-informed approach to the dam safety program, and dam safety program decisions are now be risk-informed.

Method changes

The use of the more sophisticated tools for hazard analyses used in the IPET and for the design of the HSDRRS demonstrated the value of new approaches, and the risk analysis accomplished for pre-Katrina and post-Katrina situations demonstrated the value of risk as a measure of the return on investment of such information. The Corps adapted some of their primary analytical tools to integrate risk assessment and analysis into their standard methodologies. An example is the incorporation of risk into the Hydrologic Engineering Center program HEC-FDA, Flood Damage Assessment.

One of the key bases for a risk-informed decision, and prioritization of the work, is a consideration of tolerable risk. Tolerable risk is the risk remaining after risk reduction measures have been implemented. The establishment of tolerable risk standards and methods was a key enabler for using risk to inform decisions by providing a gage for determining the effectiveness of proposed or completed risk mitigation measures. This goal remains a work in progress beyond the Dam Safety Program.

Levee safety program

The USACE Levee Portfolio Report shares the current understanding of the flood risks and benefits associated with the portfolio of levee systems within the USACE Levee Safety Program. The USACE Levee Portfolio Report is organized around risk (e.g., the flood risk associated with levees) to describe the magnitude of risk, key drivers of risk, sources of uncertainty in the understanding of risk, and distinct factors of risk within the USACE levee portfolio.

USACE levee portfolio

The USACE Levee Safety Program has conducted a comprehensive inventory, inspection, and risk assessment effort for the entire USACE levee portfolio. This provides a more comprehensive understanding of the portfolio than previously known: where the levees are (inventory); their condition (inspection); and the flood risk associated with each levee (risk assessment). The USACE levee portfolio includes about 2,220 levee systems totaling approximately 14,150 miles in length. Over 1,200 levee sponsors operate and maintain roughly 2,000 of these levee systems, spanning roughly 70% of the length of the entire portfolio. The remaining almost 200 levee systems are operated and maintained by USACE.

Risk assessments within the Levee Safety Program provide a systematic, evidence-based approach for estimating and describing the likelihood and consequences of existing and future risk associated with levee systems. Risk assessments consider what can go wrong, how it can happen, the consequences if it happens, and how likely it is to happen. To support decisions in the management of the portfolio, a levee safety action classification (LSAC) is assigned as a final step in developing a risk characterization for each levee system. LSACs range from very high risk (immediate action recommended) to very low risk (maintain routine activities). LSAC assignments are used by USACE to prioritize resources across the portfolio and to organize widespread levee-related risk information into reasonably commensurate groupings for action (Levee Portfolio Report, 2018).

National levee database

The national levee database (NLD) includes 9,272 levee systems, 29,635 miles of levees that include 47,184 levee structures. It is maintained by the Corps of Engineers in partnership with FEMA.

The NLD is a congressionally authorized database that documents levees in the USA. The NLD is maintained and published by the USACE. It recently underwent a refresh making more tools available to data managers to keep information updated and provides an improved dashboard that makes finding and understanding levee information easier than ever. NLD information includes the location, general condition, and risks associated with the levees.

The NLD contains information about the condition and risk information for approximately 2,000 levee systems (approximately 15,000 miles/mostly levees affiliated with USACE programs. An additional 6,000 levee systems – approximately 15,000 miles – have location information, but little to no information about condition and risk. One of the goals for the NLD is to include data about levees owned and operated by all other federal agencies, tribes, states, municipalities, levee boards, and private entities. This information will be added as it becomes available.

The database includes attributes of levees and floodwalls relevant to flood fighting, design, construction, operation, maintenance, repair, and inspection. Because the location and characteristics of levee systems can be viewed on a map with real-time data from other sources, such as stream gauges and weather radar, it is a useful tool for a variety of public agencies and individuals including flood plain managers, emergency management agencies, levee system sponsors, and citizens who live or work behind a levee.

The NLD information is presented in a convenient dashboard and includes the ability to search on specific areas of interest or geographically. The database is available at https://levees.sec.usace.army.mil.

Federal dam safety program

For 30 years, the federal government has used the National Dam Safety Program (NDSP) to protect Americans from dam failure. The NDSP is a partnership of the states, federal agencies, and other stakeholders that encourages and promotes the establishment and maintenance of effective federal and state dam safety programs to reduce the risks to human life, property, and the environment from dam-related hazards. The program is administered by FEMA and provides the following:

Information needs for dam safety: Under FEMA's leadership, state assistance funds have enabled all participating states to better their programs through increased inspections, emergency action planning, and the purchase of needed equipment.

Dam safety training: FEMA has expanded existing training programs and initiated new training programs to enhance the sharing of expertise between the federal and state sectors.

Grant assistance to the states: Grant assistance provides vital support for the improvement of state dam safety programs that regulate most of the 87,000 dams in the USA.

Dam safety research: A national research program in dam safety focuses on priorities and produces products for both the layperson and the expert and develops technological tools that drive data collection and analysis toward a better understanding of risk and remediation needs.

NDSP partners: To encourage individual and community responsibility for dam safety, FEMA coordinates partnerships through two federal organizations, the National Dam Safety Review Board and the Interagency Committee on Dam Safety.

Partners include the Association of Engineering Geologists, the Association of State Dam Safety Officials, the Bureau of Reclamation, the Federal Emergency Management Agency, the Federal Energy Regulatory Commission, the Institute for Water Resources, and the US Society of Dams.

USACE dam safety program

The USACE operates and maintains approximately 700 dams nationwide and in Puerto Rico that provide significant, multiple benefits to the nation – its people, businesses, critical infrastructure, and the environment. These benefits include flood risk management, navigation, water supply, hydropower, environmental stewardship, fish and wildlife conservation, and recreation.

USACE's dams are part of our nation's landscape, integral to many communities, and critical to watershed management. Our dam safety professionals carry out a dam safety program to make sure these projects deliver their intended benefits while reducing risks to people, property, and the environment through continuous assessment, communication, and management. (By comparison, there are more than 87,000 dams in the National Inventory of Dams that are federally, state, locally, and privately operated and maintained.)

ENGINEERING GUIDANCE CHANGES

The Corps has accomplished a broad spectrum of changes to their engineering guidance and practice documents. Without this comprehensive umbrella of new and updated engineer regulations (ER), engineering circulars (EC), and EMs, governance of the processes and application of the principles of risk and resilience would be futile. This includes the modernization and continued evolution of processes that are facilitated through Engineering and Construction Bulletins (ECB) and Engineering Technical Letters (ETL). The following is a sample of the key updated or new governance documents that provide policy and guidance for the application of risk and resilience in Corps practice. This includes a few draft documents that are still in development and formalization stages.

Engineer Regulations: www.publications.usace.army.mil/USACE-Publications/Engineer-Regulations/

1. ER 1110-2-1156 Safety of Dams Policy, 3/31/2014: This regulation prescribes the guiding principles, policy, organization, responsibilities, and procedures for the implementation of risk-informed dam safety program activities and a dam safety portfolio risk management process within the USACE.
2. ER 10-1-51 Roles and Responsibilities – Dam Safety Modification Mandatory Center of Expertise 9/28/2012.
3. ER 10-1-55 Organization and Functions – Roles and Responsibilities Risk Management Center, 6/30/2013.
4. ER 1100-2-8162 Incorporating Sea-level Changes in Civil Works Programs, 12/31/2013:
   This Regulation provides USACE guidance for incorporating the direct and indirect physical effects of projected future sea-level change across the project life cycle in managing, planning, engineering, designing, constructing, operating, and maintaining USACE projects and systems of projects.
5. ER 1105-2-101 Risk Analysis for Flood Risk Management Studies, 7/17/2017:
   This regulation provides guidance on risk assessment requirements for flood management studies including but not limited to feasibility studies, post-authorization changes, general reevaluation studies, dam and levee safety studies, and major rehabilitation studies. This regulation is jointly promulgated by planning and engineering. The risk framework is a decision-making process that comprises three tasks: risk assessment, risk communication, and risk management.
6. ER 1110-2-8156 Policies, Guidance, and Requirements for Geospatial Data and Systems, September 2012: General criteria for the use and development of geospatial technologies throughout the USACE. This includes, but is not limited to, geospatial system development and utilization as well as the acquisition, processing, storage, distribution, and utilization of geospatial data and systems. This ER provides a framework that promotes interoperability through the use of data standards and an open platform to move data seamlessly throughout the organization.

7. ER 1110-2-8160 Policies for Referencing Project Evaluation Grades to Nationwide Vertical Datums, March 2009: This regulation establishes USACE policies for referencing project elevation grades to nationwide vertical datums established and maintained by the US Department of Commerce. Its purpose is to ensure that controlling elevations and local datums on USACE projects are properly and accurately referenced to nationwide spatial reference systems used by other federal, state, and local agencies responsible for flood forecasting, inundation modeling, flood insurance rate maps, navigation charting, and topographic mapping.

Engineering Manuals: www.publications.usace.army.mil/USACE-Publications/Engineer-Manuals/

1. EM 1110-2-1413 Hydrologic Analysis of Interior Areas, 8/24/2018
2. EM 1110-2-1420 Engineering and Design Hydrologic Engineering Requirements for Reservoirs, 9/24/2018
3. EM 1110-2-6056 Standards and Procedures for Referencing Project Evaluation Grades to Nationwide Vertical Datums
4. EM 1110-2-3600 Management of Water Control Systems, 10/10/2017

Engineer Circulars: www.publications.usace.army.mil/USACE-Publications/Engineer-Circulars/

1. EC 1110-2-6065, Comprehensive Evaluation of Project Datums, July 2007: This document provides guidance on the proper application of vertical datums used to reference protection elevations on flood control structures or excavated depths in navigation projects – hereinafter referred to as the Comprehensive Evaluation of Project Datums (CEPD) project. It describes specific procedural actions immediately required to evaluate the accuracy and adequacy of existing flood protection elevations or controlling navigation depths relative to federal datums established by the Department of Commerce and prescribed for government-wide use by the Federal Geographic Data Committee (FGDC). This guidance implements lessons learned from the IPET study conducted after Hurricane Katrina, as identified in Volume II (Geodetic Vertical and Water Level Datums) of the Final IPET Report.
2. EC 1165-2-218, Levee Safety Policy and Procedures, DRAFT, 3/10/2018: This includes LSAC Table.
3. EC 1165-2-217, Review Policy for Civil Works, 2/20/2018: The purpose of this EC is to establish general policy and guidance for Corps of Engineers (Corps) implementation of Executive Order (EO) 11988, Floodplain Management, as amended by EO 13690, Establishing a Federal Flood Risk Management Standard and a Process for Further Soliciting and Considering Stakeholder Input1 for most Corps actions, except for those taken under the Corps Regulatory Program.
4. EC 1110-2-6067 USACE Process for the National Flood Insurance Program (NFIP) Levee System Evaluation, August 2010: The purpose is to provide a consolidated document that will guide USACE procedures for levee system evaluations in support of the NFIP as administered by the FEMA.
5. EC 1165-2-218 Policies and Procedural Guidance for Processing Requests to Alter US Army Corps of Engineers Civil Works Projects, September 2017.
6. EC 1165-XX2-XXXX, USACE Enterprise Risk Management System, Principles and Terminology (Draft): This EC begins a process to establish an Enterprise Risk Management (ERM) System within the Civil Works Program of the USACE.
7. EC 1110-2-6072 Comprehensive Guidance on Levee Safety (DRAFT), July 2015.
8. EC 1165-2-212 Sea-Level Rise Considerations for Civil Works Programs, October 2011: This circular provides USACE guidance for incorporating the direct and indirect physical effects of projected future sea-level change across the project life cycle in managing, planning, engineering, designing, constructing, operating, and maintaining USACE projects and systems of projects.

Engineering and Construction Bulletins: http://www.wbdg.org/ffc/dod/engineering-and-construction-bulletins-ecb

1. ECB 2018-14 Guidance for Incorporating Climate Change Impacts to Inland Hydrology in Civil Works Studies, Designs, and Projects 09-10-2018: This ECB reissues and updates the policy in ECB 2016-25, Guidance for Incorporating Climate Change Impacts to Inland Hydrology in Civil Works Studies, Designs, and Projects. This policy requires consideration of climate change in all current and future studies to reduce vulnerabilities and enhance the resilience of communities.
2. ECB 2018-2 Implementation of Resilience Principles in the Engineering & Construction Community of Practice, January 2018: This ECB provides the policy and guidance for applying the USACE principles of resilience – Prepare, Absorb, Recover and Adapt (PARA) – to Engineering & Construction (E&C) Community of Practice (CoP) efforts.
3. ECB 2017-15 Managed Overtopping of Levee Systems, 7/14/2017: The USACE will use this directive and guidance for commands responsible for planning, design, construction, and operation and maintenance of civil work projects. The level of detail in the overtopping analysis will depend on the study phase. This document is applicable for all USACE riverine levee and floodwall systems. This ECB provides an interim update to expired technical guidance Engineer Technical Letter (ETL) 1110-2-299 and provides a methodology for configuring the engineered capacity exceedance related to flood overtopping at a specific location or locations along the levee system.
4. ECB 2017-3 Design and Evaluation of I-Walls Including Sheet Pile Walls, 1/27/2017: This bulletin transmits specific guidance to be used for the design and evaluation of I-Walls and Sheet Pile Walls with links to the supporting documentation located on the USACE Official Publications site and/or the Technical Excellence Network (TEN) site.
5. ECB 2016-08 Interim Risk Reduction Measures (IRRMs) for Levee Safety, 2/22/2016: The USACE will use this directive and guidance for all levee systems in the USACE Levee Safety Program, which includes those (1) USACE operated and/or maintained; (2) federally authorized, typically USACE constructed, and locally operated and maintained; and (3) locally constructed and locally operated and maintained, but have been accepted into the USACE Rehabilitation and Inspection Program (RIP). Content of this ECB will be incorporated into the future Levee Safety Program EC.
6. ECB 2018-3 Using Non-NOAA Tide Gauges for Computing Relative Sea Level, February 2018.
7. ECB 2018-7 Advanced Modeling Requirements on USACE Projects, 6/6/2018: This directive renews and updates the requirements set forth in ECB 2016-3, 'Advanced Modeling Requirements on USACE Projects' (ref b). The use of advanced modeling processes and related technologies for design and construction is required as described herein.
8. ECB No. 20-xxx, Draft, Interim Approach for Risk-Informed Designs for Dam and Levee Projects.

Engineer Technical Letters: www.publications.usace.army.mil/USACE-Publications/Engineer-Technical-Letters

1. ETL 1100-2-1 Procedures to Evaluate Sea Level Change: Impacts, Responses, and Adaptation, June 2014: This technical letter provides guidance for understanding the direct and indirect physical and ecological effects of projected future sea-level change on USACE projects and systems of projects and considerations for adapting to those effects.
2. ETL 1100-2-2 Appropriate Application of Paleo-Flood Information for Hydrology and Hydraulics Decisions, October 2014: This Engineer Technical Letter establishes guidance for the appropriate use of paleo-flood analyses and information to support USACE Hydrology and Hydraulics (H&H) decision-making.

INTERAGENCY DOCUMENTS

1. FEMA P-1025, January 2015, Federal Guidelines on Dam Safety Risk Management: This addresses common understanding of risk management processes, commonly recognized standards for safety and tolerable risk, consistency in standards within the dam safety community, and technical tools and approaches related to risk analysis that can be mutually shared and jointly developed.
2. Best Practices in Dam and Levee Safety Risk Analysis (joint with USACE, USBR, and FERC), July 2005: This training manual attempts to help fill the need for practical reference material detailing risk analysis methodology for dam and levee safety applications. It contains what is considered the 'Best Practices' currently in use for estimating dam and levee safety risks at the Bureau of Reclamation and the USACE.

CONCLUDING REMARKS: WHAT NOW?

This discussion has highlighted what has been an all-to-often path to a major change in addressing national water management issues and particularly flooding. It takes a disaster or two to stimulate change and when change is achieved, it takes an excruciatingly long time to spread that change from the realm of policy to practice and an even longer period of time to implement the change on the ground nationwide. While change is relatively easy to accomplish in terms of new construction and unfortunately, for reconstruction after a disaster, it is a virtual conundrum for those areas that are living with unacceptable (intolerable) risks but have done nothing to mitigate that risk. It seems that as a nation we continue to fix the problem one disaster at a time. Breaking this cycle is one of the most difficult yet high payoff challenges we face.

Katrina provides one of the more recent examples of this cycle, especially when you consider the more recent disastrous flood events in New York/New Jersey (Superstorm Sandy), Houston (Hurricane Ike, Harvey), Puerto Rico (Hurricane Maria), and Florida (Hurricane Joaquin, Irma, Michael). Part of this enigma results from the fact that as a nation we spend 90% of our flood-related investments on recovery and only 10% on mitigation to reduce the need for recovery. Another component is the seemingly impossible task of reversing over a century of poor land-use and management decisions, both in where we develop and how we develop the land. A risk analysis for many coastal or inland areas vulnerable to flooding is too often viewed as sophisticated bad news that reveals the obvious. While it can significantly assist in prioritizing how to incrementally reduce risk, there is always the challenge of resources (unless, of course, you have a disaster).

This is not to diminish the highly important progress that is being made in the development and application of risk tools and risk-informed decision-making. These advances are essential and provide a much more productive path forward to systematically reduce the magnitude and extent of losses suffered from flooding. Examples of such progress in developing an impressive package of policy, guidance, and practice leading to a significant culture change in a large organization in a relatively short time frame are few and far between. This is an essential step forward and provides a viable platform to prepare for the next big step, dealing head-on with non-stationarity.

All of the demonstrated increased sophistication in the application of probabilistic risk analysis and systems approach to optimizing flood risk reduction does little to deal with the climate non-stationarity that is pervasive in hydrologic analysis. Non-stationarity has often been associated with the large uncertainties represented by climate change (Milly *et al.*, 2008), yet nationally we have been dealing with other forms of it indirectly and insufficiently with our traditional tools and concepts that depend on the assumptions of stationarity for their application. Most prominent of these with respect to risk are land use, population, and property value changes. By using relatively short time spans, we can 'adjust' runoff, vulnerability, and consequence estimates to make a data set 'uniform'. This, however, precludes effective use of the wealth of information available from the paleo record that can dramatically extend our 'modern data sets' to provide a richer understanding of the occurrence of major hazard events.

The larger issue is the difficulty in estimating the recurrence interval for a large event. The IPET uncertainty analyses showed that by far (71%) the greatest source of uncertainty in flooding estimates comes from the uncertainty in estimating the storm surge and wave conditions. Add to that the fact that we have so few actual records of major surge events at any one location (surge is very dependent on local physical geometry) from which to estimate recurrence intervals of such events. Now add the issue of likely climate non-stationarity in the processes themselves, which will alter the frequency and severity of individual events. The quantification of this total uncertainty paints a picture that makes the application of probabilistic risk analysis less attractive.

The current best practice to deal with this dilemma is the application of scenario analysis. Defining scenarios that represent a broad spectrum of possible change, not just with respect to natural processes, but also the socio-economics of the area of interest, provides the background within which one can begin to understand the sensitivity of existing water management capabilities and societal capabilities to cope with different types and degrees of change (without having to forecast when those changes might occur, or how often). This provides a platform for developing incremental adaptive strategies to deal with escalating change. The benefits of this approach are substantial. In addition to avoiding the minefield of uncertainty in estimating the recurrence of major events and having to decide whether to protect against the worst-case or accept less, it is possible to develop plans for incremental risk reduction as the changes unfold with time. This is both much more affordable and practical since you avoid over-investing for conditions that may not happen. Perhaps even more important, you can adapt as you learn more both with respect to what changes are most likely and how you will choose to deal with those changes.

The most recent comprehensive example of taking this approach is the Netherland's Delta Programme. The Delta Programme is an adaptive scenario-based approach to deal with the implications of climate change on water management in the Netherlands. It started with the Delta Model effort, an initiative to assemble a set of analytical models that would easily work together to understand not only change in the water regime, but also the effects of the water regime changes on the major functions of the nation (water supply, agriculture, commerce, environment, and, of course, flood safety).

The Delta Model package was used to examine a range of scenarios representing changes in water regime (e.g., sea-level rise and flow conditions on inland rivers) and changes in the socio-economics of the nation (e.g., land-use change and population shifts). The results provided knowledge of the sensitivity of current water management capabilities to those changes and their ability to meet the nation's needs. This knowledge became the basis for a planning exercise, again using the model package as a tool, to examine alternative approaches to deal with the water regime changes when they exceeded the current water management capabilities. The underlying concept was to develop strategies that allow incremental steps to adapt. This effort led to the Delta Plan which is now their long-term strategy for dealing with climate change in concert with possible socio-economic change.

The wisdom of this approach is hard to miss. Adaptation in steps, flexibility to adjust as change unfolds, the avoidance of massive resource expenditures based on highly uncertain analyses, and the opportunity to reevaluate as knowledge and analytical tools improve. Not the perfect world, but closer.

DATA AVAILABILITY STATEMENT

All relevant data are included in the paper or its Supplementary Information (https://biotech.law.lsu.edu/katrina/ipet/Volume%20I%20FINAL%2023Jun09%20mh.pdf).

REFERENCES

Halpin & Escuder (2015). Smart governance of infrastructure programs: facing the next generation of challenges and succeeding. In: *Proceedings of the 25 ICOLD Congress*, Stavanger, Norway.

IPET Final Report (2009). *Performance Evaluation of the New Orleans and Southeast Louisiana Hurricane Protection System*, Final Report of the Interagency Performance Evaluation Task Force, USACE; Volume I – Executive Summary and Overview, June 2009; Volume II – Geodetic Vertical and Water Level Datums, March 2007; Volume III – The Hurricane Protection System, June 2009; Volume IV – The Storm, March 2007; Volume V – The Performance – Levees and Floodwalls, June 2007; Volume VI – The Performance – Interior Drainage and Pumping, June 2007; Volume VII – The Consequences, March 2007; Volume VIII – Engineering and Operational Risk and Reliability Analysis, June 2009; Volume IX – General Appendices.

IPET Supplemental Report (2009). *A General Description of Vulnerability to Flooding and Risk for New Orleans and Vicinity: Past, Present, and Future, Supplemental Report of the Interagency Performance Evaluation Task Force*. US Army Corps of Engineers.

IWR Report. (1992). *Guidelines for Risk and Uncertainty Analysis in Water Resources Planning, Volume I, Principles, with Technical Appendices*, IWR REPORT 92-R-1.

Levee Portfolio Report (2018). A Summary of Risks and Benefits Associated with the USACE Levee Portfolio. Available from: https://www.usace.army.mil/Missions/Civil-Works/Levee-Safety-Program/.

Milly, P. C. D., Betancourt, J., Falkenmark, M., Hirsh, R. M., Kundzewicz, Z. W., Lettenmaier, D. P. & Stoofer, R. J. (2008). Stationarity is dead: Hither water management? *Science 319* (5863). doi.10.1126/Science.1151915.

New Orleans District (2012). *Hurricane and Storm Damage Risk Reduction System Design Guidelines, INTERIM, New Orleans District Engineering Division with Revisions Through.*

Resio, D. T. & Westerink, J. J. (2008). Modeling the physics of storm surges. *Phys. Today*. doi:10.1063/1.2982120.

Resio, D. T., Irish, J. L. & Cialone, M. A. (2008). A surge response function approach to coastal hazard assessment: Part 1, basic concepts. *Nat. Hazards*. doi:10.1007/s11069-009-9379-y.

Resio, D. T., Irish, J. L., Westerink, J. J. & Powell, N. J. (2013). The effect of uncertainty on estimates of hurricane surge hazards. *Nat. Hazards*. doi:10.1007/s11069-012-0315-1.

Resio, D. T. & Irish, J. L. (2015). Tropical cyclone storm surge risk. *Curr. Clim. Change Rep. 1*, 74–84. doi:10.1007/s40641-015-0011-9.

USACE. (2012). *Hurricane and Storm Damage Risk Reduction System Design Elevation Report*. US Army Corps of Engineers, New Orleans District.

First received 15 November 2021; accepted in revised form 15 November 2021. Available online 29 November 2021

Water Infrastructure Planning, MDCU doi: 10.2166/wp.2021.242

Infrastructure capacity planning for reducing risks of future hydrologic extremes

John J. Boland[a,*] and Daniel Peter Loucks[b]

[a] Department of Environmental Health and Engineering, Johns Hopkins University, 3801 Canterbury Road #1010, Baltimore, MD 21218, USA
[b] School of Civil and Environmental Engineering and School of Public Policy, Cornell University, Ithaca, NY 14850, USA
*Corresponding author. E-mail: jboland@jhu.edu

ABSTRACT

Floods and droughts and their associated economic, environmental, and social losses or damages are increasing in severity and frequency. Measures taken to reduce these losses or damages stemming from extreme events typically depend on how effective they are in reducing the consequences of having either too much or too little water and for longer periods of time. To identify trade-offs between the annual estimated loss or damage reduction, i.e., the benefits, however measured, and the average annual cost of various damage reduction measures, one can perform risk–cost analyses. Because of climate change, the likelihoods of future hydrologic extremes are both changing and uncertain. Also uncertain are any estimates of future damages that would occur given any specific extreme event. In addition, one cannot be certain of the future costs or benefits of damage reduction measures. This paper outlines a range of practical approaches for identifying these trade-offs, taking into account the uncertainties associated with future damages resulting from any specific flood or drought event, the changing uncertainties of future flood and drought events, and the uncertainty of future damage mitigation costs.

Key words: Damage reduction, Droughts, Floods, Infrastructure planning, Risk, Uncertainty

HIGHLIGHTS

- Explaining the impact of deep hydrologic uncertainty on water infrastructure planning.
- Discussing how the uncertainty of future hydrologic extremes influences extreme events plus safety factor approaches, scenario planning, and adaptive planning.
- Presenting a conceptual risk–cost model for infrastructure planning.
- Emphasizing the importance of determining the influence of assumptions regarding the future on current infrastructure capacity decisions.

INTRODUCTION

Rising global temperatures have resulted in more severe storms and droughts across much of Africa, Asia, Europe, and North America. Sea levels around the world are rising at rates greater than previously expected. The total human and environmental costs of these trends are often difficult to estimate, but insurance companies can and do measure their financial losses. These losses are both significant and increasing. The International Monetary Fund (IMF) estimated that worldwide climate change-induced damages cost insurance companies $80 billion in 2018, double the inflation-adjusted average for the previous 30 years. This annual damage cost exceeds $10 per year for each person living in the world today. In 2017, IMF estimates that additional uninsured losses reached $200 billion.

In some of the countries most exposed to climate change – Bangladesh, Egypt, India, Indonesia, Nigeria, the Philippines, and Vietnam – less than 1% of the population is insured against the consequences of extreme hydrologic events (Cantelmo *et al.*, 2019; Carney, 2019). These statistics, if nothing else, should motivate appropriate public agencies to develop and maintain methods to identify, evaluate, and manage these risks and to make plans to increase attention to these issues due to obvious increases in damage costs. Water infrastructure, in particular, is of critical importance to communities and economies. At a minimum, public agencies should be able to perform regular assessments of infrastructure capacity and resilience.

The role of infrastructure capacity planning in reducing the damage from floods, droughts, and water pollution is not a new subject. Approaches for defining the risks of exceeding specific values (such as flows, durations, or pollutant loads) of such events have been widely used and are typically based on historical data. The challenge today is that information derived from these historical data no longer appears representative of what might be observed in the future.

Although some sources treat the term 'risk' as interchangeable with probability, it is used here in its more widely accepted sense: to describe a combination of the probability and the consequence of some harmful event (USACE, 2019b). Because the consequences of a harmful event depend on many factors – including exposure, resilience, vulnerability, etc. – risk is a substantially more complex concept than mere probability. For this reason, planning approaches that account for probability alone fail to address many factors contributing to risks. Still, any discussion of risk begins with the consideration of probability.

The probabilities of extreme hydrologic events are changing, and the extent of those changes is uncertain. However, the direction of the changes is becoming increasingly apparent. Current calculations of the probability of exceeding any particular drought duration or any particular flood flow or stage based on historical data are likely to underestimate the probabilities of the same events in the future. Estimates of future damages resulting from any given hazardous event are also uncertain and changing over time. Similar adverse changes seem to apply to water pollution events that result from changes in population, in how land and water resources are used, managed, and developed, and in the growth of urbanization and industrialization. Again, these changes are uncertain as are their consequences. In short, future probabilities of damaging events will not resemble past or current estimates, nor will the damages associated with them (Changnon *et al.*, 2000; USGCRP, 2017; NOAA, 2018; Hill *et al.*, 2019).

Hence, it is not clear how to protect against a 100-year chance event in the future when the event associated with this likelihood is unknown. Events corresponding to the current 100-year return interval are becoming more likely. Hence return periods are likely decreasing, but in unknown ways. Therefore, the term 'return period' has little meaning when applied to future events (Salas *et al.*, 2018).

While it may be possible to identify events that society wishes to be protected from, the likelihood of any of them happening is uncertain. However, both the event's damage and its likelihood are factors usually considered in determining what is considered acceptable. For example, a family may be willing to spend the money needed to protect their home from a 100-year flood (that has a 26% chance of occurring at least once in the next 30 years), but not the amount of money needed to protect from a 200-year flood (that has only a 14% chance of occurring at least once in the next 30 years). But this calculus is upset if the future risk of a given flood event is unknown. Of all the options available for reducing risks, how much of each option should be implemented and paid for now, and on into the future, that will at least maintain acceptable levels of risk when the risks themselves are changing in unknown and possibly unpredictable ways? This paper addresses this question.

Rogers (1997) addressed the same question more than two decades ago, as has Hill *et al.* (2019) more recently at the Hoover Institution Rogers applied the standard capacity expansion model of operations research to a very narrow set of water infrastructure planning methods, all characterized by repeated choices of investment decisions over time. Hill *et al.* (2019), on the other hand, took a very broad view of the capacity planning problem, focusing on resilience and robustness as the goals of planning, rather than cost minimization alone. The purpose of risk analysis is to better inform decision-making, especially for the many components of a water resources management system whose performance possesses various degrees of uncertainty. Yet risk analysis itself introduces its own elements of uncertainty, often in the form of unfamiliar and arcane statistical and probabilistic methods. These are far less transparent than simply selecting a design event for protection, based on its probability being exceeded and its cost. Note that there are two distinctly different risk/reliability evaluation processes involved in the planning of a particular infrastructure (USACE, 2019a) – formulating and assessing options – and the subsequent design. The approach outlined in this paper for planning infrastructure capacity investments for the reduced likelihood of damage from extreme hydrologic events is a conceptual one, simplified here for clarity. For any site-specific project, the functional relationships described here would be based on site-specific data. These relationships assume that the specific infrastructure under consideration is known, as well as its cost and effectiveness in reducing various measures of damage, whether economic, environmental, or social. In other words, we take a deterministic approach but the estimates we use for costs and probabilities could be based on stochastic analyses of alternative futures. Or they could be based on subjective judgments that may later warrant adjustment after computing the estimated consequences of those judgments. For a specific location and assumed type of infrastructure, detailed simulation studies may have to be made to identify the relations and data needed for carrying out the described approaches.

Several different methods for incorporating risk into infrastructure planning are presented here. These differ significantly with respect to data requirements, analytical complexity, and potential contribution to improved infrastructure decisions. However, all of these methods are capable of recognizing hydrologic uncertainty, either implicitly (extreme event plus safety factor) or explicitly (risk–cost analysis). The critical role of adaptive planning is emphasized throughout. There is no attempt to contribute to the state of the art; the purpose of the paper is to present practical approaches to water infrastructure planners under uncertainty.

Considerations in the choice of a method include not only implementation cost but also the degree to which the method makes appropriate use of available knowledge and the degree to which it supports decisions that are efficient in terms of the most likely future hydrology but robust with respect to hydrologic uncertainty.

This paper focuses on, and limits itself to, the risks of floods and droughts, and how they might be reduced. Damages resulting from these extreme hydrologic events can be expressed in economic, environmental, or social terms, as applicable at specific locations. The costs of infrastructure designed to reduce both probability and consequences of extreme events are typically expressed in monetary units while recognizing that the infrastructure itself can have ecologic, environmental, and social impacts, as well as economic costs.

APPROACHES TO PLANNING WATER INFRASTRUCTURE UNDER UNCERTAINTY

The following paragraphs describe some approaches that have been used or proposed for infrastructure capacity planning under uncertainty. These approaches are not mutually exclusive. Actual project planning may encounter different types of uncertainty, ranging from statistical uncertainty that can be expressed in terms of available historical data to deep uncertainty, where little or nothing is known about the range of possible outcomes (Fletcher *et al.*, 2017, 2019). The planner may use different approaches to characterize different types of uncertainty. Also, this list of planning approaches is not exhaustive; other approaches are possible.

Extreme event plus safety factor

Knowing that probabilities of events may be uncertain and changing, is there value in defining engineering standards in terms of probabilities rather than just selecting some extreme event to protect from, and then adding a safety factor (e.g., 'freeboard') to deal with uncertainties and 'unknown unknowns'? Ultimately, the purpose of risk analysis is to find solutions that are risk–cost-effective. That is, the level of risk and the cost of preventing damages associated with the level of risk are deemed socially acceptable or 'tolerable'. For example, the Standard Project Hurricane used by the USACE is based on the concept of a maximum damaging Category 3 hurricane. The probable maximum precipitation, also used in some contexts, is based on a concatenation of a series of low probability events that cause soil saturation and intense rainfall. Measures may be taken to protect against any damage resulting from these events. The cost of protection from more extreme events might be considered too costly given the low probabilities of such events happening. But the practice of limiting protection to events not exceeding some arbitrary level may not be socially beneficial or acceptable.

There are weaknesses in this safety factor approach. While the definitions of the protected event and the amount of freeboard may be based on historical data, changing probabilities due to changes in land use and climate, e.g., and uncertain future consequences can render historical data irrelevant to future extremes. Therefore, many of the facets of a safety factor are essentially arbitrary choices. There is usually no determination of the efficiency of this approach under likely conditions. The cost of constructing infrastructure to this standard may greatly exceed the damage avoided, or not. In most cases, residual probabilities (for events more severe than the protected event) are not considered, even though they may be substantial. Changes in the consequences of various events, which are themselves changing, are not considered at all. Nothing can be said about the robustness of any resulting plan with respect to a range of extremes.

The extreme event plus safety factor approach has some advantages: it uses data that are often readily available, and it avoids any kind of complex analysis. But while it may have been acceptable when probabilities and possible damages were known with some degree of certainty and were not believed to be changing over time, it is seriously deficient in a world where probabilities of extreme events are increasingly uncertain, as are their consequences. Evidence of the limitations of this approach in practice is provided by the fact that after many years of designing projects based on fixed criteria such as Standard Project Flood, the US Army Corps of Engineers no longer employs this design target, opting instead for risk–cost modeling (USACE, 2019b).

Scenario planning

Because of its usefulness in cases of extreme uncertainty and change, scenario planning has found a wide range of applications. At its most basic, the method consists of the definition, analysis, and comparison of a number of scenarios. A scenario is defined by Peterson *et al.* (2003), as 'a structured account of a possible future'. A scenario can be entirely or partially quantitative. There is a large literature on the methods and criteria for developing scenarios (Amer *et al.*, 2012).

Generally, scenarios need to be comprehensive: they must include all demographic, economic, social, ecological, and other factors that determine the damage associated with a given hydrologic extreme. They must also be internally consistent. So long as scenarios are plausible, there is no need to assign a likelihood. Scenarios are about the possible, not the probable.

In most public sector applications, scenario planning considers the consequences of some policy or action, based on scenarios which define a range of possible external factors (demography, economic conditions, etc.). When considering future hydrologic extremes, however, some external factors can be held constant over all scenarios or may be varied for groups of scenarios. Each scenario incorporates the infrastructure plan that is to be evaluated. Then the main issue is the assumption, in each scenario, of some extreme hydrologic event along with the damage that would result from that event, given all the other elements of the scenario. The entire process can then be repeated with a different infrastructure plan scenario.

In its basic form, scenario planning of water infrastructure does not address efficiency. While a comparison of damage estimates for different scenarios may show a range of outcomes, since their probabilities are neither known nor assumed, there is no way to compare an estimated outcome to, e.g., infrastructure cost. The principal benefits of a scenario approach are to reveal unexpected outcomes (which may lead to the development of better plans), and to identify plans which are robust with respect to one or more, including hydrologic, uncertainties.

Adaptive planning

Hydrologic, economic, environmental, technological, and social uncertainties create a challenge to effective infrastructure capacity planning, especially where large-scale, long-lived facilities are under consideration. As the time scale increases, so does the importance of considering uncertainties with respect to the future demand for infrastructure services, future technological progress, long-term environmental impacts, etc. Commitments are often made to large, long-lived projects in the interest of minimizing life cycle cost as well political expediency. But these decisions raise the possibility of facilities that are inappropriately sized, have unexpected environmental impacts, and/or become technologically obsolete. Also, these facilities may be vulnerable to hydrologic events not anticipated at the time of planning.

Adaptive planning is one approach for reducing the long-term risks associated with these considerations, as well as those resulting from hydrologic uncertainty. Adaptive planning assumes that plans will be updated through an ongoing process of adaptable decision-making. Rather than following a predetermined path, decisions are adjusted as forecasted conditions and outcomes of prior management actions are better understood (National Research Council, 2004). The purpose of adaptive planning is to achieve more effective planning over time in the face of an uncertain future (Haasnoot *et al.*, 2013; Beh *et al.*, 2015; Hui *et al.*, 2018; Erfani & Harou, 2021).

The general principle of adaptive planning is to avoid plans or proposed actions that close future options. These options can be addressed later, as some of the uncertainties present in the initial plan are partially resolved and new ones emerge. In this way, risks due to a number of uncertainties, including hydrologic uncertainty, are considered. However, there are costs associated with this approach. It requires careful monitoring of the performance of the planned infrastructure throughout its life, as well as analysis and design costs whenever interim decisions must be made. It is possible that the life cycle cost for the adaptively managed facility might turn out to be less than for a one-time design, but it may also be greater.

Risk–cost modeling

Another approach to planning under uncertainty involves actively managing and modeling risk along with the cost of reducing risk. Hydrologic risk is managed by developing resilience in affected communities (thereby reducing exposure and vulnerability) and by planning policies and actions that mitigate adverse impacts, should a harmful event occur (National Research Council, 2012). Risk–cost modeling supports the development of effective mitigation while informing decision-makers of the possible exposure to risk.

Planning under deep uncertainty

In most cases of planning under uncertainty, something is known about our future. Both hydrologic variables and vulnerability to damage may have plausible upper and/or lower bounds. Future trends may be seen as uncertain deviations from past experience. A statistical representation of uncertain future conditions may be possible. Various planning approaches can be used, depending on the information available. Said differently, planning under uncertainty may involve the application of the mathematics of probability and statistics because some information/data are generally available.

In contrast, the most challenging planning problem arises where the future is characterized not just by limited information, but by true uncertainty. This means that no useful data exist, that is one 'doesn't have a clue'. True uncertainty is now often replaced by the term deep uncertainty. Deep uncertainty is defined by Lempert *et al.* (2003) as:

> '... the condition in which analysts do not know or the parties to a decision cannot agree upon (1) the appropriate models to describe interactions among a system's variables, (2) the probability distributions to represent uncertainty about key parameters in the models, and/or (3) how to value the desirability of alternative outcomes.'

The usual approach to planning problems of this kind is to focus on robustness, where outcomes are least sensitive to various realizations of the underlying variables. Recent work at the RAND Corporation (2021) has produced a guide to decision-making under deep uncertainty, combining the methods of scenario analysis and adaptive planning to achieve a robust, broadly acceptable plan.

INFRASTRUCTURE OPTIONS AND RISK REDUCTION

Various options exist for preventing or mitigating the adverse impacts associated with droughts or floods. These adverse effects lead to damages, which can be measured in monetary as well as non-monetary terms. The following subsections illustrate how risks and related damages can be conceptualized. This allows various options for risk reduction to be modeled under conditions of various assumed probabilities of hydrologic extremes.

Droughts

There are several mitigation measures that can be used to reduce the impacts caused by drought. These include any measure that reduces the demand for water, including, e.g.:

- Soil and water conservation and irrigation practices that result in less water use.
- Various urban water conservation measures, including pricing.
- Technological fixes, such as low-flow toilets, showerheads, and washing machines.
- Interconnecting smaller municipal water supply systems and augmenting with groundwater.
- Reducing demands through leak detection and repairs.

They also include measures that increase the supply of water, such as

- Desalination of seawater, water recycling, and rainwater harvesting.
- Conjunctive use of surface and groundwater reservoirs.
- Interconnections among rivers and reservoirs within and outside river basins.

Prolonged droughts can lead to adverse health, social, economic, and political impacts. Any measures taken in advance of drought events that address these potential impacts can reduce if not prevent their adverse impacts. Such measures include:

- Storage and distribution of food and medical supplies.
- Unemployment and crop failure insurance.

Provision of alternative sources of energy to mitigate any reductions due to lack of water, say for hydropower or cooling. None of these options can change the duration and severity of any drought. What they can do is reduce the damage resulting from it. Of interest here is the trade-off between the costs of drought damage reduction measures and the extent of such damage reductions. It is important to note that not all measures are equally reliable in their ability to eliminate or reduce drought-related damage. Even if one could effectively calculate these factors, there are always considerable uncertainties associated with each management measure, which often depends on the fallibility of human factors and interpretation of information.

Given a scenario that includes assumed estimates of the probabilities or return periods of various drought durations and severities and the damages resulting from them, the estimated annual damages associated with measures taken to reduce drought damage can be compared to annual implementation costs.

To outline how such risk–cost trade-offs can be estimated, the relationship between drought damages and drought likelihood needs to be defined. Again, since this estimation is itself uncertain, this damage function can be considered as a scenario. Risk can be expressed as the probability of some level of drought damage being exceeded.

Measures taken to mitigate drought impacts can be implemented in stages as the severity of drought increases. Drought management triggers can be assigned to drought durations and severities, i.e., the number of days of water shortage and the extent of that shortage (such as specified aquifer or reservoir storage levels or river flows). The probability distribution of annual drought damages can be based on historical annual drought data, but if so, this will not reflect future changes in this distribution due to climate change. These changes of course are unknown and hence will have to be based on assumed drought event scenarios. From these scenarios, the probability of drought duration exceedance functions can be estimated by subtracting the cumulative probability distribution of damage (the probability that the actual damage will not exceed some specified damage) from 1.

Assume the function $PE_D(D)$ defines the probability of a specific annual drought damage D being exceeded. The expected annual drought damage, $\mathrm{E_D}[D]$, is the area under this exceedance function, i.e., the integral of $PE_D(D)$ over all values of damages D.

$$\mathrm{E_D}[D] = \int PE_D(D)\mathrm{d}D \tag{1}$$

A changing climate is causing a change in this expected annual damage over time. In regions where droughts are lasting longer and hence drought damages are increasing, it is reasonable to take that into account when planning current drought damage mitigation measures. Letting the index t represent a future time period, we can develop scenarios of shifting probabilities of annual drought damages, $PE_{Dt}(D)$, and use them to compute the corresponding estimated annual drought damages, $E_{Dt}[D]$ in each period t. This will be useful for the next step of computing the estimated present net benefits of future drought mitigation measures.

Drought damages can be reduced in various ways, as discussed earlier in this section. Those damage reduction measures will have costs. For the purposes of this paper, assume these damage reduction measures are discrete independent of each other. Let $C_{\mathrm{it}}(\delta_{\mathrm{it}})$ be the estimated annual cost function applicable in period t for reducing drought damage D by δ_{it} using measure i. To be determined are the damage reduction values δ_{it} that maximize the present value of the damage reduction net benefits – the reduced estimated annual damages less the costs of damage reduction. Letting r be the periodic interest rate, this goal is equivalent to:

$$\text{Maximizing } \sum\nolimits_t \left\{\sum\nolimits_{\mathrm{i}} (\delta_{\mathrm{it}} - C_{\mathrm{it}}(\delta_{\mathrm{it}}))\right\}(1/(1+r)^{\mathrm{t}-1}) \text{ where } \sum\nolimits_{\mathrm{i}} \delta_{\mathrm{it}} \leq E_{Dt}[D] \tag{2}$$

This approach can apply to each type of damage, whether economic, environmental, or social, as long as such damages can be quantified and along with the values of $C_{\mathrm{it}}(\delta_{\mathrm{it}})$, expressed in comparable damage units. If not, it becomes a multi-objective trade-off between damage reduction and cost. More detail is available in the handbooks edited by Jain & Singh (2019), Singh (2017), Singh *et al.* (2007), and Mays (1996).

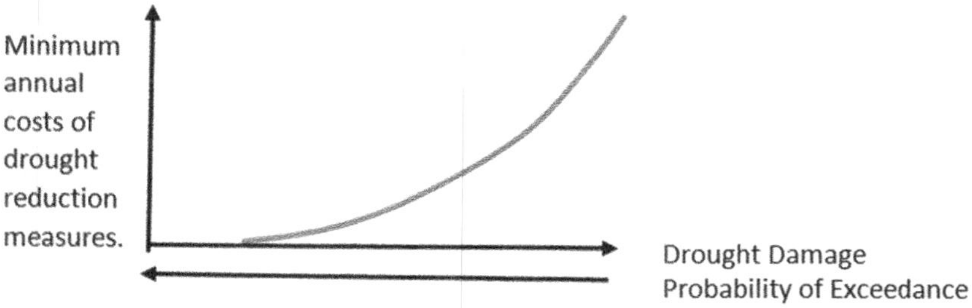

Fig. 1 | Function showing minimum total costs, $C_i(\delta_i)$, of protecting from droughts causing specified damages or having a specified probability of being exceeded. Actual cost functions, $C_i(\delta_i)$, may include fixed costs together with economies of scale.

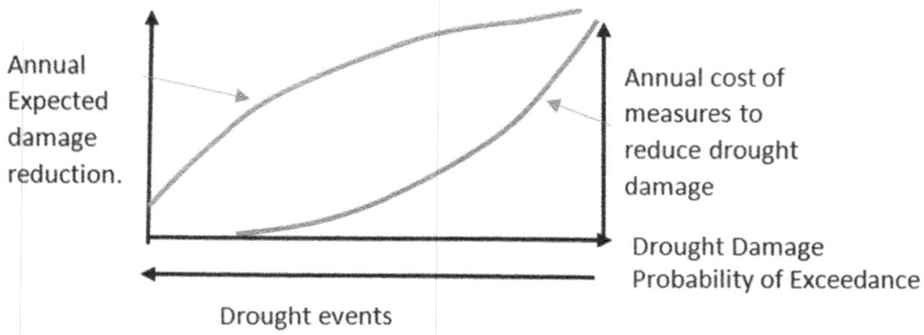

Fig. 2 | Functions showing trade-offs between minimum costs of reducing estimated annual drought damages, $\sum_i C_i(\delta_i)$, and resulting annual drought damage reduction benefits, $\sum_i \delta_i$. Actual cost functions may include fixed costs together with economies of scale that are not shown in this example.

The derivation and use of estimated annual damage reduction values as outlined above are just one approach for informing those responsible for deciding which measures to take to hedge against future droughts and their damages. Another approach is to decide what level of drought risk, $PE_D(D)$, or damage, D, to protect from, and identify the least-cost set of measures, δ_I, that will accomplish that. This trade-off is illustrated in Figure 1. That selection process could be based on other criteria besides cost, as deemed appropriate.

Obviously different drought reduction measures will result in different estimated annual damage reductions and will cost different amounts of money. Considering different sets of drought reduction measures and their annual estimated damage reductions and costs, trade-offs between those damage reductions and costs can be defined as illustrated in Figure 2. If damages are expressed in monetary terms, say dollars, the particular set of measures that maximize the estimated damage reduction less cost can be identified.

The above discussion has assumed known risks, perhaps based on historical data (Vogel & Castellarin, 2016). Of course, these return periods are changing due to many factors, including climate change. Furthermore, these changes are uncertain and this uncertainty is itself unknown (Hall *et al.*, 2014; Read & Vogel, 2015, 2016a, 2016b; Salas *et al.*, 2018). So the question of what and the extent of measures to take to accomplish future drought damage reduction targets or to protect from droughts having specified return periods is still unaddressed. These issues will be examined after a discussion of flood risk reduction measures that follow.

Floods

Unlike droughts, floods can occur relatively quickly giving less opportunities for adaptive measures as the severity of a flood increases. Strategies to reduce threats to life and property from flooding include those that:

- reduce the susceptibility to flood damage and disruption;
- reduce the adverse impacts of floods on individuals and the community; and
- alter the flood stages and/or durations themselves.

Susceptibility to flood damage and disruption can be reduced through appropriate floodplain zoning and other regulations pertaining to land-use development and drainage, disaster preparedness, assistance including flood insurance, and recovery planning and programs, flood proofing, and the development and use of flood forecasting and warning capabilities.

The adverse impact of flooding on individuals and the community can be reduced through information and education programs, flood insurance, tax adjustments, emergency and rescue services, and flood damage recovery efforts.

Flood stages and durations can be reduced through the operation of reservoirs, installation of dikes, levees, and floodwalls, channel dredging and modification, high-flow diversions, floodplain modifications, and detention storage.

Along coastal shorelines, structural measures include barrier protection, interior flood gates, road/rail elevation, levees, floodwalls, bulkheads, seawalls, revetments, beach restoration, breakwaters, and storm system drainage improvements.

Coastal zone flood management options include building retrofit (elevation and flood proofing), managed coastal retreat, emergency evacuation plans, early warning systems, public education/risk communication, flood insurance, wetland restoration, and implementing living shorelines, green storm-water management, reefs, and submerged aquatic vegetation.

Each or a combination of these flood damage reduction measures can change the relationships defining damages and the probability of exceedances associated with flood events. Following the conceptual approach taken to estimate estimated annual drought damages and how they might be reduced, consider a potential flood damage site. Let S be the stage or height of a flood at that site that causes D damages. Define the relationship between stage S and damage D as the function $D = f(S)$. Having records of the maximum flood stage in each year for a number of years allows one to define the probability of exceedance function associated with flood stages at particular locations. Let that function be designated as $PE_S(S)$. The estimated annual flood stage is the area under that probability of exceedance function:

$$\mathrm{E_S}[S] = \int PE_S(S)\mathrm{d}S \tag{3}$$

Knowing the damages and probability of exceedance associated with any specific flood stage S allows the calculation of the probability of exceedance of the corresponding specific damages:

$$PE_{\mathrm{F}}(D) = PE_S(f(S)) \text{ where } D = f(S) \tag{4}$$

and the estimated annual flood damage, $\mathrm{E_F}[D]$

$$\mathrm{E_F}[D] = \int PE_F(D)\mathrm{d}D \tag{5}$$

Any measures taken now to reduce flood damage should take into consideration the changes over time that will alter both the flood damage function, $f(S)$ (such as urban development on the floodplain) and the probability of exceedance, $PE_S(f(S))$ (such as resulting from changes in precipitation due to climate change) associated with any flood stage S. Letting the index t represents a future time period, one can create scenarios of future exceedance probabilities and damage functions to calculate future estimated annual damage functions, $E_{Ft}[D]$.

Flood damages can be reduced in various ways, either by altering the stage damage function (such as by flood proofing structures on the floodplain) or by altering the probability of exceedance distribution (such as by levees or reservoir flood storage capacity or operation during a flood period). Those damage reduction measures will have costs. Various combinations of them will have different total costs. Let $C_t(\delta_t)$ be the estimated minimum annual cost function applicable in period t for reducing the remaining estimated annual flood damage $E_{Ft}[D]$ by δ_t. To be determined are the damage reduction values δ_t that maximize the present value of the damage reduction net benefits – the reduced estimated annual damages less the costs of damage reduction. Letting r be the periodic interest rate, this goal is equivalent to:

$$\text{Maximizing} \sum\nolimits_t (\delta_t - C_t(\delta_t))\}(1/(1+r)^{t-1}) \text{ where } \delta_t \leq E_{Ft}[D] \tag{6}$$

This approach can apply to each type of damage, whether economic, environmental, or social as long as such damages can be quantified using the same units as $C_t(\delta_t)$. If not, it becomes a multi-objective trade-off between damage reduction and cost. More detail is available in the handbooks edited by Jain & Singh (2019), Singh (2017), Singh *et al.* (2007), and Mays (1996).

IMPACT OF CHANGING EXTREMES

While there are exceptions (Hirsch & Ryberg, 2012), recent observations generally show that more extremes are occurring more frequently (Doocy *et al.*, 2013; Kunkel *et al.*, 2013; Peterson *et al.*, 2013; Hall *et al.*, 2014; Blöschl *et al.*, 2015; IPCC, 2018). Hydrologically related probability distributions are changing, i.e., are non-stationary, and that change is hard to predict (Blöschl *et al.*, 2015; Hu *et al.*, 2018). From Figure 3, the translation of the event values, such as drought duration or flood stage, on the vertical axis of the upper right quadrant to the horizontal axis of the lower left quadrant is accomplished by the blue line in the upper left quadrant. Raising that linear function, such as shown in green, shortens the scale on the horizontal axis. This green function, whether linear or not, can represent the impact of extreme events due to climate change. The same events now have a higher probability of occurring and hence the estimated damages will increase.

As shown in Figure 3, increases in the magnitude of hydrologic events such as the magnitude and duration of drought restrictions or flood stages will result in increases in estimated annual damages. This will impact the trade-offs between annual estimated damage reductions and annual costs of measures taken to achieve such reductions. It will also impact the trade-offs between target return periods and the annual costs of measures taken to avoid events less than or equal to those target values. Hence, the fundamental question (addressed in the next section) is one of determining what measures should be taken today given the estimated need to increase the capacity of such measures in the future to accomplish the same, or even different, goals. Failure to consider the increasing future risks can result in infrastructure investments that will fail to function or serve society for as long as originally expected (World Energy Council, 2015; Davenport, 2018).

INFRASTRUCTURE CAPACITY PLANNING OVER TIME

Capacity planning addresses how much infrastructure capacity to build and when in order to meet the desired risk or estimated damage reduction goals over time. Whether the goal is to maximize the difference between the estimated annual damage reduction and annual infrastructure cost or to achieve protection up to a specified

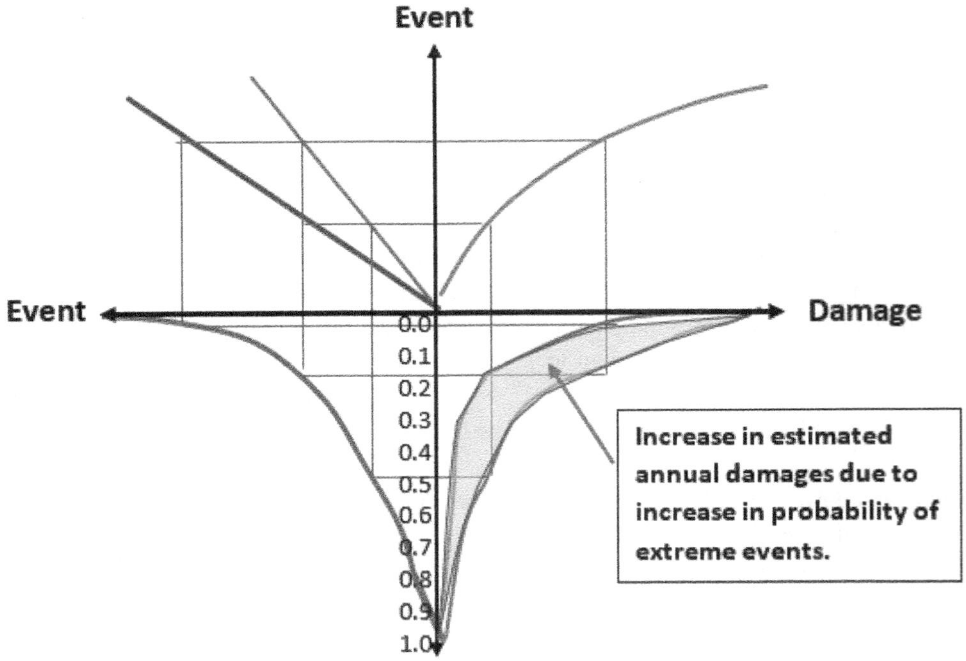

Fig. 3 | Estimating the impact of decreasing return periods associated with various hydrologic events. The magnitude of the event, however measured, increases in the direction of the arrows on the respective event axes. Please refer to the online version of this paper to see this figure in colour: http://dx.doi.org/10.2166/wp.2021.242.

return period at minimum cost, the question of what types of infrastructure and their design capacities should be built is not a trivial one. Even assuming no change in risks or potential damages over time, the fact that infrastructure costs typically include fixed as well as variable costs, and exhibit economies of scale, makes it economical to build more capacity than is needed at the time construction takes place. Add to this, the need to meet an uncertain but likely increasing future demand, the issue is how much to build and when. Here our demand is the capacity required to protect from an increasingly severe hydrological event or the capacity required to meet increasing estimated damage reduction targets.

Based on observed trends and analyses as outlined above, assume one can estimate the demand for infrastructure capacity that is needed to meet specified risk or estimated damage reduction goals over the next several decades. Also assume that the costs of new infrastructure capacity can be estimated for this period, as well as the interest rate that defines present values of costs. All these values are uncertain, and the impact of this uncertainty needs to be addressed, but first assume such data can be estimated.

Let K_t be the existing capacity of infrastructure at the beginning of time period t. Except for K_1 for the current period, $t = 1$, these values need to be defined for future periods t such that they meet or exceed the assumed known capacity demand targets, KD_t. Clearly, there are many values of K_t that will meet this constraint for each period t. Which values are best depends on the criteria chosen for evaluating alternatives. If a least-cost capacity expansion schedule is of interest, then cost estimates are needed for capacity additions, KA_t, in each period t. Assume such costs can be estimated and then discounted to the present time. Let the function $C(KA_t, K_t)$ be the present value of the cost of adding KA_t capacity to an existing capacity K_t during construction period t. Using these data, an optimization model can be developed and solved to identify the values of each KA_t variable that results in a least-cost capacity expansion schedule identifying how much capacity to add and when over a predefined time horizon.

$$\text{Minimize Total Costs} = \sum_t C(KA_t, K_t) \text{ subject to } K_t + KA_t = K_{t+1} \geq KD_t \text{ for each period } t \tag{7}$$

Whatever the values of those future capacity additions, they are based on numerous assumptions concerning time horizons, future demands, KD_t, costs, $C(KA_t, K_t)$, and interest rates, all of which are uncertain. Thus, different estimates of these parameter values should be used to test if any such assumed conditions alter the value of the variable of most interest, namely what to do now, KA_1. If not, one can be more confident of that possible decision that best satisfies the objective, in this case, to minimize the present value of the total cost. Rogers (1997) provides a more detailed discussion of this approach, although his analysis is restricted to the case where repeated investment decisions are possible over time. Also, see Hall *et al.* (2020) for an interesting case study in risk-based planning applied to the water industry in England.

When the future is not only unknown but non-stationary, i.e., there is no probabilistic basis for stating what the future may be, and as a result, any capacity expansion schedule needs to be adaptive (Kwakkel *et al.*, 2016; Pahl-Wostl, 2016; Söderholm *et al.*, 2018). Of interest is the current decision for the current time period. The capacity to add now, if any, surely needs to be based on the best guess of future demands and costs and discount rates. Any later decisions should be informed using updated estimates of all these uncertain parameter values.

SUMMARY

A changing climate is bringing with it increasing hydrologic extremes. These extremes can result in damages, the extent to which depends on measures taken in advance to reduce the damages that would otherwise occur from those extreme events. Infrastructure capacity planning cannot escape consideration of the uncertainty regarding future hydrologic extremes as well as the myriad other factors involved in infrastructure design. All of these potentially affect future flood and drought damages. This paper has outlined a number of approaches to incorporating uncertainty into risk reduction in infrastructure planning.

To estimate estimated annual damages that could result from an extreme hydrologic event one needs to know the probability of such an event being equaled or exceeded. Because of climate change, these probabilities are changing and therefore unknown. Not only is there uncertainty in the likelihood of future hydrologic extremes, but there is also uncertainty in the future damages that would occur given specific extreme events. Similarly, the future costs of damage reduction measures are uncertain.

In considering the various approaches to infrastructure planning over time and under uncertainty, emphasis has been placed on the investment in the first period. Errors due to uncertainty regarding infrastructure requirements may be less costly in later years. One reason is the opportunity for adaptive planning, an advantage that increases in value as the time between investment decisions becomes smaller. This point is illustrated by Rogers (1997). That analysis assumed a repeated choice of investment decisions over time and the result was that the first-year investment was almost invariant with respect to uncertainty regarding future conditions. As noted by Rogers, this result does not necessarily hold for the kind of large, indivisible, irreversible investments typical of water infrastructure.

All these uncertainties do not change the fact that we continue to make decisions to protect communities from future hazards. The approach to risk–cost analysis outlined in this paper is designed to assist such decision-making. Because of all these uncertainties, the approach involves using multiple scenarios, each of which includes estimates of these uncertain risks, damages, and costs. The estimated values derived from all these scenarios (or the scenario that meets a specified degree of reliability of estimated annual damage) can be used in an adaptive capacity expansion approach, where over time the analysis results can adapt to changes in future estimates of all these uncertainties.

REFERENCES

Amer, M., Daim, T. U. & Jetter, A. (2012). A review of scenario planning. *Futures 46*, 23–40.

Beh, W. H. Y., Maier, H. R. & Dandy, G. C. (2015). Adaptive, multiobjective optimal sequencing approach for urban water supply augmentation under deep uncertainty. *Water Resources Research 51*, 1529–1551.

Blöschl, G., Gaal, L., Hall, J., Kiss, A., Komma, J., Nester, T., Parajka, J., Perdigao, R. A. P., Plavcova, L., Rogger, M., Salinas, J. L. & Viglione, A. (2015). Increasing river floods: fiction or reality? *Wiley Interdisciplinary Reviews: Water*. doi:10.1002/wat2.1079.

Cantelmo, A., Melina, G. & Papageorgiou, C. (2019). *Macroeconomic Outcomes in Disaster-Prone Countries*. International Monetary Fund WP/19/217, Research Department, IMF, Washington DC.

Carney, M. (2019). *Fifty Shades of Green*. Point of View, Finance and Development, International Monetary Fund, Washington DC.

Changnon, S. A., Pielke Jr., R. A., Changnon, D., Sylves, R. T. & Pulwarty, R. (2000). Human factors explain the increased losses from weather and climate. *Bulletin of the American Meteorological Society 81*(3), 437–442.

Davenport, C. (2018). Trump's infrastructure plan may ignore climate change. It could be costly. *New York Times*, February 10.

Doocy, S., Daniels, A., Murray, S. & Kirsch, T. D. (2013). The human impact of floods: a historical review of events 1980–2009 and systematic literature review. *PLOS Currents Disasters*. doi:10.1371/currents.dis.f4deb457904936b07c09daa98ee8171a.

Erfani, T. & Harou, J. J. (2021). Adaptive water resource planning using decision-rules. *Advances in Water Research 154*, 103961.

Fletcher, S. M., Miotti, M., Swaminathan, J., Klemun, M. M., Strzepek, K. & Siddiqi, A. (2017). Water supply infrastructure planning: decision-making framework to classify multiple uncertainties and evaluate flexible design. *Journal of Water Resources Planning and Management 143*(10), 04017061. doi:10.1061/(ASCE)WR.1943-5452.0000823.

Fletcher, S., Lickley, M. & Strzepek, K. (2019). Learning about climate change uncertainty enables flexible water infrastructure planning. *Nature Communications 10*, 1782. doi:10.1038/s41467-019-09677-x.

Haasnoot, M., Kwakkel, J. H., Walker, W. E. & ter Maat, J. (2013). Dynamic adaptive policy pathways: a method for crafting robust decisions for a deeply uncertain world. *Global Environmental Change 23*(2), 485–498.

Hall, J., Arheimer, B., Borga, M., Brazdil, R., Claps, P., Kiss, A., Kjeldsen, T. R., Kriaučiūnienė, J., Kundzewicz, Z. W., Lang, M., Llasat, M. C., Macdonald, N., McIntyre, N., Mediero, L., Merz, B., Merz, R., Molnar, P., Montanari, A., Neuhold, C., Parajka, J., Perdigao, R. A. P., Plavcova, L., Rogger, M., Salinas, J. L., Sauquet, E., Schar, C., Szolgay, J., Viglione, A. & Blöschl, G. (2014). Understanding flood regime changes in Europe: a state of the art assessment. *Hydrology and Earth System Sciences 18*, 2735–2772. doi:10.5194/hess-18-2735-2014.

Hall, J. W., Mortazavi-Naeini, M., Borgomeo, E., Baker, B., Gavin, H., Gough, M., Harou, J. J., Hunt, D., Lambert, C., Piper, B., Richardson, N. & Watts, G. (2020). Risk-based water resources planning in practice: a blueprint for the water industry in England. *Water International 34*(3), 441–454.

Hill, A., Mason, D., Potter, J. R., Hellmuth, M., Ayyub, B. & Baker, J. W. (2019). *Ready for Tomorrow: Seven Strategies for Climate-Resilient Infrastructure*. Hoover Institution Press, Stanford University. Available at: https://www.hoover.org/research/readytomorrow-seven-strategies-climate-resilient-infrastructure.

Hirsch, R. M. & Ryberg, K. R. (2012). Has the magnitude of floods across the USA changed with global CO_2 levels? *Hydrological Sciences Journal 57*, 1–9. doi:10.1080/02626667.2011.621895.

Hu, Y., Liang, Z., Singh, V. P., Zhang, X., Wang, J., Li, B. & Ang, H. (2018). Concept of equivalent reliability for estimating the design flood under non-stationary conditions. *Water Resources Management 32*, 997–1011. doi:10.1007/s11269-017-1851-y.

Hui, R., Herman, J., Lund, J. & Madani, K. (2018). Adaptive water infrastructure planning for nonstationary hydrology. *Advances in Water Research 118*, 83–94.

IPCC (2018). *Summary for Policymakers*. Available at: https://archive.ipcc.ch/pdf/special-reports/sr15/sr15_spm_final.pdf.

Jain, S. K. & Singh, V. P. (2019). *Engineering Hydrology: An Introduction to Processes, Analysis and Modeling*. McGraw-Hill Education, New York, NY, USA. ISBN 978 1259641978.

Kunkel, K. E., Karl, T. R., Brooks, H., Kossin, J., Lawrimore, J. H., Arndt, D., Bosart, L., Changnon, D., Cutter, S. L., Doesken, N., Emanuel, K., Groisman, P. Y., Katz, R. W., Knutson, T., O'Brien, J., Paciorek, C. J., Peterson, T. C., Redmond, K., Robinson, D., Trapp, J., Vose, R., Weaver, S., Wehner, M., Wolter, K. & Wuebbles, D. (2013). Monitoring and understanding trends in extreme storms: state of knowledge. *Bulletin of the American Meteorological Society 94*. doi:10.1175/BAMS-D-11-00262.1.

Kwakkel, J., Walker, W. & Haasnoot, M. (2016). Coping with the wickedness of public policy problems: approaches for decision making under deep uncertainty. *Journal of Water Resources Planning and Management 142*(3), 1–5. doi:10.1061/(ASCE) WR.1943-5452.0000626.

Lempert, R. J., Popper, S. W. & Bankes, S. C. (2003). *Shaping the Next one Hundred Years: New Methods for Quantitative Long-Term Strategy Analysis (MR-1626-RPC)*. RAND Pardee Center, Santa Monica, CA, USA.

Mays, L. W. (1996). *Water Resources Handbook*. McGraw-Hill Professional, New York, NY, USA. ISBN-10 00704111500.

National Research Council (2004). *Adaptive Management for Water Resources Project Planning*. The National Academies Press, Washington, DC, USA.

National Research Council (2012). *Dam and Levee Safety and Community Resilience: A Vision for Future Practice*. The National Academies Press, Washington, DC, USA.

NOAA National Centers for Environmental Information (NCEI) (2018). *U.S. Billion-Dollar Weather and Climate Disasters*. Available at: https://www.ncdc.noaa.gov/billions/.

Pahl-Wostl, C. (2016). Water security, systemic risks and adaptive water governance and management. In: *Handbook on Water Security*. Pahl-Wostl, C., Bhaduri, A. & Gupta, J. (eds). Edward Elgar Publishing. Available at: https://www.elgaronline.com/view/9781782548003.00012.xml.

Peterson, G., Cumming, G. S. & Carpenter, S. R. (2003). Scenario planning: a tool for conservation in an uncertain world. *Conservation Biology 17*(2), 358–366.

Peterson, T. C., Heim Jr., R. K., Hirsch, R., Kaiser, D. P., Brooks, H., Diffenbaugh, N. S., Dole, R. M., Giovannittone, J. P., Guirguis, K., Karl, T. R., Katz, R. W., Kunkel, K., Lettenmaier, D., McCabe, G. J., Paciorek, C. J., Ryberg, K. R., Schubert, S., Silva, V. B. S., Stewart, B. C., Vecchia, A. V., Villarini, G., Vose, R. S., Walsh, J., Wehner, M., Wolock, D., Wolter, K., Woodhouse, C. A. & Wuebbles, D. (2013). Monitoring and understanding changes in heat waves, cold waves, floods and droughts in the United States: state of knowledge. *Bulletin of the American Meteorological Society 94*, 821–834. doi:10.1175/BAMS-D-12-00066.1.

RAND Corporation (2021). *An Interactive Guide to the Use of Methods for Decisionmaking under Deep Uncertainty (DMDU) for U.S. Bureau of Reclamation Water Resources Planning*. Available at: https://www.rand.org/pubs/tools/TL320/tool.html (accessed 20 April 2021).

Read, L. K. & Vogel, R. M. (2015). Reliability, return periods, and risk under nonstationarity. *Water Resources Research 51*. doi:10.1002/2015WR017089.

Read, L. K. & Vogel, R. M. (2016a). Hazard function analysis for flood planning under nonstationarity. *Water Resources Research 52*, 4116–4131. doi:10.1002/2015WR018370.

Read, L. K. & Vogel, R. M. (2016b). Hazard function theory for nonstationary natural hazards. *Natural Hazards Earth Systems Science 16*, 915–925. doi:10.5194/nhess-16-915-2016.

Rogers, P. (1997). Engineering design and uncertainties related to climate change. *Climatic Change 37*, 229–242.

Salas, J. D., Obeysekera, J. & Vogel, R. M. (2018). Techniques for assessing water infrastructure for nonstationary extreme events: a review. *Hydrological Sciences Journal*. doi:10.1080/02626667.2018.1426858.

Singh, V. P. (2017). *Handbook of Applied Hydrology*, 2nd ed. McGraw-Hill Education, New York, NY, USA. ISBN-10: 0071835071.

Singh, V. P., Jain, S. K. & Tyagi, A. (2007). *Risk and Reliability Analysis: A Handbook for Civil and Environmental Engineers*. ASCE Press, Reston, VA, USA. ISBN 978 07844 0891-9.

Söderholm, K., Pihlajamäki, M., Dubrovin, T., Veijalainen, N., Vehviläinen, B. & Marttunen, M. (2018). Collaborative planning in adaptive flood risk management under climate change. *Water Resources Management 32*(4), 1383–1397.

USACE (2019a). *Risk Informed Decision Making for Engineering Work During Planning Studies*. Engineering and Construction Bulletin. No. 2019-3, CECW-EC, March 6.

USACE (2019b). *Risk Assessment for Flood Risk Management Studies*. Regulation No. 1105-2-101, CECW-CP CECW-CE, July 15.

USGCRP (2017). In: *Climate Science Special Report: Fourth National Climate Assessment, Volume I*. Wuebbles, D. J., Fahey, D. W., Hibbard, K. A., Dokken, D. J., Stewart, B. C. & Maycock, T. K. (eds). U.S. Global Change Research Program, Washington, DC, USA, p. 470. doi:10.7930/J0J964J6.

Vogel, R. M. & Castellarin, A. (2016). Risk, reliability, and return periods and hydrologic design. Chapter 78. In: *Handbook of Applied Hydrology*, 2nd ed. Singh, V. P. (ed.). McGraw-Hill Education, New York, NY, pp. 78.1–78.10, p. 1440.

World Energy Council (2015). *The Road to Resilience – Managing and Financing Extreme Weather Risks*. London. Available at: http://www.oliverwyman.com/content/dam/marsh/Documents/PDF/UK-en/TheRoad to Resilience Managing and Financing Extreme Weather Risks-09-2015.pdf.

First received 1 November 2021; accepted in revised form 1 November 2021. Available online 12 November 2021

Water Infrastructure Planning, MDCU doi: 10.2166/wp.2021.269

US Flood risk management in changing times

Gregory B. Baecher* and Gerald E. Galloway

Department of Civil and Environmental Engineering, A. James Clark School of Engineering, University of Maryland, College Park, MD 20472, USA
*Corresponding author. E-mail: gbaecher@umd.edu

GBB, 0000-0002-9571-9282; GEG, 0000-0002-8596-0961

ABSTRACT

The traditional regulatory and policy approach to flood risk in the US has been the optimization of benefits and costs, broadly mandated by federal policy. However, optimization may not be the best approach to flood risk management in light of the deep uncertainties we now face. A more incremental approach using a satisficing strategy may be. Flood risk is a function of the hydrologic factors that produce a hazard and the consequences of the hazard interfacing with the people and property exposed. Regretfully, both hydrologists and climatologists seem unable to provide the clairvoyant guidance needed by the water community facing major decisions on flood risk management in the coming years. As the seminal 'Red Book' noted, two things have become second nature to policy analysts and risk managers: absolute safety is unachievable, and it is necessary to distinguish between science and policy. The forcing elements and largest unknowns in determining risk rest with understanding the hydrologic factors involved in shaping the hazard.

Key words: Benefit–cost, Flood risk, Hydrology, Stationarity, Uncertainty

HIGHLIGHTS

- Flood risk is a function of hazard, exposure, and vulnerability.
- Hydrologists and climatologists seem unable to provide guidance for coming years.
- The US operates with an inconsistent approach to dealing with hydrologic factors.
- Quantitative risk analysis may not be the best approach considering the deep uncertainties such analyses face.
- A more incremental and satisficing one may be.

INTRODUCTION

The seminal 'Red Book', *Risk assessment in the federal government: managing the process* published by the U.S. National Academies[1] (NRC, 1983) warned of two things that have become second nature to most policy analysts and risk managers. These are

- Absolute safety is unachievable, and
- It is necessary to distinguish between science and policy.

Each applies to flood risk management. Critics contend that the results of risk assessment are often seen as scientific findings by regulators and the public, whereas in fact they are based in part on other considerations.

[1] U.S. National Academies of Science, Engineering, and Medicine (NASEM).

The Committee believed that guidelines can lead to risk assessments that clearly delineate the limits of current scientific knowledge and the policy basis for choosing among inference options. 'Flood risk is a function of the hydrologic factors that produce flood waters and the consequences of the flood waters interfacing with the people and property exposed to the hazard' (Figure 1). With the extent of exposure to flood hazard, the composition of the communities affected (vulnerability), and steps taken to mitigate the hazard, the forcing elements and largest unknowns in determining risk rest with understanding the hydrologic factors involved in shaping the hazard (Kron *et al.*, 2019).

Hydrology has likely been with us since the beginnings of civilization, probably even before. Hydrological science is more recent, but still mature. Yet, our ability to characterize hydrological risks remains imperfect, even in that stationary world in which few hydrologists or risk analysts still believe (Galloway, 2011). Some hydrologists are skeptical that current data yet indicate significant changes in river parameters. Climatologists,

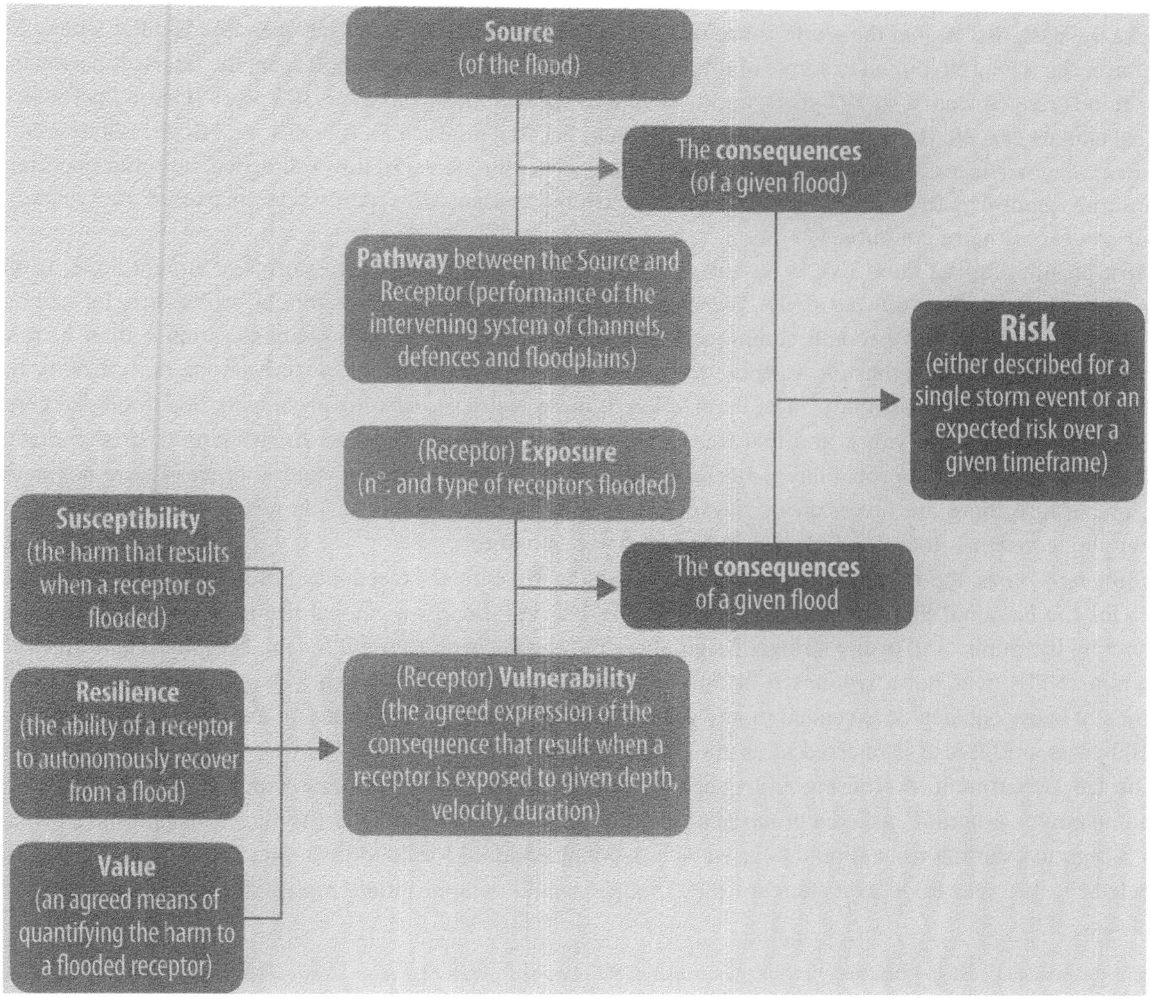

Fig. 1 | The components of risk from *Flood Risk Management: A Strategic Approach* (Sayers *et al.*, 2013).

however, predict a climate change shift away from current conditions to a more turbulent flood and drought future.

Regretfully, both hydrologists and climatologists seem unable to provide guidance to the water community facing major decisions on flood risk management for coming years. This is no one's fault; the challenge is hard. Nonetheless, there is a need for the hydrology and meteorology communities to provide analysts and officials the information they require now to operate today's water projects and to plan and design for those of tomorrow. A revised way of thinking about flood risk and hydrologic factors that influence it – a different paradigm – may be called for.

FACTORS TO CONSIDER

Flood frequency is the number of times a flood above a given discharge or stage is likely to occur over a given number of years (WMO, 1992).

Nearly, all analyses of historical flood information to estimate future flood risk assume that future climate will be like the past, that is, that the world is stationary. Even so, observed intervals between floods differ widely from their average. The well-known example of this is Passau where during the past 500 years the intervals between 50-year floods ranged from 4 to 192 years, and between 100-year floods from 37 to 192 years (Figure 2). These historical climate variations are unpredictable even at multi-decadal time scales. A home or a bridge built to serve for the next 50 years faces a risk that a new phase of long-term climate variability will arrive, and that uncertainty cannot be quantified from short historical records. Unlike Passau, US records seldom exceed 100 years, and 50 or so years is more common.

Most people think of flood risk as having to do with the frequency of severe storms or stream flows, leading to notions such as the *100-year flood*. We measure these frequencies by an annual probability, for example, the $p = 0.01$ *flood*. It has become common to talk of flood recurrence in probabilities rather than in return periods, but one suspects that most people feel more comfort with the latter, as few among our citizenry have ever taken a course in probability[2]. The implication of these metrics is that the probability of a flood of a certain size has some particular value, and that this frequency is a property of nature. In risk jargon, it is an *aleatory* uncertainty. Granted, the probability may change with time either because the natural processes are not stationary, or because there are upstream or site-specific anthropogenic changes in the watershed. Nonetheless, the underlying concept is that flood frequency is a property of nature.

In the late 1960s, the US government decided to use the 1% annual exceedance probability (AEP) flood as the basis for the National Flood Insurance Program. The 1% AEP flood was thought to be a fair balance between protecting the public and overly stringent regulation, but the decision to use it for the National Flood Insurance Program (NFIP) was not a risk-based or an economic calculation. Because the 1% AEP flood has a 1-in-100 chance of being equaled or exceeded in any year, a flood that large or larger has an average recurrence interval of 100 years, and it also is referred to as the *100-year flood*.

The US Department of Housing and Urban Development (HUD), which was responsible until 1979 for the flood insurance program[3], asked a group of academic experts, headed by Gilbert White to meet at the University of Chicago to examine what should be used as a national standard for flood insurance requirements. The group selected the 100-year flood as a standard they thought would be appropriate representing a balance between a

[2] There is substantial evidence that many of the public understand odds and pictures far more clearly than they do either probabilities or percentage frequencies (CDC, 2018).

[3] The Federal Emergency Management Agency (FEMA) assumed these responsibilities from HUD.

Fig. 2 | High-water scale (1501–2002) taken in 2012 at the tower of the Altes Rathaus, Passau, Germany, as of September 2012 (Eychaner, 2015). Reproduced with permission of the Association of State Flood Plain Managers.

flood height that would prevent substantial damages and a standard that would be economically affordable. One attendee, John Sheaffer, a former student of Gilbert White, indicated that:

> *'[...] the discussions focused on the selection of a flood frequency determination that would delineate the regulatory area. The 100-year flood emerged as the flood to be used in the flood insurance program. There was no data on 100-year floods, but it was stated that it had 'a nice sound' to it and would give an allusion [sic] of safety. As a result, all of the definitions developed at the Seminar, e.g., floodway, flood fringe, and regulatory area, were based on the 100-year flood. Also, the communities were to formulate land use regulations for the 100-year flood plain. Each locality was to regulate the extension of public facilities such as streets, sewers, gas, electricity and water in the 100-year floodplain regulatory area'* (Sheaffer, 2004).

Floods of other average recurrence intervals are defined in the same way: a 50-year flood has 1 chance in 50 (2%) of being exceeded in any 1 year, a 200-year flood has 1 chance in 200 (0.5%), and so on. The chance of an X-year flood occurring in a single year is 1 divided by X. More recently, people are talking about larger floods, such as the 500-year flood, as the citizenry's tolerance for risk has diminished and exposures of land use and development have increased. The '500-year flood' corresponds to an AEP of 0.2%, which means a flood of that size or greater has a 0.2% chance in a given year. In 2007, the California Legislature set the acceptable risk criterion on urban levees in the Central Valley of that state at a 200-year return period (SB5, 2007).

In earlier times, other standards were used. Prior to the NFIP, the Tennessee Valley Authority (TVA) used the *maximum probable flood*, and the US Army Corps of Engineers (USACE) used the *standard project flood (SPF)*. Neither of these was a probabilistic concept, and neither was particularly replicable by different engineers. According to Robinson (2004), other standards were also in use during this period. The Connecticut Resources Commission adopted 5–7 times the mean annual flood, equating to a 35- to 150-year level. The reason for this standard was given to be that there was no uniform method for determining flood frequencies. The Soil Conservation Service (SCS) used the 25-year flood in rural areas and the 100-year flood in urban areas. The USGS at the time avoided the 100-year flood because most basins had much shorter record lengths. Nonetheless, by the 1970s, federal agencies began agreeing on the 100-year flood as the standard for floodplain management that would be required by FEMA and formalized with the NFIP. As far as one can tell, this criterion was not based on economic analysis. USACE continued to apply project by project criteria including use of the SPF in their designs.

Famously, the Netherlands Ministry of Infrastructure and Water Management (Rijkswaterstaat) adopted a 10,000-year standard for its coastal flood defenses along the North Sea and a 2,000-year standard for its Rhine River levees, although their calculation of return periods differ somewhat from US practice. Unlike other regions of the world – specifically coastal America – the additional meters elevation to provide this added protection along the Dutch coast are by comparison few. The derivative of elevation to AEP is larger in the Netherlands, so extra protection is comparatively less expensive (Jorrison *et al.*, 2016)[4].

The Dutch example illustrates the point that setting risk standards is a matter of trade-offs. If the cost of increased safety is small, then why not tighten the safety standard? If the cost is large, then one has to think more carefully about what is being bought and at what cost. This is the case in New Orleans where the derivative of elevation with risk reduction is larger than in the Netherlands (i.e., the levees have to be raised a great deal higher and at much greater cost to achieve even a 1,000-year elevation). Everything depends on trade-offs.

Risk analysts separate uncertainties into at least two types: *aleatory* due to randomness in nature, and *epistemic* due to the lack of knowledge. Aleatory uncertainties are properties of nature and are irreducible, but epistemic uncertainties are properties of the mind and are, in principle, reducible to zero with enough information. The common notion in engineering and planning is that flood frequencies are aleatory uncertainties. Actually, this is mostly a convenient assumption of the way we model floods. Pierre-Simon Laplace, the great Enlightenment probabilist, for example, thought there was no such thing as randomness in nature and that all uncertainty was epistemic (Laplace, 1814). From the view of risk analysis and planning, the uncertainties of flood discharge or river height are not simple frequencies of nature but more complicated probabilities of how we know things.

From a purely statistical view, the limited historical record of floods means that the sampling error in flood discharge and river height is reasonably large. Along the Rhine or Danube that history may be as much as 500 years – on the Nile at Cairo it is perhaps 1,400 years – but for North American streams, the length of record is much shorter. This means that the aleatory natural process of rainfall, runoff, and streamflow is supplemented by epistemic uncertainty in the parameters of the frequency distributions describing those natural processes.

In the early 1990s, USACE, at its Hydraulic Engineering Center in Davis, California, began developing a 'risk & uncertainty' approach to flood hazard damage reduction studies (USACE, 2008). This resulted in a risk analysis

[4] In late 2016, the Dutch Parliament amended the national water law to include a risk-based standard of 1:100,000 years with a goal of reaching that level by 2050.

methodology for dimensioning flood levee heights that incorporates uncertainty due to both the expected AEP of river discharge (i.e., the best estimate of the aleatory frequency of flooding) and to the parameter uncertainty in those exceedance probabilities (i.e., the epistemic or statistical uncertainty due to limited data). The distinction between these two uncertainties is seen in Figure 2. The standard deviations referred to in the figure are those involving limited numbers of data from which to estimate flood frequencies.

How should these two sorts of uncertainty be combined? USACE adopted the approach of using the expected (aleatory) flood discharge with some return period, say 100 years, and adding to it some fraction of the epistemic uncertainty. This results in what has been called, a 'conditional non-exceedance probability' flood. The percentage is usually taken to be 95%. For example, in Figure 3, the best estimate of the discharge with a 100-year return period is about 6,000 cfs (170 m^3/s). The mean plus two standard deviations of parameter uncertainty is about 12,000 cfs (340 m^3/s). Presuming for illustration that the uncertainty is Normally distributed, plus two standard deviations above the mean has about a 98% chance of not being exceeded. So, one would say that the '100-year flood with a conditional non-exceedance probability of 98%' is 12,000 cfs (340 m^3/s). Many people find this confusing.

An alternative way of combining aleatory and epistemic uncertainty is by convolving them together to get a single probability. This is sometimes called an *expected probability* (NRC, 2000) or the *Bayesian predictive probability* (Mosleh & Bier, 1996), but it is just a simple probabilistic calculation, presuming the two types of uncertainty are interchangeable. Not all risk analysts accept this presumption.

The distinction between aleatory and epistemic uncertainty becomes especially important when considering the fragility term in the risk equation. Fragility describes the response of protective works to water loads, for example, levee performance. The geotechnical probabilities inherent in forecasting the performance of protective works seldom have to do with aleatory uncertainties, that is, natural frequencies in space or time. They have to do with the characterizations of geotechnical or constructed systems, with uncertainties about engineering parameters, as-built conditions, and models. That is, the uncertainties underlying fragility curves are mostly epistemic. These uncertainties are not describable as annualized values. If a levee is weak, it is not weak on so

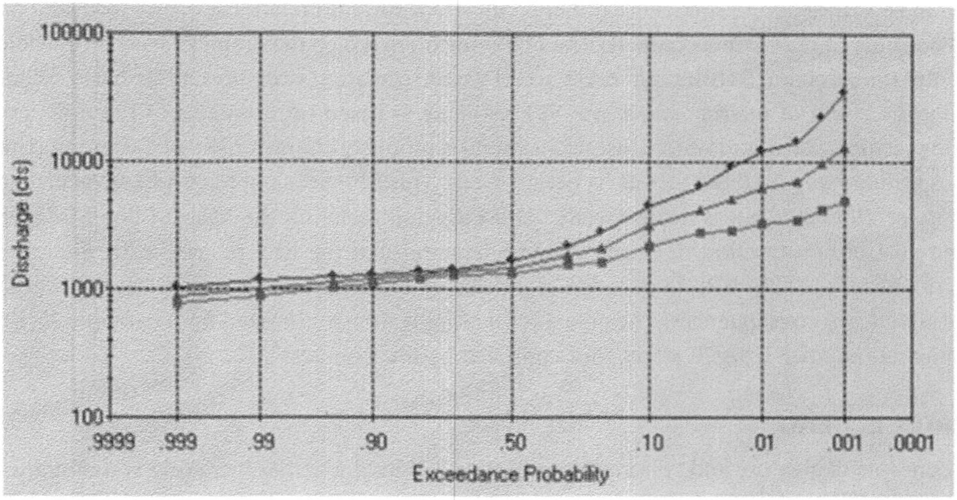

Fig. 3 | Exceedance-probability function and error limit values for the South Fork Bear Creek (NRC, 2000). The middle (red) curve is the expected value, whereas the upper (blue) and lower (green) curves are plus and minus two standard deviations, respectively. Please refer to the online version of this paper to see this figure in color: https://doi.org/10.2166/wp.2021.269.

many days per year; it is either weak or it is not, one just does not know which. The annualized probability only arises in the hazard curve.

As noted, TVA at one time used a *maximum probable flood* and USACE used a *standard project flood*, neither of which was probabilistic. These were point-estimated, deterministic flood heights used for dimensioning flood protection (aka, flood risk management) structures.

A widely used concept in the US is the *Probable Maximum Flood* (PMF). The PMF is the most severe possible flood in a particular drainage area. It is calculated by combining information about the precipitation (PMP), geography, and water management strategies of a particular area. According to the Federal Energy Regulatory Commission (FERC, 2001),

'The Probable Maximum Flood (PMF) is the flood that may be expected from the most severe combination of critical meteorological and hydrologic conditions that are reasonably possible in a particular drainage area'.

'[The] Probable Maximum Precipitation (PMP) is the greatest depth of precipitation for a given duration that is theoretically physically possible over a particular size storm area at a particular geographical location at a particular time of year'.

The SPF is the volume of streamflow expected to result from the most severe combination of meteorological and hydrologic conditions which are reasonably characteristic of the geographic region involved, excluding extremely rare combinations. It is considered to be in the range of half of the flow of the PMF.

For the Mississippi River, a *project specific* or *design flood* is used by USACE to aid in the design and execution of flood protection in the Mississippi Valley. The current project design flood, developed between 1954 and 1955 resulted from a thorough and cooperative effort by the Weather Bureau, USACE, and the Mississippi River Commission that incorporated previously unavailable data regarding the sequence, severity, and distribution of past major storms and investigated 35 different hypothetical combinations of actual storms that produced significant amounts of precipitation and runoff (Mississippi River Commission, 2007). The *standard project storm* estimate for a particular drainage area and season of the year in which snow melt is not a major consideration intends to represent the most severe flood-producing rainfall depth–area–duration relationship and isohyetal pattern of any storm that is considered reasonably characteristic of the region in which the drainage basin is located, giving consideration to the runoff characteristics and existence of water regulation structures in the basin. Where floods are predominantly the result of melting snow, the SPF estimate is based on estimates of the most critical combinations of snow, temperature, and water losses considered reasonably characteristic of the region (USACE, 1965).

USACE has traditionally used the SPF as its design event. This SPF does not have a fixed recurrence frequency such 100 years or 200 years but is a site-specific determination made on the basis of flood frequency, damage potential, and cost of construction. It is generally understood that the SPF ranges in the vicinity of a 200- to 500-year event (FEMA, 2020). SPF is a design storm based on lesser storm than PMF but where such flood would create significant consequences. The actual level of mitigation authorized for a dam or a levee is proposed by the Administration, after USACE study, and approved by the Congress[5].

STATIONARITY IS DEAD

Beyond the concepts of aleatory and epistemic uncertainty is that of whether the world is stationary. Can we rely on historical frequencies to forecast those of the future? Has the world ever been stationary? That is perhaps both

[5] Agricultural areas may receive less protection.

a scientific and a philosophical question. The pronouncement in 2008 that 'stationarity was dead' (Milly *et al.*, 2008) coupled with an increasing number of large storm events and growing recognition of the impacts of climate change, brought into question continuing reliance on the use of stationarity in assessing flood risk[6].

Predicting future flood risk from past floods assumes that the future will be like the past. Yet most geo-scientists probably assume that climate variability and change will bring a future unlike the past (IPCC, 2012). Historical data only provide estimates of historical recurrences. The extent to which flood risk will change in the future is unknown with certainty. Some argue that the end certainty is unquantifiable; others might argue that the end certainty is at least difficult to quantify.

In the world of risk analysis, such uncertainties are called 'deep' (Cox, 2012). Simply, this means that the relationship between alternatives and their consequences is unknown even to a probabilistic description. Some deep uncertainties are of the type that Secretary of Defense Rumsfeld called 'unknown-unknowns' (USDOD, 2002), or others have called 'black swans' (Aven, 2014).

In the realm of flood risk, these deep uncertainties are perhaps less enigmatic, but still difficult to forecast in probabilities. One sees this manifest in the reports of the IPCC in which projections are treated as scenarios rather than as probabilities. That is not to say that projections of climate change, increasing flood risk, or sea-level rise cannot be made probabilistically, but only that many scientists and politically prominent organizations seem reluctant to put probabilities to them. Given this widespread reluctance, there may be an implicit concurrence that quantitative risk analysis itself may not be the best approach to flood risk management in light of these deep uncertainties.

ECONOMIC IMPACT VS. SAFETY

In the US, flood insurance was established by Congress through the National Flood Insurance Act of 1968. A series of subsequent amendments to the Act made insurance mandatory for properties in identified high-hazard or 'A' zones (100-year flood plains) financed by federally backed mortgages. These amendments also regulated new construction in high hazard zones. When catastrophic floods cause damages beyond those expected and in other than high hazard zones, low interest loans from the federal government may be available for repair and recovery (*floodsmart.gov*).

Two kinds of impacts and two kinds of consequences are important here: economics and safety. Economics has to do with financial losses that the catastrophe causes. These include damages to structures, business interruption, and things that cost money. Safety has to do with morbidity and mortality losses. The more relevance of the latter is the number of lives lost. The NFIP deals only with the former. Until the calamity that was Hurricane Katrina in New Orleans, USACE also only considered economic losses of floods. Since then, and in concordance with the US Bureau of Reclamation practice since the failure of Teton dam in 1976, potential loss of life has become a dominating criterion for dam and levee safety, and increasingly has become of concern for flood risk reduction projects (Figure 4).

Economic losses in the US water resource project history have traditionally been treated using benefit–cost analysis (BCA). In this approach, consequences even if not obviously economic are nonetheless monetized to provide commensurate metrics. The discounted sum of monetized project benefits is compared to the discounted sum of monetized project costs to generate a proportionality index. If this index is sufficiently larger than 1.0, the benefits are said to outweigh the costs and the project is deemed worthy of moving forward. For many project

[6] The 2008 paper by Milly *et al.* was followed by another paper in 2015 (Milly *et al.*, 2015) indicating that the problem was far from being solved.

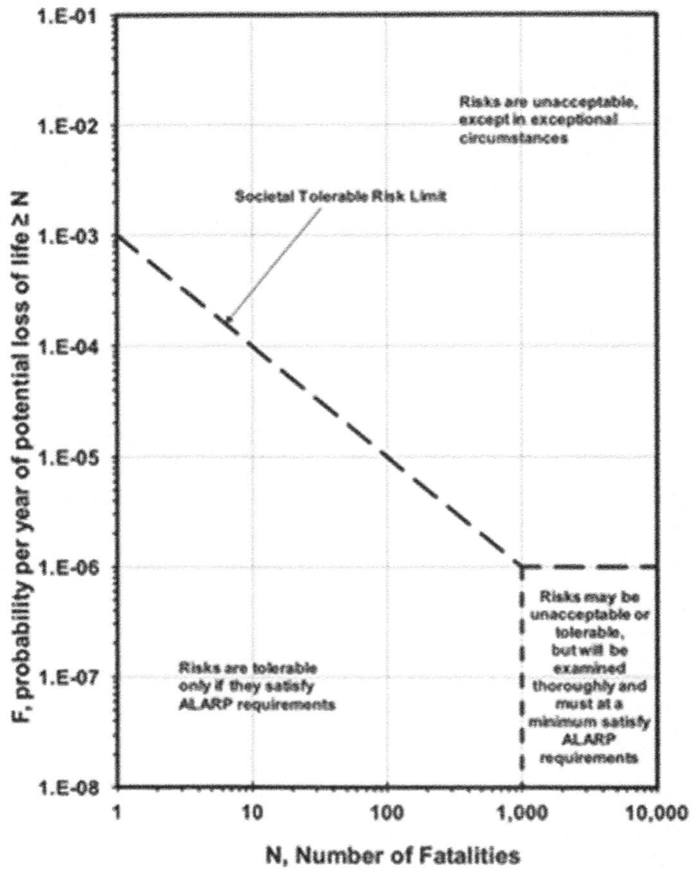

Fig. 4 | USACE societal risk guidelines for existing dams (Reclamation & USACE, 2015). ALARP means, 'as low as reasonably practicable'.

benefits and costs, there exist markets from which prices can be taken. For those for which markets do not exist or are difficult to identify, then non-market proxy methods like surveys are sometimes used (NRC, 2004b).

The willingness-to-pay (WTP) for some benefit or to avoid some risk is a common economic measure used in monetizing non-commensurate benefits and costs. The WTP is the maximum price at which a consumer will buy a unit of some product or service. There are a variety of indirect means for attempting to measure WTP and as might be expected there is always some uncertainty about the values inferred.

One approach to monetizing safety impacts is to attempt to quantify the WTP for marginal reductions in the risk to life. This is sometimes called the *value of a statistical life* (VSL). Engineers and planners are usually reluctant to quantify the proverbial 'value of life', and the VSL attempts to sidestep the issue by valuing risk reduction at the margin (Viscusi, 1993). This VSL can be inferred from decisions made by citizens in life-threatening situations, or by direct questioning. This is an approach common in regulatory use, such as the US Environmental Protection Agency (US EPA, 2014) or the US Department of Transportation (USDOT, 2012). It is not common in flood risk management or other water resource infrastructure decision making.

In dam and levee safety, several major federal agencies in flood management (USACE, Reclamation, FERC, FEMA) have adopted a 'tolerable risk criteria' approach pioneered by the UK Health and Safety Executive for nuclear power (HSE, 2001; Figure 3). The history of this approach is summarized by Baecher *et al.* (2015). In essence, tolerable risk criteria attempt to establish the equivalent of standards for risk to life safety. They do not use trade-offs or economic considerations. Such criteria are, for the most part, not based on demonstrated social preferences but on the judgments of engineers and planners. Social perceptions of risk to life are known to be influenced by a variety of factors – voluntary vs. involuntary, dread vs. usual, occupational vs. non-occupational, public vs. private, and so on – few of which are reflected in the tolerable risk curves in water resource practice (Starr, 1969; Slovic, 1987).

MUDDYING THE WATERS: 'SATISFACTORY' ALTERNATIVES

In the recent past, flood risk management has relied upon either statistical analysis of historical flood frequencies or on deterministic bounding concepts such as the PMF or design flood. In an age of non-stationarity and evolving climate neither of these seems to seize the current challenge: how to make rational decisions in this age of change and indefiniteness?

The traditional flood risk management approach, faced either from the point of view of an economist or an engineer, is optimization. How can we best balance the probabilities of adverse natural events against the associated costs of protection and consequence? This has been the standard rational utilitarian approach since the Victorian age. But does it work in an age of vagueness? Perhaps not.

It seems that what is needed is more akin to the *muddling through* approach of Charles Lindblom (1959), something that Herbert Simon (1956) would have called *satisficing*. Rather than seeking optimal solutions, perhaps our planning objective should be to find incremental, satisfactory solutions which are resilient against the vagaries of future climate, future land uses and development, and other factors which affect risk reduction and the return on investment in flood protection measures. Satisfactory plans lack the attractive property of optimization but benefit from robustness. The question is, what does it mean for a flood protection measure to be, 'satisfactory' or 'good enough' given present knowledge? That is a less obvious question to answer than is that of optimizing the trade-off between risk and cost.

This might be an approach to which Gilbert White (1961) could be sympathetic. He proposed thinking about natural resource decisions in the sense of satisficing. He cited March and Simon's (1958) book on decision making in organizations (1958) as well as Firey's, *Man, Mind, and Land* (1960) who present concepts along this line. Kates (1971) reprises White's thinking in the context of natural hazards and risk, as does the National Academy (NRC, 2004a) in its report on *Adaptive Management for Water Resources Project Planning*.

A more adaptive approach to flood risk management need to postulate no 'utility function' for the organization or citizenry, or does it require an elaborate procedure for calculating marginal rates of substitution among different costs and benefits. Natural hazard problems like flood risk are characterized by computational intractability and inadequate information, both of which inhibit the use of mathematical optimization. Simon observed in his Nobel Lecture that 'decision makers can satisfice either by finding optimum solutions for a simplified world, or by finding satisfactory solutions for a realistic world. Neither approach, in general, dominates the other and both have continued to co-exist in the world of management science'. This is similar to the approach adopted by USACE is the aftermath of Hurricane Katrina (Figure 5) called 'stepping down risk'.

To the extent we are still trying to maximize something, it is our confidence of a good enough outcome even if things go poorly. There is no particular reason to assume that the solution that is best in a utility calculation is also the solution that is most robust to error in the data and assumptions underlying that calculation.

Our uncertainty in today's flood risk management is more extreme than the uncertainty when rolling dice. Knight (1921) distinguished between probabilistic risk which can be measured and insured against, and the non-probabilistic 'true uncertainty', which is the source of entrepreneurial profit (and loss) in the natural

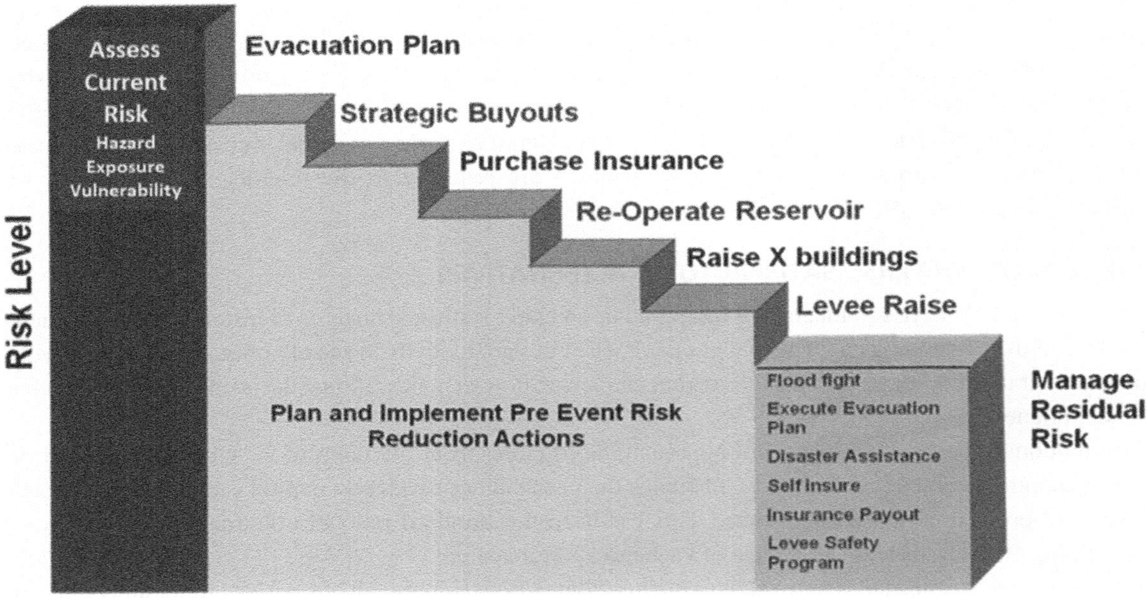

Fig. 5 | Integrated flood risk management from a satisficing approach (*Courtesy*: USACE).

world. Here is an alternative to utility maximization: it is maximizing robustness to uncertainty of a satisfactory outcome. *Robust satisficing* is appropriate when probabilities are known only vaguely. The maximizer of utility seeks the answer to a single question: which option provides the highest benefit–cost. The robust satisficer seeks the answer to two questions: what is a 'good enough' outcome; and of the options that produce a good enough outcome, which one will be robust to the widest range of possible futures?

THE PATH FORWARD

For the first two centuries of the nation's settlement of floodplains, the determination of the flood risk to be tolerated (level of protection to be provided or required) was based on the judgment of local officials responsible for carrying out the construction or controlling the development of the landscape. Decisions were made based on the knowledge available at the site and were tied to both engineering and political factors as well as statistics – what was known about the floods that had occurred and might occur in the future, the landscape, and the necessity for flood mitigation for social or economic purposes.

The 20th century brought more science into these judgments and saw the development of mathematically based methodologies for determining the recurrence interval of flood events and the use of climatology and meteorology to determine the characteristics of large floods. By the end of the century, both approaches were in play with the National Flood Insurance Program focusing its attention on the 100-year flood event as an insurance and floodplain management standard, and USACE developing risk standards relying on a combination of climatology, meteorology, and statistics.

After major climate-related disasters in the first decades of the 21st century, President Barack Obama proposed that all flood risk reduction projects undertaken by the federal government should require efforts to be made to accommodate future changes in the flood levels that might occur under climate change and physical changes in the areas being considered for projects. The federal government examined what standards should be used to prevent building new or rebuilding damaged projects whose characteristics had or would have already proven inadequate in dealing with the 21st century flood events for support of new federal projects and the repair of those damaged by floods. 'The government concludes that, in the face of the conditions previously mentioned, efforts needed to be made to accommodate the seemingly inevitable rise in flood elevations'.

In 2015, President Obama issued Executive Order 13690 promulgating a federal flood risk management standard (FFRMS) to ensure that agencies expand management from the current base flood level to a higher vertical elevation and corresponding horizontal floodplain to address current and future flood risk and ensure that projects funded with taxpayer dollars last as long as intended. Under the new standards, projects being supported by the federal government, agencies would have followed one of three options designed to adapt to future conditions:

1. Utilizing best-available, actionable data and methods that integrate current and future changes in flooding based on science.
2. Two or three feet of elevation (depending on criticality), above the 100-year, or 1%-annual-chance, flood elevation; or
3. 500-year, or 0.2%-annual-chance, flood elevation.

The FFRMS would not apply beyond projects supported by the federal government; however, the presence of such an approach would influence local and state agencies.

During the period that federal agencies were putting the standard into use, President Trump was elected and rescinded the Executive Order 13690. While there may not be a federal standard, much of the business world has been carefully examining the FFRMS and voluntarily including elements of it in their plans. In late 2019, the governor of the state of Virginia, facing significant sea-level rise along the Virginia coast and riverine climate change impacts throughout the state, issued State Executive Order 45: Floodplain Management Requirements and Planning Standards for State Agencies, Institutions, and Property, which mirrors much of the thrust of the FFRMS.

In addition, the Congress, through the 2018 National Defense Authorization Act in a section, Flood Risk Disclosure for Military Construction, directed the Secretary of Defense to require the secretaries of the military departments to:

> *'[...] when mitigating the flood risk of a major or minor military construction project within or partially within the 100-year floodplain, the Secretary concerned shall require any mitigation plan to assume an additional, two feet above the base flood elevation for non-mission critical buildings, as determined by the Secretary; and three feet above the base flood elevation for mission-critical buildings, as determined by the Secretary'.*

On January 20, 2021, on his first day in office, President Joe Biden issued Executive Order 13990, Protecting Public Health and the Environment and Restoring Science To Tackle the Climate Crisis, which among other things rescinded Executive Order 13807, August 15, 2017 (Establishing Discipline and Accountability in the Environmental Review and Permitting Process for Infrastructure Projects), which rescinded the previously issued Executive Order establishing the FFRMS and efforts are underway in federal agencies to establish agency guidelines for implementation of the standard.

Even within the *federal* role, the US operates with an inconsistent approach to flood risk. There is no consistent approach across federal agencies or among states and localities. No federal agency exercises direction or even oversight as the world moves into a rapidly changing climate and weather situation. As noted, optimizing risk may not be the best approach to flood risk management in light of the deep uncertainties we now face, but a more incremental and satisficing one may be.

FUNDING

Funding for this research was provided by the Water Center of the University of Maryland, College Park.

DATA AVAILABILITY STATEMENT

All relevant data are included in the paper or its Supplementary Information.

REFERENCES

Aven, T. (2014). *Risk, Surprises and Black Swans: Fundamental Ideas and Concepts in Risk Assessment and Risk Management*. Routledge, Abingdon, Oxon. New York, NY.

Baecher, G. B., Abedinisohi, F. & Patev, R. C. (2015). *Societal Risk Criteria for Loss of Life Concepts, History, and Mathematics*. University of Maryland, College Park, MD.

CDC (2018). *If the Material Uses Numeric Probability to Describe Risk, Is the Probability also Explained with Words or a Visual? | The CDC Clear Communication Index | Centers for Disease Control and Prevention*. Center for Disease Control. Available at: https://www.cdc.gov/ccindex/tool/page-20.html (accessed January 11, 2020).

Cox, L. A. (2012). Confronting deep uncertainties in risk analysis. *Risk Analysis 32*(10), 1607–1629.

Eychaner, J. H. (2015). *Lessons from a 500-Year Record of Flood Elevations*. Technical Report 7, Association of State Floodplain Managers, Madison, p. 25.

FEMA (2020). *Myths and Facts about the National Flood Insurance Program*. Federal Emergency Management Agency. Available at: https://www.fema.gov/news-release/2011/10/13/myths-and-facts-about-national-flood-insurance-program (accessed January 11, 2020).

FERC (2001). Determination of the probable maximum flood (Chapter VIII). In: *Engineering Guidelines for the Evaluation of Hydropower Projects*. Prepared by the Office of Energy Projects (OEP). United States Federal Energy Regulatory Commission, Washington DC.

Firey, W. I. (1960). *Man, Mind, and Land: A Theory of Resource Use*. Free Press, Glencoe, Ill.

Galloway, G. E. (2011). If stationarity is dead, what do we do now? *JAWRA Journal of the American Water Resources Association 47*(3), 563–570.

HSE (2001). *Reducing Risks, Protecting People – HSE's Decision Making Process*. UK Health and Safety Executive, HMSO, London.

IPCC (2012). *Managing the Risks of Extreme Events and Disasters to Advance Climate Change Adaption: Special Report of the Intergovernmental Panel on Climate Change*. C. B. Field and Intergovernmental Panel on Climate Change (eds.). Cambridge University Press, New York, NY.

Jorissen, R., Kraaij, E. & Tromp, E. (2016). Dutch flood protection policy and measures based on risk assessment. In: *FLOODrisk 2016 – 3rd European Conference on Flood Risk Management*. E3S Web of Conferences 7, 20016.

Kates, R. W. (1971). Natural hazard in human ecological perspective: hypotheses and models. *Economic Geography 47*(3), 438.

Knight, F. H. (1921). *Risk, Uncertainty, and Profit*. Hart, Schaffner & Marx; Houghton Mifflin Company, Boston, MA.

Kron, W., Eichner, J. & Kundzewicz, Z. W. (2019). Reduction of flood risk in Europe - Reflections from a reinsurance perspective. *Journal of Hydrology 576*, 197–209.

Laplace, P. S. (1814). *Philosophical Essay on Probabilities*. Dover Publications, Inc, New York.

Lindblom, C. E. (1959). The science of 'muddling through'. *Public Administration Review 19*(2), 79–88.

March, J. G. & Simon, H. A. (1958). *Organizations*. Wiley, New York.

Milly, P. C. D., Betancourt, J., Falkenmark, M., Hirsch, R. M., Kundzewicz, Z. W., Lettenmaier, D. P. & Stouffer, R. J. (2008). Stationarity is dead: Whither water management? *Science 319*, 573–574.

Milly, P. C. D., Betancourt, J., Falkenmark, M., Hirsch, R. M., Kundzewicz, Z. W., Lettenmaier, D. P., Stouffer, R. J., Dettinger, M. D. & Krysanova, V. (2015). On critiques of 'stationarity is dead': Whither water management? *Water Resources Research 51*(9), 7785–7778.

Mississippi River Commission (MRC) (2007). *The Mississippi River and Tributaries Project: Controlling the Project Flood*. MRC, Vicksburg, MS.

Mosleh, A. & Bier, V. M. (1996). Uncertainty about probability: a recociliation with the subjectivist viewpoint. *IEEE Transactions on Systems, Man, and Cybernetics – Part A: Systems and Humans 26*(3), 1083–1090.

National Research Council (1983). *Risk Assessment in the Federal Government: Managing the Process*. National Academy Press, Washington, DC.
National Research Council (2004a). *Adaptive Management for Water Resources Project Planning*. National Academies Press, Washington, DC.
National Research Council (2004b). *Valuing Ecosystem Services: Toward Better Environmental Decision-Making*. National Academies Press, Washington, DC.
NRC (2000). *Risk Analysis and Uncertainty in Flood Damage Reduction Studies*. National Academies Press, Washington, DC.
Reclamation and USACE (2015). Best Practices and Risk Methodology. US Bureau of Reclamation, Denver, p. 20.
Robinson, M. F. (2004). History of the 1% annual chance flood standard. In: Olinger, L. W., Plasencia, D. & Galloway, G. E. (eds). *Reducing Flood Losses: Is the 1% Chance (100-year) Flood Standard Sufficient?* Association of State Floodplain Managers, Madison, WI, pp. 1–8.
Sayers, P., Li, Y., Galloway, G., Penning-Rowsell, E., Shen, F., Wen, K., Chen, Y. & Le Quesne, T. (2013). *Flood Risk Management: A Strategic Approach*. UNESCO, Paris.
SB5 (Senate Bill 5) State of California (2007). Sacramento.
Sheaffer, J. (2004). The evolution of the 100-Year flood standard. In: Olinger, L. W., Plasencia, D. & Galloway, G. E. (eds). *Reducing Flood Losses: Is the 1% Chance (100-year) Flood Standard Sufficient?* Association of State Floodplain Managers, Madison, WI, pp 9-11.
Simon, H. A. (1956). Rational choice and the structure of the environment. *Psychological Review 63*(2), 129–138.
Slovic, P. (1987). Perception of risk. *Science 236*(4799), 280–285.
Starr, C. (1969). Social benefit versus technological risk. *Science 165*, 1232–1238.
USACE (1965). *Standard Project Flood Determination*. US Army Corps of Engineers (USACE), Washington, DC.
USACE (2008). *HEC-FDA Flood Damage Reduction Analysis: Users Manual CPD-72, Version 1.2.4*. US Army Corps of Engineers, Hydrologic Engineering Center, Davis, CA.
USDOD (2002). *Defense.gov Transcript: DoD News Briefing – Secretary Rumsfeld and Gen. Myers*. DoD News Briefing Secretary Rumsfeld and Gen. Myers, USDOE News Transcript Press Operations. Available at: https://archive.defense.gov/Transcripts/Transcript.aspx?TranscriptID=2636 (June 14, 2019).
USDOT (2012). *Economic Values Used in Analyses*. US Department of Transportation. Available at: https://www.transportation.gov/regulations/economic-values-used-in-analysis (December 10, 2019).
USEPA (2014). *Mortality Risk Valuation*. US EPA, Overviews and Factsheets. Available at: https://www.epa.gov/environmental-economics/mortality-risk-valuation (accessed December 10, 2019).
Viscusi, W. K. (1993). The value of risks to life and health. *Journal of Economic Literature 31*(4), 1912–1946.
White, G. F. (1961). The choice of use in resource management. *Natural Resources Journal 1*(1), 23–40.
WMO (1992). *International Glossary of Hydrology*. World Meteorological Organization, Geneva.

First received 18 November 2021; accepted in revised form 29 November 2021. Available online 1 December 2021

Water Infrastructure Planning, MDCU doi: 10.2166/wp.2021.244

On the need for streamflow drought frequency guidelines in the U.S.

Richard M. Vogel[a,*] and Charles N. Kroll[b]

[a] Department of Civil and Environmental Engineering, Tufts University, Medford, MA 02155, USA
[b] Environmental Resources Engineering, SUNY ESF, Syracuse, NY, USA
*Corresponding author. E-mail: richard.vogel@tufts.edu

ABSTRACT

Extreme drought and resulting low streamflows occur throughout the U.S., causing billions of dollars in annual losses, detrimentally impacting ecosystems, as well as agricultural, hydropower, navigation, water supply, recreation, and a myriad of other water resource systems, leading to reductions in both the effectiveness and resiliency of our water resource infrastructure. Since 1966, with the introduction of Bulletin 13 titled 'Methods of Flow Frequency Analysis', the U.S. adopted uniform guidelines for performing flood flow frequency analysis to ensure and enable all federal agencies concerned with water resource design, planning, and management under flood conditions to obtain sensible, consistent, and reproducible estimators of flood flow statistics. Remarkably, over one-half century later, no uniform national U.S. guidelines for hydrologic drought streamflow frequency analysis exist, and the various assorted guidelines that do exist are not reliable because (1) they are based on methods developed for floods, which are distinctly different than low streamflows and (2) the methods do not take advantage of the myriad of advances in flood and low streamflow frequency analyses over the last 50 years. We provide a justification for the need for developing national guidelines for streamflow drought frequency analysis as an analog to the existing national guidelines for flood frequency analysis. Those guidelines should result in improved water resources design, planning, operations, and management under low streamflow conditions throughout the U.S. and could prove useful elsewhere.

Key words: 7Q10, Baseflow, Hydrologic design, Low flow frequency analysis, Low streamflow, Precipitation

HIGHLIGHTS

- Since 1966, the U.S. has adopted uniform guidelines for performing flood flow frequency analysis.
- Over one-half century later, no analogous uniform national guidelines exist for hydrologic drought streamflow frequency analysis.
- Existing low streamflow frequency methods are based mostly on flood frequency analysis and do not take advantage of the myriad of advances in both areas over the last 50 years.
- Numerous innovations in low flow frequency analysis are described, using analogies to recent advances in flood frequency analysis.
- A justification is provided for the need to develop national guidelines for streamflow drought frequency analysis.
- Such guidelines would enable agencies concerned with water resource design, planning, and management under drought conditions to obtain sensible, consistent, and reproducible estimators of low streamflow statistics.

1. INTRODUCTION AND BACKGROUND

Recently, the U.S. has undergone a series of extreme drought events, where losses to crops, livestock, and the overall gross domestic product are estimated to be in the billions of dollars (e.g., Fannin, 2012; Folger *et al.*, 2012; Howitt *et al.*, 2014; NOAA, 2019). For example, the 2012 drought/heatwave was the most extensive drought to affect the U.S. since the 1930s with over 33.9 billion dollars of damage and 123 deaths (NOAA, 2019). Of all weather-related disasters in the U.S., droughts alone accounted for $144 billion (41.2%) of the estimated $349 billion total cost of all weather-related disasters over the period 1980–2003 (Ross & Lott, 2003). In economic terms alone, droughts are the costliest natural disasters to strike the U.S. (Cook *et al.*, 2007). However, put in context, Stakhiv *et al.* (2016) argued that due to 'continuous adaptation by the various economic sectors associated with agriculture, as well as the growing suite of drought mitigation and compensation policies and programs of various governmental institutions', annual losses due to drought constitute only a miniscule percentage of the overall U.S. economy and only about 5% of agricultural sector production of the U.S. in 2015. Still, on a global scale, of all the 20th century natural hazards, droughts had the greatest detrimental impact (Bruce, 1994; Obasi, 1994), in part, due to the highly interrelated nature of water availability, agriculture, food supply, and the environment. Shifts in climate, water availability, and water demand make the estimation of low-flow statistics even more challenging.

As we describe here, there are many useful analogies between flood and low streamflow frequency analyses within the context of the design, planning, operations, and management of water resource systems. First and foremost, flood and low streamflows are both extreme events, and there may be considerable benefits from drawing upon analogies in flood frequency analysis (FFA) that are useful in drought streamflow frequency analysis. In a recent review article on drought, Brunner *et al.* (2021) suggest that 'droughts and floods should be studied in a joint framework' and that 'uniting droughts and floods in one framework requires stochastic continuous models and an improved joint representation of both types of extremes in hydrological models. This can be obtained by better representing processes important for both types of extremes in model structure and by developing calibration strategies specifically targeting both droughts and floods. Tackling these challenges will allow us to derive more reliable flood and drought predictions and ultimately help to minimize the negative impacts of extreme events.'

Unfortunately, to date, there has been little effort to integrate our knowledge of flood and drought streamflow frequency analyses. Since the introduction of the first U.S. federal guidelines for flood flow frequency analysis in 1966 (Bulletin 13, 'Methods of Flow Frequency Analysis'), there has been a relatively steady transfer and application of improvements and developments in flood flow frequency analysis resulting in subsequent Bulletins 15, 17, 17B, and 17C, so that the existing guidelines for flood frequency analysis in the U.S. (England *et al.*, 2018) represent the current state of the art. Remarkably during that same period of over a half century, there has been no effort analogous to those referenced national guidelines on flood frequency to ensure and enable coordination between all U.S. federal agencies concerned with water resource design, planning, and management under low streamflow conditions to obtain sensible, consistent, and reproducible estimators of low-flow statistics and indices.

Our goal is to document the need for guidelines for drought streamflow frequency analysis in the U.S., analogous to the Bulletin 13 effort, which at the time it was developed in 1966 was simply a description of the current state of the art of FFA within federal agencies, with evaluation and testing of those methods left to subsequent efforts resulting in Bulletins 15, 17, 17B, and 17C. Naturally, any such guidelines recommended here would also prove useful for other countries, regions, and even continents of the world, because as we describe below, the U.S. is not alone on this matter. We begin by providing the necessary introduction and background needed to describe the various aspects associated with the development of national guidelines for low-flow frequency analysis within the context of water resource system design, operations, planning, and management.

1.1. On the need for national standards for flood and drought streamflow frequency analyses

Engineering standards are documents that specify the characteristics and technical details that must be met by the procedures employed by an engineer within the domain that the standards cover. Engineering standards are normally published documents that establish specifications and procedures designed to ensure the reliability, consistency, and effectiveness of the methods applied by engineers. By developing and applying national standards for engineering methods, government agencies can help to ensure that the approaches to be employed are consistent, compatible, safe, and effective.

Many national programs are based on and consist of the accepted and uniform methods for deriving flood and drought streamflow frequencies. For example, the FEMA flood insurance program is based on standard and uniform methods of flood flow frequency analysis outlined in Bulletin 17C (England *et al.*, 2018). Similarly, the U.S. Army Corps of Engineers and the U.S. Bureau of Reclamation employ standardized methods for project justification and selection based on the computation of expected annual damages to determine the design which maximizes net benefits. While some national standards, such as those for FFA (England *et al.*, 2018), have undergone a steady evolution and development over the years in an effort to include the most up to date advances in associated research and technology, other national standards either do not exist, as in the case of low-flow frequency analysis, or are subject to considerable criticism, as in the case of federal water quality standards based on the geometric mean, which have been employed by the U.S. E.P.A. for decades and are subject to continued highly critical concerns (see Vogel (2020) for a summary).

Another important dimension associated with engineering standards relating to droughts involves the implementation of federal disaster assistance programs (U.S. Department of Agriculture, 2017) as well as numerous other federal financial assistance programs for drought emergencies. Numerous federal agencies offer loans or grants to assist with drought-related emergencies, and all such programs require uniform and reliable statistical approaches to evaluate damage costs associated with such drought emergencies. For example, programs for farming and ranching operations are administered by the U.S. Department of Agriculture through the USDA Farm Service Agency, the USDA Rural Development, and the USDA Natural Resources Conservation Service. Similarly, the U.S. Small Business Administration provides loans to businesses, including nonprofit organizations, and the U.S. Bureau of Reclamation offers grants for drought resiliency planning and project implementation.

The development and application of national standards for FFA has enabled federal agencies to adopt and apply consistent, reproducible, sensible, and state-of-the-art methods. Importantly, those same national guidelines and standards for FFA are also normally adopted by state agencies, as well as by engineering practitioners in private practice. Without such standards for either flood or drought streamflow frequency analysis, there is little assurance that results in project justifications, designs, and operations will perform as expected, or that applied standards will be reproducible in practice.

1.2. On the definition of drought

Interestingly, in spite of the extraordinary societal importance of drought impacts, there is no universal definition of drought. Dracup *et al.* (1980), Wilhite & Glantz (1985), Tallaksen & Van Lanen (2004), Mishra & Singh (2010), Sheffield & Wood (2011), Van Loon (2015), and the World Meteorological Organization and Global Water Partnership (2016) provided reviews of the wide variety of drought definitions. Lloyd-Hughes (2014) argued that due to the myriad views of droughts, corresponding to the disparate viewpoints of scientists, water managers, water users, and others, it is unlikely that any single method of assessing and describing drought severity will ever be suitable for all circumstances and users. Lloyd-Hughes (2014) argued that any suitable universal description of drought would require reference to water supply, demand, and management, in addition to the usual focus on meteorological and/or hydrologic conditions.

Mishra & Singh (2010) summarized the commonly employed four categories of drought: hydrological, meteorological, agricultural, and socio-economic, coined earlier by Wilhite & Glantz (1985). Lloyd-Hughes (2014) stressed that the influence of human intervention through water management is intrinsic to hydrological, agricultural, and socio-economic drought, and can only be ignored in the case of purely meteorological drought. We focus on the water management aspects of drought as related to the operations of water resource systems under drought conditions, and thus, our work does not attempt to classify, assess, or define drought. Rather, we are concerned with the development and application of methods for quantifying the frequency and magnitude of low streamflow conditions in rivers and water resource systems under drought conditions.

1.3. On the distinction between drought indices and low streamflow statistics

Mishra & Singh (2010), Heim (2012), Svoboda *et al.* (2015), Bachmair *et al.* (2016), the World Meteorological Organization and Global Water Partnership (2016), and Hasan *et al.* (2019) reviewed the myriad of hydroclimatic indices that have been introduced for classifying, monitoring, assessing, and managing various categories of drought. For example, the U.S. national drought atlas developed by Werick *et al.* (1994a, 1994b) summarized the Palmer Drought Severity Index, which is a meteorological drought index for characterizing the severity of drought. Svoboda *et al.* (2015) reviewed a number of more recent national drought studies. The U.S. Drought Monitor integrates streamflow, precipitation, and additional drought-related information, as well as expert knowledge, in generating maps across the U.S. to characterize current drought conditions (Svoboda *et al.*, 2002). Elsewhere, large-scale national or even multinational drought monitoring and early warning systems are more commonly based on meteorological (precipitation-based) drought indicators (Bachmair *et al.*, 2016).

On the one hand, the large number of available drought indices highlights both the challenge and the importance of developing an objective drought definition. On the other hand, rather than clarifying the definition of drought, the plethora of indices creates further confusion regarding the selection of an appropriate metric for a particular setting. While our discussions are closely related to previous work on drought indices for use in classification, monitoring, and management, our argument appears to be unique due to its focus on the characterization of hydrological drought within the context of low streamflow frequency analysis, for the sole purpose of improving water resource system design, operation, planning, and management activities. Thus, we do not focus on the myriad of hydroclimatic indices useful for describing meteorological, agricultural, and socio-economic drought, and instead focus on the probabilistic behavior of streamflow under drought conditions.

Our focus on drought is quite different from the previously mentioned studies, because we only concentrate upon drought-based statistics computed from low streamflows which are specifically useful for a wide range of water resource design, operations, planning, and management applications. Such streamflow drought statistics are distinct from many of the traditional meteorological drought statistics, because they are useful for evaluating river conditions during drought, and may not be useful for assessing, monitoring, and communicating the severity and duration of droughts, as is the case for other indices such as those reported by Werick *et al.* (1994a, 1994b), Svoboda *et al.* (2015), Bachmair *et al.* (2016), Tijdeman *et al.* (2018), and many others.

1.4. On the importance and need for drought and low-flow statistics

Low streamflow conditions have important societal impacts including, but not limited to, the need for municipal, industrial, agricultural, residential, recreational, and other water use restrictions; reductions in water quality, hydropower generation and power plant cooling efficiency; disruptions to navigation; agricultural/food system losses and stress; harm to drought sensitive species; ecological (instream) flow reductions; increased fire risk; and increased groundwater depletion.

Design, operations, planning, and management for the prevention of hydrologic drought is often accomplished via the computation of various low streamflow statistics, such as the 7-day, 10-year low streamflow (7Q10), or the daily streamflow with an annual exceedance probability of 95% (Q95). Such low-flow statistics are required in a wide range of water resource design, planning, operations, and management settings (Smakhtin, 2001; Ouarda *et al.*, 2008), as well as for the maintenance of environmental flows. Ensuring adequate streamflow under drought conditions is extremely important for protecting a wide range of aquatic species as evidenced from reviews by Dewson *et al.* (2007), Ouarda *et al.* (2008), Hulley *et al.* (2014), and Warren *et al.* (2015).

Hydrologic drought streamflow statistics are used routinely for a wide range of water management activities. For example, low-flow statistics may be used to determine the quantity of water that can be safely withdrawn from a particular stream or reservoir without causing harm to aquatic species, to ensure adequate streamflow at downstream locations, and for preparing drought management plans. They may also be used for water quality purposes, such as to determine design discharges to rivers to ensure acceptable waste-load allocations from point sources (such as for National Pollution Discharge Elimination System permits) and for the development of total maximum daily loads for streams and watersheds.

Our focus on water availability and streamflow under drought conditions is quite different from the numerous previous and ongoing studies that have focused on drought monitoring, modeling, and assessment. Such studies have generally reported meteorological, hydrological, and agricultural drought indices to enable stakeholders to better understand and plan for the socio-economic consequences of drought. Instead, our primary goal is to encourage the exploitation of developments over the past half century in flood and low-flow frequency analysis for the purpose of developing standardized, consistent, and proven approaches for estimating low streamflow statistics for use in water resource design, operations, planning, and management activities across all federal, state, and local agencies within the U.S. It is our hope that such guidelines would find use in many other settings elsewhere around the world.

2. DEVELOPMENT OF GUIDELINES FOR DETERMINING DROUGHT AND LOW-FLOW FREQUENCY

The behavior of extreme low streamflows is governed by completely different processes than that of floods. The dominant processes that govern the behavior of low streamflows include watershed hydrogeology, long-term spatial and temporal climatic variability, the lack of precipitation, and anthropogenic influences relating to both ground and surface water withdrawals and return flows. One expects distinct differences between the frequency analysis of flood flows and low streamflows. Estimators of low streamflow statistics are known to have extremely high relative errors (even higher than for floods) due, in part, to a lack of data, measurement errors, complex hydrologic processes, the inadequate or improper characterization of watershed hydrogeology, and roundoff errors due to USGS reporting standards (Archfield & Vogel, 2009). The development of frequency analysis procedures that are specifically suited for low streamflows is not only needed but also long overdue. The following sections describe the various aspects, developments, and methodologies which form the basis of guidelines for performing streamflow drought frequency analysis (DFA) in the U.S. and elsewhere.

2.1. Existing reviews, guidelines, and software for low-flow frequency analysis in the U.S. and elsewhere at gauged sites

In comparison to guidelines for FFA, there is remarkably little guidance on how to perform low-flow frequency analysis in the U.S. and elsewhere. The only manuals or comprehensive studies we are aware of on low-flow frequency analysis in the U.S. are now largely outdated (Riggs, 1972; ASCE Committee, 1980), although the U.S. Geological Survey maintains an active website titled 'Hydrologic Drought and Low-Flow Frequency' (https://water.usgs.gov/osw/techniques/lowflow.html), which contains various valuable sources of information in the form of memos, training documents, software, and references. Another comprehensive, yet largely outdated manual of practice on hydrologic frequency analysis published by the U.S. Army Corps of Engineers (1993) contains only a single chapter of length 3 pages devoted to low-flow frequency analysis.

Perhaps, the greatest attention to the development of national guidelines for low-flow frequency analysis was developed for the U.K. by Gustard *et al.* (1992). This was a significant effort which began with the comprehensive study by the Institute of Hydrology (1980) along with seven other large low-flow studies summarized in Table 1.3 of Gustard *et al.* (1992) for various regions in the U.K. To date, the most comprehensive set of guidelines on low-flow frequency analysis that we are aware of is the report titled 'Manual on Low-Flow Estimation and Prediction' authored by Alan Gustard and Siegfried Demuth and published by the World Meteorological Organization in 2008. The purpose of that manual was 'to publish state-of-the-art analytical procedures for estimating, predicting and forecasting low river flows at all sites, regardless of the availability of observational data.' The intent of the manual by Gustard & Demuth (2008) was to enable agencies to 'predict and forecast low flows for a wide range of applications, including national and regional water resource planning, abstraction management, public water supply design, instream flow determination, effluent dilution estimates, navigation, the design of run of river hydropower schemes, the design of irrigation schemes, and water resources management during low-flow conditions.' Our goal is to emphasize the need for guidelines analogous to those reported by Gustard & Demuth (2008) which are suited to conditions in the U.S. There are several other manuals of low-flow frequency analysis which are not nearly as comprehensive as Gustard & Demuth (2008) as well as software for their implementation, as is described below in the section on regional methods. Another valuable source of research and guidance on methods for the estimation of low-flow statistics at gauged sites are the reviews by Smakhtin (2001) and Ouarda *et al.* (2008) as well as the summary articles by Burn *et al.* (2008) and Hulley *et al.* (2014) on the behavior and challenges associated with characterizing low flows in Canada.

2.2. Existing software for low-flow frequency analysis

Numerous software packages exist in the U.S. for the estimation of low streamflow statistics including the U.S. Geological Survey software packages known as STREAMSTATS, GLSNET, SWToolBox, SWSTAT, the U.S. E.P.A. program DFLOW, and the U.S. Army Corps of Engineers package HEC-SSP. In addition, numerous software packages exist for other regions of the world, such as the package LFSTAT that implements all of the methods outlined in the manual of practice by Gustard & Demuth (2008).

Regional methods of frequency analysis are now widely used in many regions of the world for both flood discharges and precipitation and are generally preferred over at-site methods (Ouarda, 2017). By comparison to FFA, the use of regional methods in low-flow frequency analysis is in its infancy (Ouarda *et al.*, 2008, 2018). Some examples of readily available software for the implementation of regional low-flow frequency analysis includes the Low Flows 2000 package (Young *et al.*, 2003), which has been adopted by the Scottish Environment Protection Agency for estimating low-flow statistics in ungauged catchments. Low Flows 2000 allows the user to estimate annual and monthly flow duration curves and to carry out a complete regional frequency analysis. Another software package known as REGIONS (Ouarda *et al.*, 2002) represents a general tool for the regional frequency analysis of extreme low and flood events. In the province of Quebec, the software package ARIDE (Ouarda *et al.*, 2005) combines hydrological, statistical, and GIS tools to enable a regional low-flow frequency analysis at any site in the region. Ouarda *et al.* (2018) recommend the use of readily available Generalized Additive Modeling (GAM) software for the implementation of regional low-flow frequency methods. An exhaustive review of the existing software for DFA is needed to determine common and promising features for use in developing guidelines for the U.S. and elsewhere.

While drought streamflow statistics are sorely needed, existing methods for their estimation in the U.S. are still based generally on recommendations for FFA. For instance, the U.S. EPA uses the software program DFLOW (Rossman, 1990; EPA, 2021) to estimate design streamflows for low-flow analysis and water quality standards. In DFLOW, the estimation of the 7Q10 is based on fitting a log Pearson Type III (LP3) distribution with parameters estimated using the at-site method of moments. The same method is used by the USGS software program SWSTAT (https://water.usgs.gov/software/SWSTAT/). Both of these estimation procedures use the LP3 distribution, a distribution recommended for FFA (Rossman, 1990), yet both approaches ignore advances in FFA (e.g., regional estimators of skew, regional hydrologic regression, outlier detection, etc.) that have been recommended in practice for FFA for over 35 years (IAC, 1982), as well as regional methods of FFA based on L-moments (Hosking & Wallis, 2005), which were used in the first national drought atlas (Werick *et al.*, 1994a, 1994b), and most importantly the four recent innovations associated with Bulletin 17C completed in 2018 (Griffis & Stedinger, 2007; England *et al.*, 2018). The four innovations include the adoption of a generalized representation of flood data that includes interval data and the recognition that floods can either be higher or lower than the level of certain infrastructure, such as bridge decks; the use of the expected moments algorithm, which incorporates historical and 'paleoflood' information (data from floods in the distant past); a generalized approach to the identification of low outliers in flood data; and an improved method for computing confidence intervals. A natural next step to Bulletin 17C (England *et al.*, 2018) would be to evaluate the suitability of the recent innovations in that Bulletin, with suitable revisions, for use in the estimation of low streamflow statistics.

2.3. Characterization of the probability distribution of low streamflow

Given the distinct differences in processes leading to extreme low and flood streamflows, one expects their probability distribution to behave differently. Matalas (1963), Vogel & Kroll (1989), Kroll & Vogel (2002), Bhatti *et al.* (2019), and others have provided recommendations for suitable probability distributions for low streamflow series (LSS) in the U.S. Kroll & Vogel (2002) and Bhatti *et al.* (2019) performed national assessments of the probability distribution of LSS across the conterminous U.S. Bhatti *et al.* (2019) found that the log Pearson Type III (LP3) distribution was generally favored for both intermittent and perennial sites, whereas the Pearson Type III (P3) and the 3-parameter lognormal distributions (LN3) also performed well at perennial sites. In addition, Bhatti *et al.* (2019) found that the generalized extreme value (GEV), LP3, and LN3 distributions all performed well for modeling the distribution of the inverse of LSS at perennial sites. Interestingly, using very different goodness-of-fit criteria than either Bhatti *et al.* (2019) or Kroll & Vogel (2002), Matalas (1963) also recommended the P3 distribution for use in models of LSS in the U.S. The findings of Kroll & Vogel (2002) and Bhatti *et al.* (2019) are also consistent with more recent assessments elsewhere. For example, in a national study for the U.K., Zaidman *et al.* (2002) recommended the use of the P3 distribution over all other alternative distributions considered. In a subsequent analysis, Zaidman *et al.* (2003) recommended the use of both the GEV and P3 distributions for low-flow frequency analysis in the U.K. In Canada, Yue & Pilon (2005) recommended the P3 distribution and also identified the GEV and LP3 as acceptable distributions. A thorough analysis of appropriate probability distributions for LSS using the most current data sets and methods is warranted. An alternative to assigning a particular probability distribution to the series of annual minimum d-day LSS is through the use of nonparametric empirical probability distributions. Takeuchi (1988) employed empirical distributions, termed flood and low-flow duration curves, for use in mapping the regional behavior and persistence of both floods and droughts, an alternative approach which may improve low-flow frequency analysis.

2.4. Regional methods of low-flow frequency analysis

Regional statistical methods introduced by Hosking & Wallis (2005) and many others are now in widespread use in flood and precipitation frequency analyses and should be useful for low-flow frequency analysis. This section summarizes numerous innovations in regional FFA, which should prove useful for developing guidelines for drought flow frequency analysis.

2.4.1. Regional estimators of skewness and GEV shape parameter

The use of regional estimators of skewness is commonplace in FFA (Griffis & Stedinger, 2009). While such regional estimators have been shown to improve FFA, their use in DFA has never been explored, and the simpler at-site skew estimators, which generally have poor statistical properties (Stedinger *et al.*, 1993; Ouarda, 2017; Stedinger, 2017), are commonly employed for DFA (Rossman, 1990). Procedures for estimating regional skewness for the LP3 distribution, which are now widely used in FFA, could be adapted for use in DFA for estimating regional skewness for the P3 or LP3 distributions at intermittent sites. Likewise, it is now common practice to employ regional estimators of the shape parameter of the GEV distribution (Hosking & Wallis, 2005), which has found an increasing application in FFA outside the U.S. (see, for example, Tables 3 and 4 in Vogel & Wilson, 1996; Salinas *et al.*, 2014). Regional estimators of the GEV model shape parameter are critical to obtain reliable estimators of flood quantiles (Lima *et al.*, 2016), and we anticipate that similar conclusions should apply to quantiles of LSS at perennial sites where the GEV model has been found to provide a good approximation to LSS (Kroll & Vogel, 2002; Zaidman *et al.*, 2003).

2.4.2. Innovations in the estimation of low-flow statistics at ungauged sites

Estimating low streamflow statistics at ungauged river sites continues to be a very challenging problem as evidenced by workshops and special issues on the subject (see Spence *et al.*, 2008, 2013). It is now common knowledge in FFA that regional methods are more attractive than at-site methods (Hosking & Wallis, 2005). Ouarda *et al.* (2008) provides a brief review of regional methods for low-flow frequency analysis. Regional regression models, perhaps the most common approach for the estimation of low streamflow statistics and flow duration curve quantiles at ungauged river sites, have been shown to perform much worse for low flows than for floods (e.g., Thomas & Benson, 1970; Smakhtin, 2001; Kroll *et al.*, 2004). Vogel & Kroll (1992) document that inclusion of hydrogeologic indices and physically based information in regional regression models for DFA could lead to considerable improvements. Recent advances in hydrologic regionalization for both FFA and DFA should prove useful for improving low streamflow prediction at ungauged sites, including ways to improve hydrogeologic characterization (Kroll *et al.*, 2004; Eng *et al.*, 2011; Stagnitta *et al.*, 2017), the use of the map correlation method (Archfield & Vogel, 2010), methods related to the region of influence (Tasker *et al.*, 1996), and most importantly recent advances in the application of both Generalized Additive Modeling (GAM) (Ouarda *et al.*, 2018) and Bayesian generalized least-squares hydrologic regression methods (Reis *et al.*, 2020).

2.4.3. Index drought analogy to the index flood method

There is a good deal of research that documents that the regional flood frequency method, termed the index flood method introduced by Dalrymple (1960), is one of the most efficient regional estimation methods (Cunnane, 1988; Potter & Lettenmaier, 1990). Stedinger (2017) and Ouarda (2017) review the background, advantages, and implementation of the index flood method based on the method of L-moments. Pilon (1990) extended the index flood method, used in regional FFA, to low-flow frequency analysis when the regional distribution is assumed to be a three-parameter Weibull. The performance of the approach was demonstrated by its application to the southern part of the province of Ontario in Canada. The extension of the index flood method using regional L-moments and the P3 and GEV models to the index drought problem is a natural and straightforward application and could be accomplished with minimal effort. There is clearly a great deal of value to extending the index flood method to create an index drought approach implemented using regional L-moments for use with both the P3 and GEV models for low-flow frequency analysis.

2.4.4. Estimation of low streamflow statistics at partially gauged sites

Previous research has introduced methods for the incorporation of short or spot measurements at the site of interest to improve estimators of low streamflow statistics. Stagnitta *et al.* (2017) indicated that baseflow correlation procedures (Stedinger & Thomas, 1985; Zhang & Kroll, 2007a) produce improved low-flow estimators compared with the use of regional regression when only four baseflow measurements were available. Stagnitta *et al.* (2017) also explored methods for estimating hydrogeologic indices (baseflow recession and aquifer time constants) from spot measurements and their use in regional regression. While this method was never superior to the use of baseflow correlation, it did show promise when compared with regional regression without hydrogeologic indices. Another use of spot measurements involves the estimation of flow duration curves at ungauged sites with the quantile–probability–probability–quantile (QPPQ) daily streamflow simulator (Archfield & Vogel, 2010). It is also possible to transfer information from multiple sites in a region (Zhang & Kroll, 2007b). A summary and evaluation of existing approaches for the integration of spot measurements into low-flow frequency analysis is sorely needed.

3. NONSTATIONARY LOW-FLOW FREQUENCY ANALYSIS

As evidenced from numerous recent review articles, research on approaches to nonstationary FFA is burgeoning due to the need to capture impacts of changes in climate, land-use, water-use, and other anthropogenic impacts on water management (Madsen *et al.*, 2013; Hall *et al.*, 2014; Bayazit, 2015; Salas *et al.*, 2018; Villarini *et al.*, 2018; Francois *et al.*, 2019). National trend studies have found widespread upward trends in annual minimum daily streamflows across broad regions of the U.S. (Lins & Slack, 1999, 2005; Douglas *et al.*, 2000). In fact, those studies found much greater and widespread evidence of increasing trends in annual minimum streamflows than in annual maximum flood flows. In spite of the widespread evidence of the nonstationary behavior of LSS, few U.S. studies exist (see recent discussions by Liu *et al.* (2015); Ahn & Palmer (2016); Kiem *et al.* (2021)). Nonstationary trends in LSS and associated DFA could have a critical impact on water resource management, yet most current low streamflow frequency analysis procedures assume that future conditions will be similar to those observed in the past, a very poor assumption.

Numerous procedures summarized in the above review articles on nonstationary FFA are now available and easily adapted for performing nonstationary low-flow frequency analysis. Promising approaches include a nonparametric approach tested by Blum *et al.* (2019) on LSS and further analyzed by Vogel & Kroll (2020), and a parsimonious nonstationary approach introduced by Serago & Vogel (2018) for FFA, which could be easily adapted for use with LSS. Blum *et al.* (2019) tested a new approach to nonstationary low-flow frequency analysis that applies a stationary nonparametric quantile estimator to the most recent subset of the historical streamflow record, which is expected to exhibit stationary behavior. That approach was found to result in improvements in accuracy and reductions in bias over the standard approach of using the entire historical record. Vogel & Kroll (2020) generalized the results of Blum *et al.* (2019) and derived expressions for computing the optimal length of the recent subset record resulting in minimum variance estimators of the low-flow statistic. An alternative parametric yet parsimonious approach introduced by Serago & Vogel (2018) for nonstationary FFA could easily be adapted for use with LSS and a wide range of possible probability distributions. The approach introduced by Serago & Vogel (2018) is unique because it only requires a single additional model parameter, so that only three or four parameters are needed for a full nonstationary frequency analysis, compared with most of the methods summarized in the above review articles which often require five to eight parameters for nonstationary frequency analysis. Another attractive approach to handling nonstationary behavior is the seasonal stochastic streamflow modeling approach introduced by Kiem *et al.* (2021) for evaluating water supply reliability.

3.1. Impact of stochastic persistence on low-flow frequency analysis

A unique feature of LSS is that, unlike flood series, they tend to exhibit stochastic persistence resulting from groundwater and other storage processes, which have a critical impact on baseflow processes and thus LSS. For example, in a national assessment of the existence of stochastic persistence and deterministic trends on LSS, Douglas *et al.* (2000) found statistically significant serial correlation in annual minimum seven-day low flows at approximately 25% of the 1,474 river stations analyzed across the U.S. Douglas *et al.* (2002) and others have shown that the integration of the serial correlation of LSS is of critical importance when estimating the expected return period associated with streamflow drought events. The effect of stochastic persistence of LSS on engineering design quantiles should, therefore, be integrated into guidelines for low-flow frequency analysis.

Both stationary and nonstationary series of annual minimum low streamflows are likely to exhibit stochastic persistence, and it is very difficult to distinguish between trends and persistence as was so clearly demonstrated by Cohn & Lins (2005). For example, the existence of serial correlation is known to confound our ability to detect trends (Douglas *et al.*, 2000), though numerous procedures are available for removing the impact of serial correlation when performing trend detection. Since we expect LSS to exhibit both trends and persistence, an effort should be made to account for both of these factors in any resulting low streamflow frequency analysis.

3.2. Deterministic watershed models for estimating nonstationary hydrologic drought statistics

Deterministic watershed models (DWMs) provide tremendous promise for simulating future, possibly nonstationary, hydrologic conditions in a watershed due to their ability to account for the integrated impacts of future changes in climate, land use, water withdrawals, and other factors. DWMs provide a very useful tool for modeling future streamflow traces, which can then be subject to low-flow frequency analysis analogous to the way in which DWMs are increasingly being applied for FFA (Blazkova & Beven, 2009; Steinschneider *et al.*, 2014; Francois *et al.*, 2019).

There are dozens of DWMs to choose from as evidenced by the variety of textbooks which summarize numerous software packages for their implementation (Beven, 2000; Wagener *et al.*, 2004; Singh & Frevert, 2006; Singh, 2012) including the highly sophisticated recently released National Water Model (Maidment, 2017) that can provide near real-time, high spatial resolution streamflow forecasts for the entire U.S. Farmer & Vogel (2016) highlight that if a DWM is to be used for estimating either a low-flow statistic, such as the 7Q10 or Q95, or a flood flow statistic, such as the 100-year flood, it is necessary to implement the deterministic model in a stochastic mode to avoid bias in corresponding streamflow statistics. To enable the implementation of DWMs in a stochastic mode, Vogel (2017) advanced a new generation of DWMs, which he terms stochastic watershed models (SWMs), which are useful for generating streamflow ensembles as input to nearly all modern risk-based water resource decision-making approaches. SWMs are simply DWMs implemented using stochastic meteorological series, stochastic model parameters, and stochastic model errors to generate ensembles of streamflow traces that represent the variability in possible future streamflows (see Steinschneider *et al.* (2014) for a good example and Francois *et al.* (2019) for a recent review). Each SWM model trace or ensemble can be used to estimate a LSS that can then be used to obtain estimators of low-flow statistics such as the 7Q10 or Q95. Such an analysis results in an ensemble of estimates of 7Q10 and/or Q95, which can then be used to construct confidence intervals for the corresponding population value of those statistics.

While the use of DWMs for modeling average and high flows is now commonplace, there are fewer examples of studies that have evaluated the ability of DWMs for simulating low streamflow conditions. In a review of the use of DWMs for characterizing low streamflow, Davison & van der Kamp (2008) conclude that in spite of the plethora of DWMs available for simulating low streamflow, there are only a very limited number of studies that have evaluated the ability of DWMs to adequately characterize baseflow processes. Van Loon (2015) reviews numerous studies that have shown that low streamflows are usually poorly captured by DWM simulations, which is no surprise given the numerous challenges associated with modeling low flows summarized by Smakhtin (2001), Ouarda *et al.* (2008), and others.

Several authors have introduced new approaches to evaluate the goodness of fit of DWM simulations when one's primary interest is in the ability of such models to capture the behavior of LSS. For example, Pushpalatha *et al.* (2012) and Garcia *et al.* (2017) have suggested examining the efficiency of the inverse of LSS, rather than computing the efficiency of the streamflows or their natural logarithms as is common practice when evaluating the ability of DWMs for simulating average and flood flows. Farmer *et al.* (2015a) provides examples and reasoning for the importance of examining the goodness of fit of the inverse of daily streamflow when one's interest is in low streamflow. There is a need for more comparative assessments of the ability of various types of models for reproducing the behavior of low streamflows, analogous to the recent study by Farmer *et al.* (2015b).

4. SUMMARY AND CONCLUSIONS

On a global scale, of all the 20th century natural hazards, droughts had the greatest detrimental impact (Bruce, 1994; Obasi, 1994) in part due to the highly interrelated nature of water availability, agriculture, food supply, and the environment. In Section 1.4, we reviewed the myriad of societal consequences that stem from low streamflow conditions that occur during drought events. Shifts and changes in climate, water availability, and water demand will tend to aggravate societal consequences associated with drought and make planning for future low streamflow events even more challenging.

Since the introduction of the first U.S. federal guidelines for FFA in 1966 (Bulletin 13, 'Methods of Flow Frequency Analysis'), there has been a relatively steady transfer and application of improvements and developments in FFA resulting in subsequent Bulletins 15, 17, 17B and 17C, so that the existing guidelines for FFA in the U.S. represent the current state of the art (England *et al.*, 2018). Remarkably during that same period of over a half century, there has been no analogous effort to ensure and enable all federal agencies concerned with water resource design, planning, and management under drought conditions to obtain sensible, consistent, and reproducible estimators of low streamflow statistics. Interestingly, this situation is remarkably similar to the European Union (EU) which has a detailed plan for flood risk management and improving societal resilience, but neglects drought risk management despite high drought damages (Priest *et al.*, 2016). Clearly many of the issues raised in this commentary also pertain to the EU and elsewhere.

Our primary goal is to set the stage to exploit developments over the past half century in flood and low-flow frequency analysis for the purpose of developing standardized, consistent and proven approaches for estimating low streamflow statistics for use in water resource design, operations, planning, and management activities across all federal, state, and local agencies in the U.S. While our discussions are closely related to previous work on drought indices for use in classification, monitoring, and management of drought events, our recommendations distinguish themselves by their focus on the characterization of hydrological drought for the purpose of improving water resource system design, operations, planning, and management activities.

We have outlined numerous potentially fruitful areas of research related to the development of U.S. national guidelines for streamflow DFA including:

1. Review of environmental streamflow requirements and compilation of a list of estimators of low streamflow statistics and their use in ecological and water resource design, operations, planning, and management applications.
2. Review of the existing software for DFA and a gap analysis to identify needed features for use in developing national guidelines for the U.S.

3. Evaluating the suitability of the recent innovations in FFA outlined in Bulletin 17C, with revisions suited to LSS, for use in the estimation of low streamflow statistics at both gauged and ungauged sites.
4. Exploring the use of alternative probability distributions to describe LSS at intermittent and perennial sites with special attention given to modeling of inverse streamflows found to be quite attractive in low-flow frequency analysis; see, for example, Pushpalatha *et al.* (2012), Farmer *et al.* (2015a), Garcia *et al.* (2017), and Bhatti *et al.* (2019).
5. Investigating recent developments in the regional estimation of streamflow statistics and their application to gauged and ungauged sites, and partially gauged sites.
6. Research related to nonstationary low streamflow frequency analysis, with particular emphasis on the impact of changes in climatic, land-use, and water demand conditions.

DATA AVAILABILITY STATEMENT

All relevant data are included in the paper or its Supplementary Information.

REFERENCES

Ahn, K.-H. & Palmer, R. N. (2016). Use of a nonstationary copula to predict future bivariate low flow frequency in the Connecticut River basin. *Hydrological Processes 30*, 3518–3532. doi:10.1002/hyp.10876.

Archfield, S. A. & Vogel, R. M. (2009). The implications of discretizing continuous random variables: an example using the U.S. Geological Survey reporting standards for streamflow data. In: *ASCE-EWRI, Environmental and Water Resources Congress, 2009*. Kansas City.

Archfield, S. A. & Vogel, R. M. (2010). Map correlation method: selection of a reference streamgage to estimate daily streamflow at ungauged catchments. *Water Resources Research 46*. doi: 10.1029/2009WR008481.

ASCE Task Committee on Low-Flow Evaluation, Methods, and Needs of the Committee on Surface-Water Hydrology of the Hydraulics Division (1980). Characteristics of low flows. *ASCE Journal of Hydraulics 106*(HY5), 7.

Bachmair, S., Stahl, K., Collins, K., Hannaford, J., Acreman, M., Svoboda, M., Knutson, C., Smith, K. H., Wall, N., Fuchs, B., Crossman, N. D. & Overton, I. C. (2016). Drought indicators revisited: the need for a wider consideration of environment and society. *Wiley Interdisciplinary Reviews: Water 3*(4), 516–536. https://doi.org/10.1002/wat2.1154.

Bayazit, M. (2015). Nonstationarity of hydrological records and recent trends in trend analysis: a state-of-the-art review. *Environmental Processes 2*, 527–542.

Beven, K. J. (2000). *Rainfall-Runoff Modeling: the Primer*. John Wiley and Sons Ltd, New York.

Bhatti, S. J., Kroll, C. N. & Vogel, R. M. (2019). Another look at the probability distribution of low streamflow series in the US. *Journal of Hydrologic Engineering 24*(10), 04019043.

Blazkova, J. S. & Beven, K. (2009). A limits of acceptability approach to model evaluation and uncertainty estimation in flood frequency estimation by continuous simulation: Skalka Catchment, Czech Republic. *Water Resources Research 45*, W00B16. http://dx.doi.org/10.1029/2007WR006726.

Blum, A. G., Archfield, S. A., Hirsch, R. M., Vogel, R. M, Kiang, J. E. & Dudley, R. W. (2019). Updating estimates of low streamflow statistics to account for possible trends. *Hydrological Sciences Journal*. doi.10.1080/02626667.2019.1655148.

Bruce, J. P. (1994). Natural disaster reduction and global change. *Bulletin of the American Meteorological Society 75*, 1831–1835.

Brunner, M. I., Slater, L., Tallaksen, L. M. & Clark, M. (2021). Challenges in modeling and predicting floods and droughts: a review. *Wiley Interdisciplinary Reviews – Water*. doi: 10.1002/wat2.1520.

Burn, D. H., Buttle, J. M., Caissie, D., MacCulloch, G., Spence, C. & Stahl, K. (2008). The processes, patterns and impacts of low flows across Canada. *Canadian Water Resources Journal 33*(2), 107–124.

Cohn, T. A. & Lins, H. F. (2005). Nature's style: Naturally trendy. *Geophys. Res. Lett. 32*, L23402, doi:10.1029/2005GL024476.

Cook, E. R., Seager, R., Cane, M. A. & Stahle, D. W. (2007). North American drought: reconstructions, causes, and consequences. *Earth-Science Reviews 81*, 93–134.

Cunnane, C. (1988). Methods and merits of regional flood frequency analysis. *Journal of Hydrology 100*(1–3), 269–290.

Dalrymple, T. (1960). Flood frequency methods. *U.S. Geological Survey, WaterSupply Paper 1543A*, 11–51.

Davison, B. & van der Kamp, G. (2008). Low-flows in deterministic modelling: a brief review. *Canadian Water Resources Journal 33*(3), 181–193.

Dewson, Z. S., James, A. B. W. & Death, R. G. (2007). A review of the consequences of decreased flow for instream habitat and macroinvertebrates. *Journal of the North American Benthological Society 26*(3), 401–415.

Douglas, E. M., Vogel, R. M. & Kroll, C. N. (2000). Trends in flood and low flows in the United States. *Journal of Hydrology 240*(1–2), 90–105.

Douglas, E. M., Vogel, R. M. & Kroll, C. N. (2002). Impact of streamflow persistence on hydrologic design. *Journal of Hydrologic Engineering, ASCE 7*(3), 220–227.

Dracup, J. A., Seong, L. K. & Paulson Jr., E. G. (1980). On the definition of droughts. *Water Resources Research 16*, 297–302.

Eng, K., Kiang, J. E., Chen, Y. Y., Carlisle, D. M. & Granato, G. E. (2011). Causes of systematic over- or underestimation of low streamflows by use of index-streamgage approaches in the United States. *Hydrological Processes 25*, 2211–2220.

England Jr., J. F., Cohn, T. A., Faber, B. A., Stedinger, J. R., Thomas Jr., W. O., Veilleux, A. G., Kiang, J. E. & Mason Jr., R. R., (2018). *Guidelines for Determining Flood Flow Frequency – Bulletin 17C: U.S. Geological Survey Techniques and Methods, Book 4, Chap. B5.* p. 148. https://doi.org/10.3133/tm4B5.

Environmental Protection Agency (EPA) (2021). *Environmental Modeling Community of Practice: DFLOW.* Available at: https://www.epa.gov/ceam/dflow (accessed 23 April 2021).

Fannin, B. (2012). *Texas Agricultural Drought Losses Total $7.62 Billion, Texas A&M AgriLife Today.* Available at: http://today.agrilife.org/2012/03/21/updated-2011-texas-agricultural-drought-losses-total-7-62-billion/.

Farmer, W. H. & Vogel, R. M. (2016). On the deterministic and stochastic use of hydrologic models. *Water Resources Research 52*. doi:10.1002/2016WR019129.

Farmer, W. H., Over, T. M. & Vogel, R. M. (2015a). Multiple regression and inverse moments improve the characterization of the spatial scaling behavior of daily streamflows in the Southeast United States. *Water Resources Research 51*. doi:10.1002/2014WR015924.

Farmer, W. H., Archfield, S. A., Over, T. M., Hay, L. E., LaFontaine, J. H. & Kiang, J. E. (2015b). *A Comparison of Methods to Predict Historical Daily Streamflow Time Series in the Southeastern United States, USGS Scientific Investigations Report, 2014-5231.*

Folger, P., Cody, B. A. & Carter, N. T. (2012). Drought in the United States: causes and issues for congress. In: *Congressional Research Service, 7-5700, RL34580.*

Francois, B., Schlef, K. E., Wi, S. & Brown, C. M. (2019). Design considerations for riverine floods in a changing climate – a review. *Journal of Hydrology 574*(2019), 557–573.

Garcia, F., Folton, N. & Oudin, L. (2017). Which objective function to calibrate rainfall–runoff models for low-flow index simulations? *Hydrological Sciences Journal 2*(7), 1149–1166. doi.10.1080/02626667.2017.1308511.

Griffis, V. W. & Stedinger, J. R. (2007). Evolution of flood frequency analysis with Bulletin 17. *Journal of Hydrologic Engineering 12*(3), 283–297.

Griffis, V. W. & Stedinger, J. R. (2009). The log-Pearson type 3 distribution and its application in flood frequency analysis, 3. Sample Skew and Weighted Skew Estimators. *Journal of Hydrologic Engineering 14*(2), 209–212.

Gustard, A., Bullock, A. & Dixon, J. M. (1992). *Low Fow Estimation in the United Kingdom (IH Report No. 108).* Institute of Hydrology, Wallingford, Oxon, p. 88.

Gustard, A. & Demuth, S. (2008). *Manual on Low-Flow Estimation and Prediction, Operational Hydrology Report No. 50.* WMO-No. 1029, World Meteorological Organization, Geneva, Switzerland, p. 136.

Hall, J., Arheimer, B., Borga, M., Brázdil, R., Claps, P., Kiss, A., Kjeldsen, T., Kriauciuniene, J., Kundzewicz, Z. & Lang, M. (2014). Understanding flood regime changes in Europe: a state of the art assessment. *Hydrology and Earth System Sciences 18*, 2735–2772.

Hasan, H. H., Razali, S. F. M., Muhammad, N. S. & Ahmad, A. (2019). Research trends of hydrological drought: a systematic review. *Water 11*(11), 2252. doi:10.3390/w11112252.

Heim, R. R. (2012). A review of twentieth century drought indices used in the United States. *Bull. Am. Meteorol. Soc. 83*, 1149–1165. https://doi.org/10.1175/1520-0477-83.8.1149.

Hosking, J. R. M. & Wallis, J. R. (2005). *Regional Frequency Analysis: An Approach Based on L-Moments.* Cambridge University Press, Cambridge, UK.

Howitt, R. E., Medellin-Azuara, J., MacEwan, D., Lund, J. R. & Sumner, D. A. (2014). *Economic Analysis of the 2014 Drought for California Agriculture.* Center for Watershed Sciences, University of California, Davis, CA, p. 20. Available at: http://watershed.ucdavis.edu.

Hulley, M., Clarke, C. & Watt, E. (2014). Occurrence and magnitude of low flows for Canadian rivers: an ecozone approach. *Canadian Journal of Civil Engineering 41*(1), 1–8. doi:10.1139/cjce-2013-0300.
Institute of Hydrology (1980). *Low Flow Studies*. Institute of Hydrology, Wallingford, Oxon, UK. In four volumes.
Interagency Advisory Council (IAC) on Water Data (1982). *Guidelines for Determining Flood Flow Frequency, Bulletin 17B*. U.S. Department of the Interior Geological Survey Office of Water Data Coordination, Reston, VA, p. 194.
Kiem, A. S., Kuczera, G., Kozarovski, P., Zhang, L. & Willgoose, G. (2021). Stochastic generation of future hydroclimate using temperature as a climate change covariate. *Water Resources Research 56*, 2020WR027331. https://doi.org/10.1029/2020WR027331.
Kroll, C. N. & Vogel, R. M. (2002). The probability distribution of low streamflow series in the United States. *Journal of Hydrologic Engineering, ASCE 7*(2), 137–146.
Kroll, C., Luz, J., Allen, B. & Vogel, R. M. (2004). Developing a watershed characteristics database to improve low streamflow prediction. *Journal of Hydrologic Engineering 9*, 116–125.
Lima, C. H., Lall, R. U. & Troy, T. (2016). A hierarchical Bayesian GEV model for improving local and regional flood quantile estimates. *Journal of Hydrology 541*, 816–823.
Lins, H. F. & Slack, J. R. (1999). Streamflow trends in the United States. *Geophysical Research Letters 26*(2), 227–230.
Lins, H. F. & Slack, J. R. (2005). Seasonal and regional characteristics of US streamflow trends in the United States from 1940 to 1999. *Physical Geography 26*(6), 489–501.
Liu, D. D., Guo, S. L., Lian, Y. Q., Xiong, L. H. & Chen, X. H. (2015). Climate-informed low-flow frequency analysis using nonstationary modelling. *Hydrological Processes 29*(9), 2112–2124.
Lloyd-Hughes, B. (2014). The impracticality of a universal drought definition. *Theoretical and Applied Climatology 117*, 607–611. doi:10.1007/s00704-013-1025-7.
Madsen, H., Lawrence, D., Lang, M., Martinkova, M. & Kjeldsen, T. (2013). *A review of applied methods in Europe for flood-frequency analysis in a changing environment*. NERC/Centre for Ecology & Hydrology, Edinburgh.
Maidment, D. R. (2017). Conceptual framework for the national flood interoperability experiment. *Journal of the American Water Resources Association (JAWRA) 53*(2), 245–257. doi: 10.1111/1752-1688.12474.
Matalas, N. C. (1963). *Probability Distribution of Low Flows*. Professional Paper 434-A, U.S. Geological Survey, Washington, DC.
Mishra, A. K. & Singh, V. P. (2010). A review of drought concepts. *Journal of Hydrology 391*, 202–216.
National Oceanic and Atmospheric Administration (2019). *Billion-Dollar Weather and Climate Disasters: Table of Events*. NOAA National centers for Environmental Information. Available at: https://www.ncdc.noaa.gov/billions/events/US/1980-2019.
Obasi, G. O. P. (1994). WMO's role in the international decade for natural disaster reduction. *Bulletin of the American Meteorological Society 75*(9), 1655–1661.
Ouarda, T. B. M. J., Gingras, H., Kouider, A., Rudolf, Z. R., Hache, M., Barbe, M. & Bobee, B. (2002). Regions, a general and automatic model for flood and drought regional estimation at ungauged basins. In: *1st Biennial Workshop of the Statistical Hydrology Committee of the CGU-HS, 2002 Annual Scientific Meeting of the CGU, Ban., AB, May 18–21, 2002.*
Ouarda, T. B. M. J., Jourdain, V., Gignac, N., Gingras, H., Herrera, H. & Bobee, B. (2005). *Development of a Low-Flow Regional Estimation Model for the Inhabited Part of Quebec, the ARIDE Model*. INRS-ETE, Research Report No. R-684-f1, p. 174 (in French).
Ouarda, T. B. M. J., Charron, C. & St-Hilaire, A. (2008). Statistical models and the estimation of low flows. *Canadian Water Resources Journal* 33(2), 195–206. doi:10.4296/cwrj3302195.
Ouarda, T. B. M. J. (2017). Regional flood frequency modeling. In: *Handbook of Applied Hydrology*, Chapter 77, 2nd edn. Singh, V. P. (ed.). McGraw Hill, New York.
Ouarda, T. B. M. J., Charron, C., Hundecha, Y., St-Hilairea, A. & Chebana, F. (2018). Introduction of the GAM model for regional low-flow frequency analysis at ungauged basins and comparison with commonly used approaches. *Environmental Modelling and Software 109*, 256–271.
Pilon, P. J. (1990). The Weibull Distribution applied to regional low flow frequency analysis. In: *Regionalisation in Hydrology*. (M.A. Beran, M. Brilly, A. Becker & O. Bonacci, eds). IAHS Publication No. 191, IAHS Press, Institute of Hydrology, Wallingford, UK., pp. 227–237.
Potter, K. W. & Lettenmaier, D. P. (1990). A comparison of regional flood frequency estimation methods using a resampling method. *Water Resources Research 26*(3), 415–424.
Priest, S. J., Suykens, C., Van Rijswick, H. F. M. W., Schellenberger, T., Goytia, S. B., Kundzewicz, Z. W., Van Doorn-Hoekveld, W. J., Beyers, J. -C. & Homewood, S. (2016). The European Union approach to flood risk management and improving

societal resilience: lessons from the implementation of the Floods Directive in six European countries. *Ecology and Society 21*(4), 50. https://doi.org/10.5751/ES-08913-210450.

Pushpalatha, R., Perrin, C. & LeMoine, N. (2012). A review of efficiency criteria suitable for evaluating low-flow simulations. *Journal of Hydrology 420*, 171–182.

Reis Jr., D. S., Veilleux, A. G., Lamontagne, J. R., Stedinger, J. R. & Martins, E. S. (2020). Operational Bayesian GLS regression for regional hydrologic analyses. *Water Resources Research 56*, e2019WR026940. https://doi.org/10.1029/2019WR026940.

Riggs, H. C. (1972). Low-flow investigations, chapter B1. In: *Techniques of Water Resources Investigations of the United States Geological Survey, Book 4, Hydrologic Analysis and Interpretation*. U.S. Government Printing Office, Washington, DC.

Ross, T. & Lott, N. (2003). *A Climatology of 1980–2003 Extreme Weather and Climate Events, National Climatic Data Center Technical Report No. 2003-01*. NOAA/NESDIS, National Climatic Data Center, Asheville, NC.

Rossman, L. A. (1990). *DFLOW User's Manual, US EPA Risk Reduction Engineering Laboratory*. Office of Research and Development, Cincinnati, OH, p. 31.

Salas, J. D., Obeysekera, J. & Vogel, R. M. (2018). Techniques for assessing water infrastructure for nonstationary extreme events: a review. *Hydrological Sciences Journal*. https://doi.org/10.1080/02626667.2018.1426858.

Salinas, J. L., Castellarin, A., Viglione, A., Kohnova, S. & Kjeldsen, T. (2014). Regional parent flood frequency distributions in Europe – Part 1: is the GEV model suitable as a pan-European parent? *Hydrology and Earth Systems Science 18*, 4381–4389.

Serago, J. & Vogel, R. M. (2018). Parsimonious nonstationary flood frequency analysis. *Advances in Water Resources 112*, 1–16.

Sheffield, J. & Wood, E. F. (2011). *Drought: Past Problems and Future Scenarios*. Earthscan, London and Washington, DC.

Singh, V. P. (2012). *Computer Models of Watershed Hydrology*. Water Resources Publications, Golden, Co. ISBN-13: 978-1887201742.

Singh, V. P. & Frevert, D. K. (2006). *Watershed Models*. CRC Press, Boca Raton, FL.

Smakhtin, V. U. (2001). Low-flow hydrology: a review. *Journal of Hydrology 240*(3–4), 147–186.

Spence, C., Whitfield, P. H. & Ouarda, T. B. M. J. (2008). Introduction to the special issue on low-flow Prediction in Ungauged Basins (PUB) In Canada. *Canadian Water Resources Journal 33*(2), 103–106. doi:10.4296/cwrj3302103.

Spence, C., Whitfield, P. H., Pomeroy, J. W., Pietroniro, A., Burn, D. H., Peters, D. L. & St-Hilaire, A. (2013). A review of the Prediction in Ungauged Basins (PUB) decade in Canada. *Canadian Water Resources Journal 38*(4), 253–262. doi:10.1080/07011784.2013.843867.

Stagnitta, T. J., Kroll, C. N. & Zhang, Z. (2017). A comparison of methods for low streamflow estimation from spot measurements. *Hydrological Processes*. doi:10.1002/hyp.11426.

Stakhiv, E. Z., Werick, W. & Brumbaugh, R. W. (2016). Evolution of drought management policies and practices in the United States. *Water Policy 18*, 122–152.

Stedinger, J. R. (2017). Flood frequency analysis. In: *Handbook of Applied Hydrology*, Chapter 76, 2nd edn. Singh, V. P. (ed.). McGraw Hill, New York.

Stedinger, J. R. & Thomas, W. (1985). Low-flow frequency estimation using base-flow measurements. U.S. Geological Survey Open-File Report 85-95, U.S. Geological Survey, 22 pp, doi:10.3133/ofr8595.

Stedinger, J. R., Vogel, R. M. & Foufoula-Georgiou, E. (1993). Frequency analysis of extreme events, chapter 18. In: *Handbook of Hydrology*. Maidment, D. R. (Editor-in-Chief). McGraw-Hill Book Company, New York, NY.

Steinschneider, S., Wi, S. & Brown, C. (2014). The integrated effects of climate and hydrologic uncertainty on future flood risk assessments. *Hydrological Processes*. http://dx.doi.org/10.1002/hyp.10409.

Svoboda, M. D., LeComte, D., Hayes, M., Heim, R., Gleason, K., Angel, J., Rippey, B., Tinker, R., Palecki, M., Stooksbury, D., Miskus, D. & Stephens, S. (2002). The drought monitor. *Bulletin of the American Meteorology Society 83*, 1181–1190.

Svoboda, M. D., Fuchs, B. A., Poulsen, C. C. & Nothwehr, J. R. (2015). The drought risk atlas: enhancing decision support for drought risk management in the United States. *Journal of Hydrology 526*, 274–286.

Takeuchi, K. (1988). Hydrological persistence characteristics of floods and droughts. *Interregional comparisons. Journal of Hydrology 102*, 49–67.

Tallaksen, L. M. & Van Lanen, H. A. J. (eds) (2004). Hydrological drought: processes and estimation methods for streamflow and groundwater. In: *Developments in Water Science*, Vol. 48, Elsevier Science B.V., Amsterdam, The Netherlands..

Tasker, G. D., Hodge, S. A. & Barks, C. S. (1996). Region of influence regression for estimating the 50-year flood at ungauged sites. *Water Resources Bulletin 32*(1), 163–170.

Thomas, D. M. & Benson, M. A. (1970). Generalization of streamflow characteristics from drainage-basin characteristics. U.S. Geological Survey Water-Supply Paper 1975, U.S. Geological Survey, Reston, VA.

Tijdeman, E., Barker, L. J., Svoboda, M. D. & Stahl, K. (2018). Natural and human influences on the link between meteorological and hydrological drought indices for a large set of catchments in the contiguous United States. *Water Resources Research 54*, 6005–6023. https://doi.org/10.1029/ 2017WR022412.

U.S. Army Corps of Engineers (1993) Hydrologic Frequency Analysis, EM 1110-2-1415, see chapter 4 titled: Low Flow Frequency Analysis.

U.S. Department of Agriculture (2017). *Farm Service Agency Disaster Assistance Programs at a Glance, Fact Sheet.* p. 2.

Van Loon, A. F. (2015). Hydrological drought explained. *Wiley Interdisciplinary Reviews – Water* 2(4), 359–392.

Villarini, G., Taylor, S., Wobus, C., Vogel, R. M., Hecht, J., White, K. D., Baker, B., Gilroy, K., Olsen, J. R. & Raff, D. (2018). *Floods and Nonstationarity: A Review, CWTS 2018-01.* U.S. Army Corps of Engineers, Washington, DC.

Vogel, R. M. & Kroll, C. N. (1989). Low-flow frequency analysis using probability plot correlation coefficients. *Journal of Water Resources Planning and Management, ASCE 115*(3), 338–357.

Vogel, R. M. & Kroll, C. N. (1992). Regional geohydrologic-geomorphic relationships for the estimation of low-flow statistics. *Water Resources Research 38*(9), 2451–2458.

Vogel, R. M. & Wilson, I. (1996). The probability distribution of annual maximum, minimum and average streamflow in the United States. *Journal of Hydrologic Engineering, ASCE 1*(2), 69–76.

Vogel, R. M. (2017). Stochastic watershed models for hydrologic risk management. *Water Security*. http://dx.doi.org/10.1016/j.wasec.2017.06.001.

Vogel, R. M. (2020). The geometric mean? *Communications in Statistics: Theory and Methods*. doi:10.1080/03610926.2020.1743313.

Vogel, R. M. & Kroll, C. N. (2020). A comparison of estimators of the conditional mean under non-stationary conditions. *Advances in Water Resources*. https://doi.org/10.1016/j.advwatres.2020.103672, 2020.

Wagener, T., Wheater, H. S. & Gupta, H. V. (2004). *Rainfall-Runoff Modelling in Gauged and Ungauged Catchments*. Imperial College Press, London, p. 332.

Warren, M., Dunbar, M. J. & Smith, C. (2015). River flow as a determinant of salmonid distribution and abundance: a review. *Environmental Biology of Fishes 98*(6), 1695–1717.

Werick, W. J., Willeke, G. E., Guttman, N. B., Hosking, J. R. M. & Wallis, J. R. (1994a). National Drought Atlas Developed, EOS, AGU, Vol. 75, p. 89–90. doi:10.1029/94EO00706.

Werick, W. J., Willeke, G. E., Guttman, N. B., Hosking, J. R. M. & Wallis, J. R. (1994b). The National Drought Atlas, Geophysics News 1993, p. 8–10.

Wilhite, D. A. & Glantz, M. H. (1985). Understanding the drought phenomenon: the role of definitions. *Water International 10*(3), 111–120.

World Meteorological Organization and Global Water Partnership (2016). Handbook of drought indicators and indices. In: *Integrated Drought Management Programme, Integrated Drought Management Tools and Guidelines Series 2*. Svoboda, M. & Fuchs, B. A., (eds). World Meteorological Organization, Geneva, Switzerland, p. 45.

Young, A. R., Grew, R. & Holmes, M. G. R. (2003). Low flows 2000: a national water resources assessment and decision support tool. *Water Science & Technology 48*(10), 119–126.

Yue, S. & Pilon, P. (2005). Probability distribution type of Canadian annual minimum streamflow. *Hydrological Sciences Journal 50*(3), 438. doi:10.1623/hysj.50.3.427.65021.

Zaidman, M. D., Keller, V. & Young, A. R. (2002). *Low Flow Frequency Analysis: Guidelines for Best Practice.* Centre for Ecology & Hydrology, Environment Agency, UK, p. 33.

Zaidman, M. D., Keller, V., Young, A. R. & Cadman, D. (2003). Flow-duration-frequency behavior of British rivers based on annual minimum data. *Journal of Hydrology 277*, 195–213.

Zhang, Z. & Kroll, C. (2007a). Closer look at the baseflow correlation method. *Journal of Hydrologic Engineering 12*, 190–196.

Zhang, Z. & Kroll, C. (2007b). The baseflow correlation method with multiple gauged sites. *Journal of Hydrology 347*, 371–380.

First received 1 November 2021; accepted in revised form 1 November 2021. Available online 12 November 2021

© 2022 The Authors

Water Infrastructure Planning, MDCU doi: 10.2166/wp.2021.267

Incorporating decadal climate variability information in the operation and design of water infrastructure

J. Rolf Olsen[a,*], Vikram M. Mehta[b] and Harvey Hill[c]

[a] Charlottesville, VA, USA
[b] Center for Research on the Changing Earth System (CRCES), Clarksville, MD, USA
[c] Johnson Shoyama Graduate School of Public Policy, University of Saskatchewan, Saskatoon, Saskatchewan, Canada
*Corresponding author. E-mail: j.rolf.olsen@gmail.com

ABSTRACT

The high thermal and mechanical inertia of the oceans results in slow changes in sea surface temperatures (SSTs). Changes in SSTs, in turn, can impact atmospheric circulation including water vapor transport, precipitation, and temperatures throughout the world. The Pacific Decadal Oscillation (PDO), the tropical Atlantic SST gradient variability, and the West Pacific Warm Pool are patterns of natural climate variability that tend to persist over decadal time periods. There are current efforts to produce decadal climate predictions, but there is limited understanding if this information can be used in water resources management. Understanding the current state of decadal climate variability (DCV) phenomena and the probability of persisting in that state may be useful information for water managers. This information could improve forecasts that aid operations and short-term planning for reservoir management, domestic and industrial water supplies, flood risk management, energy production, recreation, inland navigation, and irrigation. If conditions indicate a higher likelihood of drought, reservoir managers could reduce flood storage space and increase storage for conservation purposes. Improved forecasts for irrigation could result in greater efficiencies by shifting crops and rotational crop patterns. The potential benefits of using a forecast must be balanced against the risk of damages if the forecast is wrong. Seasonal forecasts using DCV information could also be used to inform drought triggers. If DCV indices indicate that the climate has a higher probability of dry conditions, drought contingency plans could be triggered earlier. Understanding of DCV phenomena could also improve long-range water resources planning. DCV can manifest itself in relatively short-term hydrologic records as linear trends that complicate hydrologic frequency analysis, which has traditionally assumed that hydrologic records are stationary.

Key words: Decadal climate variability, Drought management, Nonstationary flood frequency analysis, Reservoir management

HIGHLIGHTS

- Knowledge of the current phase of decadal climate variability (DCV) indices and the probability of persisting in that state may be useful information for water managers for seasonal forecasts.
- DCV information could inform the allocation of flood storage and conservation storage spaces in reservoirs, but the potential use of DCV information depends on the risk if the forecasts are wrong.
- Analysis of DCV phenomena could help explain trends and breakpoints in nonstationary flood frequency analysis.

INTRODUCTION

Climate is a major driver of hydrologic conditions and varies on multiple time scales from sub-seasonal to decadal and centennial. This paper focuses on decadal climate variability (DCV) and its impacts on water resources management. DCV can be defined as quasi-oscillatory variability that occurs at approximately 8- to 20-year periods (Mehta, 2017, 2020). There are many potential water sectors affected by DCV, including reservoir management, flood and drought risk management, domestic and industrial water supplies, irrigation, energy production, recreation, and inland navigation (Mehta, 2017). DCV's manifestation in relatively short-term hydrologic records as linear trends can confound hydrologic frequency analysis, which has traditionally assumed that hydrologic records are statistically stationary. DCV and its hydrologic effects also confound the attribution of observed anomalies to natural variability or climate change, which can confuse adaptation planning in the short, medium, and long term. Therefore, it is very important to understand and, if possible, predict DCV effects on impacted sectors to inform water resources management in the short, medium, and long term. Empirical data analysis and simulations with climate models in the last 20–25 years have improved the understanding of DCV and have also stimulated decadal climate prediction efforts (see Mehta (2020) for a comprehensive review of the state of the DCV science and prediction). The goals of this paper are to familiarize the water resources management community with DCV and to evaluate the potential use of DCV information for the operation and design of water infrastructure. Although the examples illustrated here are from the United States, they are potentially applicable to other parts of the world.

BACKGROUND ON DCV

Natural processes in the Earth's climate system can cause internally generated climate variations at a variety of regular or irregular periods due to interactions among atmosphere, oceans, land surface, and snow and ice. Heat from the oceans is the major source of atmospheric motion. The specific heat capacity of sea water is approximately four times larger than air, and the density of sea water is approximately 1,000 times greater than air. Therefore, the oceans can absorb and retain much more heat than the atmosphere because the total heat capacity is the product of mass (density × volume) and specific heat capacity. It is estimated that the upper two meters of the world's oceans contain as much heat as the entire global atmosphere. Therefore, the ocean to atmosphere heat transfer is the major source of heat for the atmosphere.

Changes in sea surface temperatures (SSTs) can impact atmospheric circulation including water vapor transport, precipitation, and temperatures throughout the world. SST changes can cause changes in jet stream location and strength, and thereby cause changes in storminess and precipitation (rain, snow) over the planet's oceans and land areas. The high thermal (heat capacity of sea water) and mechanical (density of sea water) inertia of the oceans cause very slow currents and waves, and very slow changes in temperatures, including SSTs. This slow change is in marked contrast to much faster atmospheric winds, waves, and temperature changes. In this way, SST changes over a few seasons to a few decades can cause changes in water vapor and heat (temperature) transports, and storminess over continents.

El Niño, for example, is a periodic warming of SSTs in the eastern tropical Pacific Ocean. The Southern Oscillation refers to an oscillation in surface air pressure between eastern and western Pacific regions. The two together are referred to as El Niño-Southern Oscillation (ENSO). La Niña is a periodic cooling of the eastern tropical Pacific Ocean (Philander, 1990; Glantz *et al.*, 1991; Rasmusson, 1991; Glantz, 2001). ENSO is the major pattern of natural climate variability on seasonal to interannual time scales. ENSO events happen irregularly at around 2- to 7-year intervals. Figure 1 shows a time series of SST in the Niño 3.4 region, with warmer SSTs being El Niño events. ENSO attributes, such as the frequency and intensity of El Niño and La Niña events, undergo decadal variability.

Other SST variability patterns – the Pacific Decadal Oscillation (PDO), the Tropical Atlantic SST gradient (TAG for brevity) variability, the West Pacific Warm Pool (WPWP) SST variability – have approximately decadal (8–20 years) periods. The PDO is a pattern of decadal–multidecadal variations that span the Pacific Ocean such

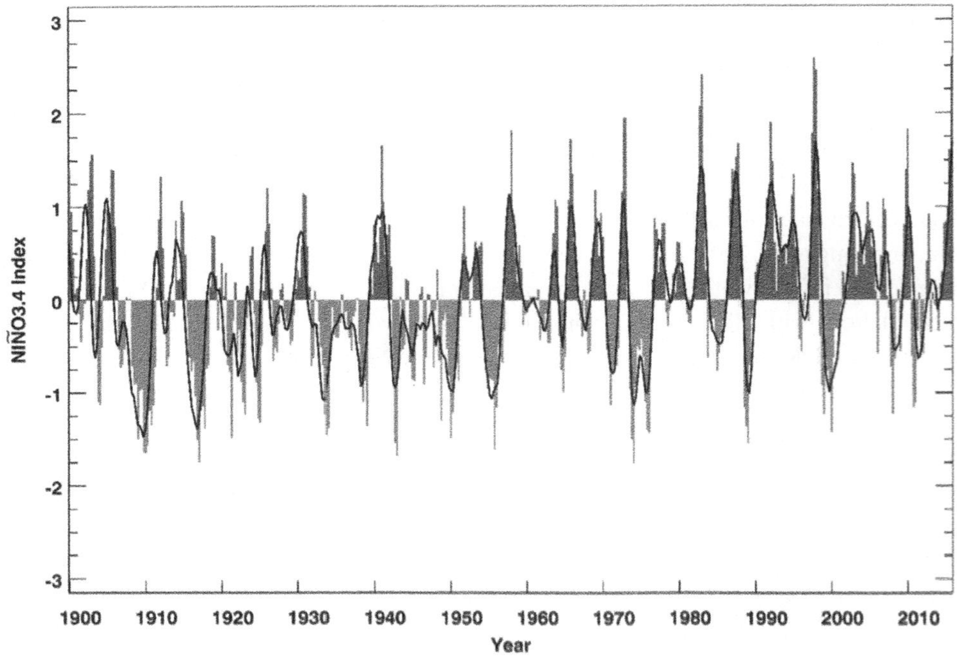

Fig. 1 | The 3-month average (Dec–Jan–Feb, Mar–Apr–May, Jun–Jul–Aug, Sep–Oct–Nov) Niño3.4 SST index (bars) from January 1900 to November 2015 and its 4-season running-average smoothed version (line).

that there are SST anomalies of one sign (positive or negative) in the tropical–subtropical Pacific and of the opposite sign in the mid-latitude Pacific (Mantua *et al.*, 1997). The PDO index time series is shown in Figure 2. In the positive or warm PDO phase, the SSTs in the tropical–subtropical Pacific and along the North and South American coasts are warmer than average, and those in the mid-latitude central and western Pacific are cooler than average. In the negative or cool PDO phase, the SSTs in the tropical–subtropical Pacific and along the North and South American coasts are cooler than average, and those in the mid-latitude central and western Pacific are warmer than average. Figure 2 shows that the PDO index from 1,900 to 2015 has undergone variability at a variety of periods from interannual to multidecadal, but the smoothed index clearly shows the prominence of decadal–multidecadal variability.

The TAG index is derived by averaging SST anomalies in a tropical North Atlantic box (5°–20°N, 30°–60°W) and a tropical South Atlantic box (0°–20°S, 30°W–10°E) and subtracting the latter from the former (Mehta, 1998; Rajagopalan *et al.*, 1998). The time series of the TAG index is shown in Figure 3. Empirical orthogonal function–principal component analysis of tropical Atlantic SST anomalies also reveals this SST pattern as the dominant pattern of tropical Atlantic SST variability. The TAG index varies at interannual to decadal and longer timescales as shown in Figure 3. The Atlantic Multidecadal Oscillation (AMO) Index is defined as a coherent pattern of variability in North Atlantic SSTs and is formed by averaging SST anomalies from the Equator to 80°N. The AMO index shows dominant variability at a period of 60–80 years and minor variability at a period of 10–12 years (Folland *et al.*, 1986, 2001; Schlesinger & Ramankutty, 1994).

The WPWP SST variability is yet another DCV phenomenon (Mehta, 2020). The western waters of the equatorial Pacific and the eastern Indian Ocean have the warmest seawaters in the world. The WPWP index from 1900 to 2015 has a clear warming trend, but the de-trended index shows variability over decadal time scales.

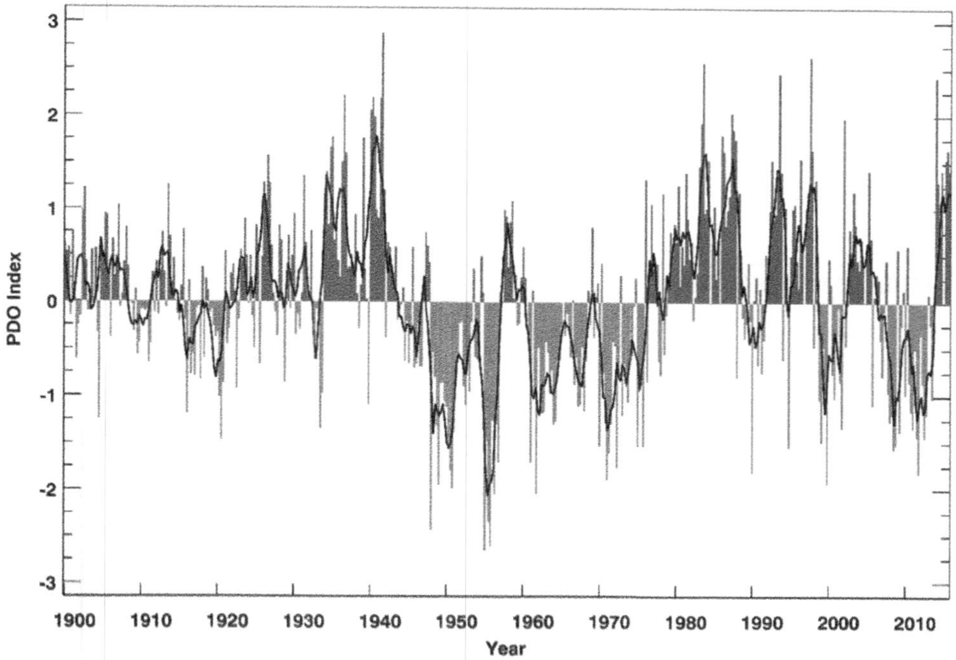

Fig. 2 | The 3-month average (Dec–Jan–Feb, Mar–Apr–May, Jun–Jul–Aug, Sep–Oct–Nov) PDO Index.

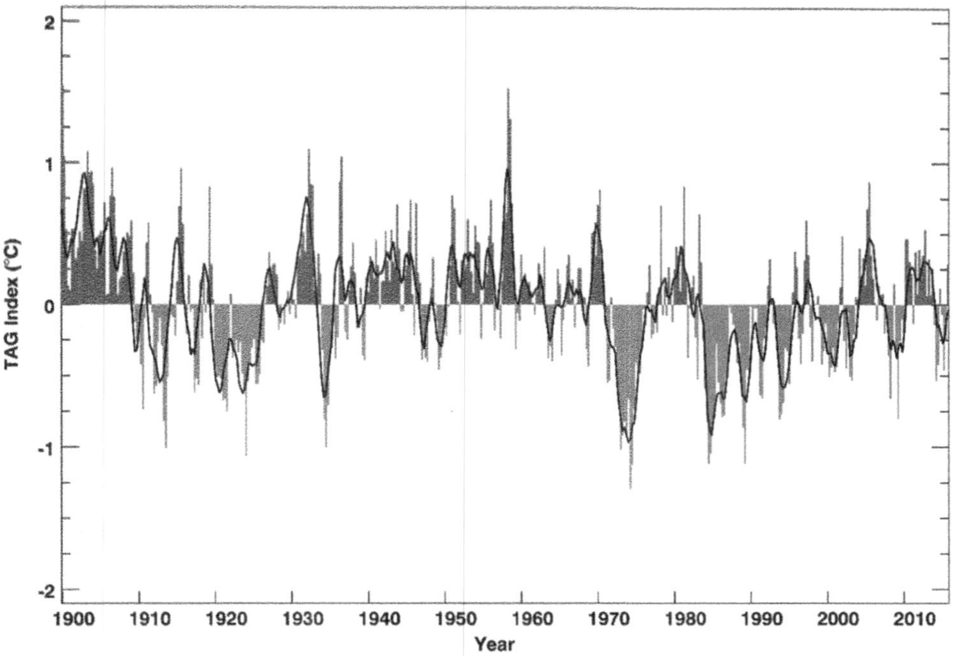

Fig. 3 | The 3-month average (Dec–Jan–Feb, Mar–Apr–May, Jun–Jul–Aug, Sep–Oct–Nov) TAG.

These DCV phenomena influence variability in hydrology throughout the world. Different phases of these DCV indices have been associated with wet and dry periods on land. In North America, for example, different phases of the PDO and AMO have been associated with dry or wet periods (Enfield *et al.*, 2001; McCabe *et al.*, 2004, 2008; Schubert *et al.*, 2004a, 2004b; Mo *et al.*, 2009). Water managers may be able to use the phases of these DCV indices and their observed correlations with continental hydrometeorology to support their decision-making, as, for example, in reservoir management and the operation of irrigation systems that depend on reservoir systems.

DECADAL CLIMATE PREDICTABILITY

The thermal and mechanical inertia of the oceans is a potential source of climate predictability. There are empirical and dynamic modeling methods to estimate the probability of future climate conditions. Empirical methods use observed data such as SSTs. The observed data are a relatively short time series compared to the decadal time scale and is a shortcoming of empirical methods (Mehta, 2020). Dynamic methods use Earth System Models (ESMs) to simulate future climate.

Coupled Model Intercomparison Projects Phase 5 (Taylor *et al.*, 2012) and Phase 6 (Eyring *et al.*, 2016), conducted under the World Climate Research Program in the last 10 years, included coordinated decadal climate hindcasts (retrospective forecast of past decades) and forecasts by approximately 30 complex climate models from many countries. Also, the Near-Term Climate Prediction Project of the World Meteorological Organization is now operationally producing worldwide decadal climate forecasts available from the U.K. Meteorological Office. The results showed some skill in predicting North Atlantic temperature. There is less prediction skill in the North Pacific temperatures (Kirtman *et al.*, 2013; Meehl *et al.*, 2014). One source of predictability is volcanic eruptions that influence SSTs and are one of the sources of decadal SST hindcast skill in the year(s) following significantly large eruptions (Mehta *et al.*, 2013, 2019).

Some of the major problems with decadal climate prediction include the following: (1) relatively short time series of instrument-based global ocean observations, especially sub-surface observations, for understanding, model initialization, and comparison with prediction; (2) an insufficient understanding of fundamental physics of DCV; (3) an insufficient theoretical understanding of possible behaviors of geographically varying, complex and non-linear dynamical systems with mixed initial and boundary values; and (4) global climate models displaying less than satisfactory skill in simulating climate in general and DCV in particular (Meehl *et al.*, 2009, 2014; Mehta *et al.*, 2011).

Models of non-linear dynamical systems are sensitive to initial conditions. The initial conditions for the global climate cannot be known with certainty. Ensembles of multiple simulations with different initial conditions are one method to handle this uncertainty. Multi-model ensembles using predictions from different ESMs are an approach to the initial condition and model uncertainty (Kirtman *et al.*, 2013).

Decadal climate patterns such as the PDO and TAG tend to persist over time. This persistence has the potential to provide information about the climate in the next season or year. Many of the studies involving the impacts of DCV on hydrometeorology are based on the phase of the DCV phenomena, whether positive or negative. These phases tend to persist from year to year. The transition probabilities can be estimated from the observed record by whether the index remains in the current state or transitions to an alternative state. These probabilities are uncertain since they are estimated from a limited sample size and assume stationarity. Table 1 shows the observed transition probabilities for the PDO from 1 year to the next. Table 2 shows the transition probabilities for the TAG index. The PDO shows more persistence than the TAG. The negative or positive state of the index may be associated with wet or dry local conditions, and the transition probabilities may provide some guidance if conditions will continue into the next year.

Table 1 | Phase transition probabilities for the PDO from 1 year to the next year.

Phase	Positive	Negative
Positive	0.72	0.28
Negative	0.20	0.80

DCV AND WATER MANAGEMENT OPERATIONAL DECISIONS

Use of forecasts in reservoir management

Water resources managers must often balance competing uses of water, such as municipal and industrial water supply, irrigation for agriculture, support of navigation on rivers, energy production, recreation, water quality, and support for the aquatic environment. This section will examine the potential use of DCV information for reservoir management. Reservoir management decisions include the allocation of storage between flood and conservation purposes and the allocation of water among multiple conservation purposes. The section will also look at the policy for using climate forecasts by the U.S. Army Corps of Engineers (USACE), a major water manager in the United States.

The flood control pool is used for flood storage to store water during floods to reduce flood damages downstream. The conservation pool, or multipurpose pool, is for additional seasonal flood storage and for the storage of multiple purposes including hydroelectric power, irrigation, navigation, municipal and industrial (M&I) water supply, downstream water quality, recreation, and support for the environment including fish and wildlife. The inactive capacity is used to maintain a minimum pool and to store sediment. The reservoir operator must allocate reservoir storage into flood storage space and conservation storage and to balance the multiple purposes.

The capacities of the flood storage and conservation storage may vary during the year depending on the flood season. Flood storage may increase in the anticipation of seasons with more runoff and flooding, while conservation storage may increase for additional water for other purposes during the summer. Reservoir rule curves show how reservoir storage should be divided between conservation storage and flood storage throughout the year. An example of a reservoir rule curve for Oroville Dam in California is given in Figure 4. Major floods occur in winter and spring due to snow melt and rain on snow events. Water levels are reduced in fall to prepare for the winter flood season. For Oroville, the size of the flood storage space is based on the observed precipitation that has occurred in the watershed during the autumn and the winter. The flood storage space is increased or decreased based on the accumulation of snow in the basin and the anticipated volume of spring runoff, but the refill begins on the same date each year. Figure 5 shows another reservoir rule curve for Shasta Dam in California. The volume of the flood storage space is the same at the beginning of the flood season. The refill rate for the reservoir and beginning time can vary based on the observed inflows to the reservoir.

Information on DCV along with the current and predicted ENSO state may indicate that there is an increased likelihood of dry or wet conditions currently and in the next couple seasons. Information that there is a higher

Table 2 | Phase transition probabilities for the TAG index from 1 year to the next year.

Phase	Positive	Negative
Positive	0.55	0.45
Negative	0.53	0.47

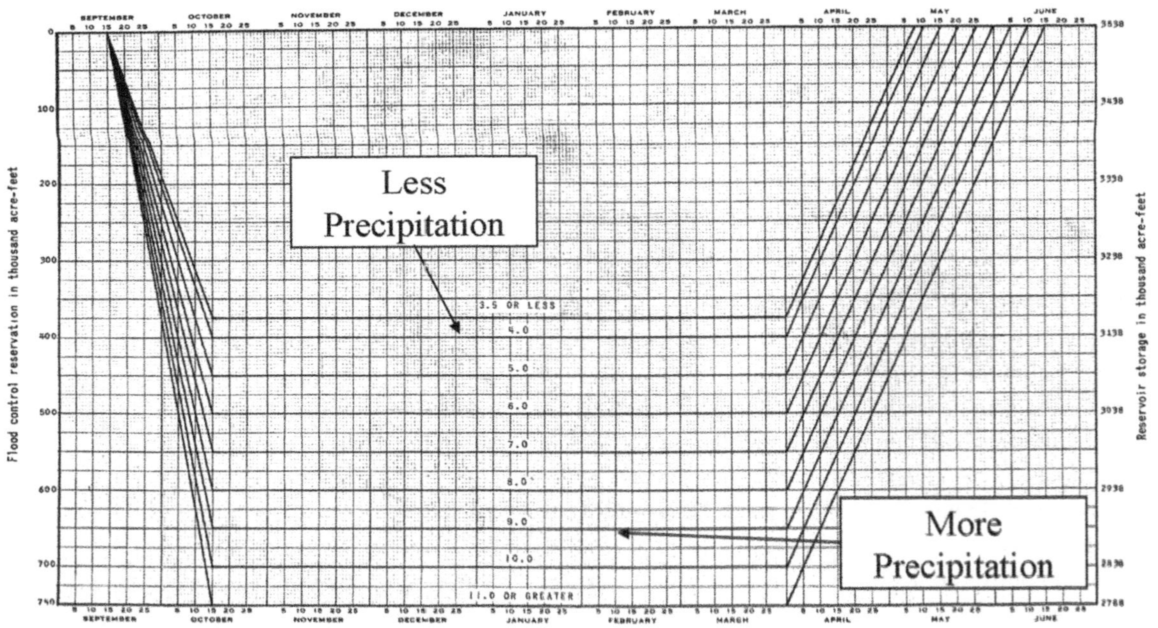

Fig. 4 | Reservoir rule curve for Oroville Dam in California (USACE, 2005).

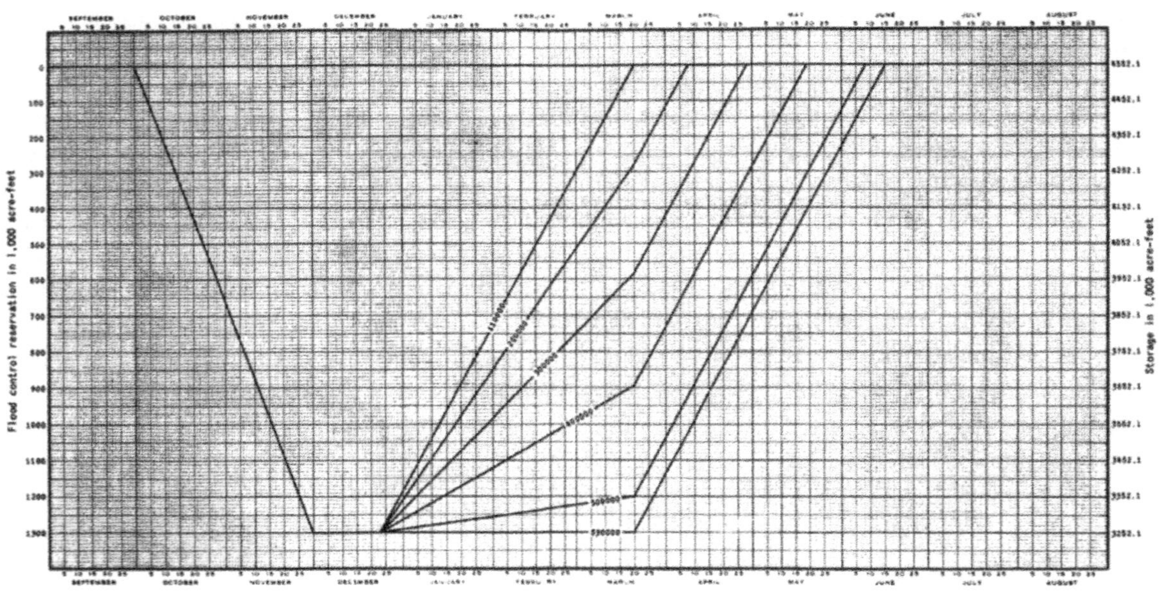

Fig. 5 | Reservoir rule curve for Shasta Dam in California (USACE, 1977).

probability of dry conditions could possibly be used to decrease flood storage to provide more water for conservation purposes or refill the reservoir early to ensure a full supply during summer. Hamlet & Lettenmaier (1999) said that the use of PDO and ENSO information could extend the lead time of streamflow forecasts for the

Columbia River basin by about 6 months over current forecasting practices based on snowpack in the basin. Hamlet *et al.* (2002) stated the use of these forecasts during conditions associated with higher streamflows would provide greater flexibility for fall hydropower generation without threatening other uses.

The U.S. Army Corps of Engineers Engineer Manual on Management of Water Control Systems (EM 1110-2-3600), released in October 2017, specifically addresses the use of climate forecasts in reservoir operations. The Engineer Manual (EM) acknowledges some usefulness of climate forecasts.

> 6.3.3.1. Long-term (monthly, seasonal, or annual) weather forecasts can be useful for planning purposes and for developing reservoir runoff forecasts several months in advance.

However, the EM says the patterns associated with DCV are limited in their usefulness for long-range forecasts.

> 6.3.3.2. Some of the commonly discussed large-scale weather patterns are limited in their ability to be incorporated into long-range forecasts, such as the Arctic Oscillation, PDO, and AMO.

The EM acknowledges that ENSO can have impacts on climate in the United States.

> 6.3.3.3. The ENSO cycle, however, has been demonstrated to having distinct impacts on various regions of the continental United States based on both the phase (warm versus cold) and intensity. These correlations are greatest from late fall into early spring with the general correlation during an El Nino event being cooler and wetter along the southern tier states and warmer and drier from the Pacific Northwest into the Northern Rockies. The opposite impacts are often observed within these regions during a La Nina episode.

The bottom line of the EM is a warning not to reduce flood storage space based on seasonal climate forecasts or climate indices:

> 6.3.3.4. However, it should be noted that not every event within the ENSO cycle has the same characteristics or intensity, so the water manager should be cautioned not to reduce the level of flood risk management based solely on long-range weather forecasts or climatic indexes. These forecasts can inform the water manager of the possibility of those conditions; however, an unjustified use of such forecasts may, in the long run, result in inaccurate regulation of project facilities, and under extreme conditions, impact the projects' authorized functional use. (USACE, 2017a)

A major concern is the uncertainty of the climate system and the impacts in the basin and the potential consequences if conditions turn out different than forecast. The potential benefits of using climate information and the reliability of the information will need to be demonstrated and evaluated before they will be adopted by USACE reservoir managers.

There are efforts underway in USACE and other water management agencies to evaluate the potential use of forecasts in reservoir management. For example, a project is underway in California to evaluate forecast-informed reservoir operations (FIRO) for Lake Mendocino. The preliminary viability assessment of FIRO was concerned with the following questions.

- If FIRO is implemented, will operation improve reliability in meeting water management objectives and ability to meet environmental flow requirements, and to what extent?
- If FIRO is implemented, will operation adversely affect flood risk management in the system? If so, where and to what extent can that be mitigated?
- What meteorological and hydrological forecast skill is required to enable FIRO to be implemented? Is current forecast skill adequate to support FIRO, and what improvements would be needed to enable the full implementation of FIRO? (FIRO, 2017)

These questions could also serve as a guide for evaluating the use of climate forecasts, including DCV information, in reservoir management.

Case study 1: drought contingency planning and Apalachicola–Chattahoochee–Flint River Basin

Another potential application of DCV information is informing drought management. A drought contingency plan is triggered when certain conditions are met. Some indicators of drought are below normal precipitation, lower than average reservoir inflows, and low reservoir levels. The potential use of DCV information for reservoir drought contingency planning will be considered in this section. The case study will be the Apalachicola–Chattahoochee–Flint River Basin, which is located in the southeastern United States in Georgia, Alabama, and Florida. The USACE operates several dams and reservoirs in the basin:

- Buford Dam and Lake Sidney Lanier
- West Point Dam and Lake
- Walter F. George Lock and Dam and Lake
- George W. Andrews Lock and Dam and Lake George W. Andrews
- Jim Woodruff Lock and Dam and Lake Seminole

The reservoirs serve multiple purposes. Lake Sidney Lanier and West Point Lake contain flood storage space to store flood waters. Generally, water is stored during high-flow periods in the winter and spring and is released during the dry summer and fall. Hydropower generation occurs at all the dams except George W. Andrews Lock and Dam. In addition, there are other hydropower dams on the Chattahoochee River operated by private companies. There is a navigation channel on the Apalachicola and Chattahoochee Rivers. The ACF basin provides water supply to many municipalities, including the city of Atlanta. Lake Lanier and downstream of Buford Dam are major sources of M&I water supply. The USACE projects also release minimum flows that provide water quality benefits. In addition, USACE dams release flows to benefit the aquatic environment, fish, and wildlife and to support fish spawning. Finally, the lakes provide significant recreation benefits, which are degraded when reservoir elevation falls below certain levels. During normal flow periods, the ACF basin can meet these multiple needs, but during drought not all water management objectives can be met.

The water control plan for the ACF must balance the needs of these multiple purposes. The ACF projects are operated as a coordinated system to the maximum extent possible rather than as independent, individual projects. Figure 6 shows the composite storage for the basin and the action zones within the conservation storage. The zones are described as follows:

Zone 1: If all the lakes are in Zone 1 or above, the river system would operate in a fairly normal manner. Releases can be made for hydroelectric power, water supply, and water quality. If system composite conservation storage is in Zone 1, releases can be made in support of a navigation season (January to April or May). Drought contingency operations cease when levels return to the composite action Zone 1 in accordance with the Drought Contingency Plan.

Zone 2: Hydroelectric power generation is supported at the same or a reduced level. Water supply and water quality releases are met. Minimum flow targets are met. If system composite conservation storage is in Zone 2, releases can be made in support of a navigation season (January to April or May).

Zone 3: Hydroelectric power generation is supported at a reduced level. Water supply and water quality releases are met. Minimum flow targets are met. If system composite conservation storage is in Zone 3, navigation is not supported. Drought contingency operations are triggered when levels drop to Zone 3.

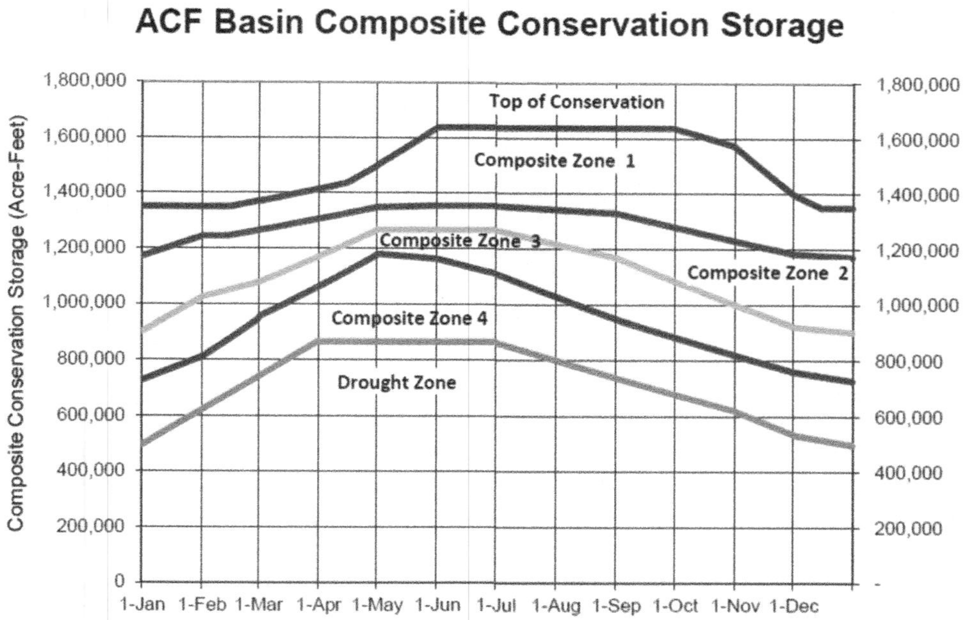

Fig. 6 | ACF basin composite conservation storage (USACE-SAM, 2017).

Zone 4: Hydroelectric power demands will be met at a minimum level and might occur for concurrent uses only. Water supply and water quality releases are met. Minimum flow targets are met. If system composite conservation storage is in Zone 4, navigation is not supported.

Drought Zone: Hydroelectric power will only be met as a result of meeting other project purposes. Water supply and water quality releases are met. Minimum flow targets are met but are reduced to their lowest level. If system composite conservation storage is in the Drought Zone, navigation is not supported and the emergency drought operations are triggered. (USACE-SAM, 2017)

In the Southeastern United States Gulf Region, La Niña conditions are associated with below normal precipitation (Ropelewski & Halpert, 1989) and below normal streamflow (Dracup & Kahya, 1994). In addition, several studies have found a link between decadal climate patterns and streamflow in the region. Tootle *et al.* (2005) showed that the AMO may influence La Niña impacts on streamflow in the Southeast United States. Stevens & Ruscher (2014) examined precipitation in the ACF basin and found that the southern part of the basin is primarily influenced by ENSO conditions, while the northern part of the basin is influenced more by the AMO and the PDO. Johnson *et al.* (2013) analyzed stream flow in the ACF basin and determined that the phase of the AMO modulated the impacts of La Niña. During La Niña years, the streamflow was about 50% lower during the positive phase of the AMO compared with the negative phase of the AMO. Singh *et al.* (2015) examined the baseflows of the Flint River. They found that during positive phases of AMO and PDO, La Niña conditions were associated with severe drought and a greater decrease in baseflow levels in the Flint River. This research has shown that DCV influences drought in the ACF basin. The current DCV conditions could be used to influence when the drought zone is activated, but a thorough assessment of using the DCV information in reservoir operations would first need to be conducted. The assessment would determine the benefits and costs of using the DCV information on each purpose under scenarios both when the DCV information provides a timelier drought trigger or not. This assessment must then be reviewed with stakeholders to determine if the change in operations is acceptable.

The use of DCV information in stakeholder decisions

In addition to reservoir managers, the stakeholders who use this water may also be able to use DCV information to improve their decision-making. Electrical energy production affected by river flows includes hydropower and thermal energy plants cooled by river flows. Knowledge of current decadal climate patterns, such as whether there is a higher probability of dry or wet conditions, could help schedule hydropower production. In the Columbia River basin, forecasts using ENSO and PDO information could support increased autumn hydropower generation during wet conditions associated with higher streamflows (Hamlet *et al.*, 2002). Potential power transfers from the Pacific Northwest to California could be predictable based on forecasts using ENSO and PDO (Voisin *et al.*, 2006).

Agriculture is another potential application. Climate information may allow farmers to adapt management decisions such as fertilizer applications and crop selection to upcoming weather conditions (Mjelde & Hill, 1999; Hill & Mjelde, 2002; Meza *et al.*, 2008). A study of farmers in the Missouri River Basin on the use of DCV information in crop selection and irrigation decisions found that the benefit of using DCV information could be substantial (Fernandez *et al.*, 2016).

DCV information could support inland navigation and shipment decisions. Firms may ship their products either domestically or for export using inland navigation. They need to decide when and where to ship and what mode of transportation to use. During low-flow periods and channel restrictions, barge spot rates increase. During severe droughts, the channel may close. Drought forecasts using climate information may improve shipping decisions. Other firms include barge companies and their competitors. Barge companies may be required to reduce loads or change barge configurations during low-flow periods leading to higher unit costs and shipping rates (Olsen *et al.*, 2005). The competitors to the barge industry include railroad and trucking companies. During channel closures or restrictions, the volume of their freight may increase as shippers move to alternatives. Drought forecasts could help these firms deploy their assets.

In addition, forecasts of dry conditions could help channel operators plan their dredging resources in advance. Drought forecasts may inform dredging decisions. Dredging is often performed during droughts to maintain channel depths. Channel operators could plan for increased dredging volumes by staging dredges or awarding additional dredging contracts in the anticipation of drought conditions.

DCV, NONSTATIONARITY, AND WATER RESOURCES PLANNING

Another potential use of DCV information is in water resources infrastructure planning. Water managers often use future economic benefits to evaluate the future performance of projects. In the United States, for example, 'the Federal objective of water and related land resources project planning is to contribute to national economic development consistent with protecting the Nation's environment, pursuant to national environmental statutes' (USWRC P&G, 1983). Future economic benefits are estimated using hydrologic frequency distributions fitted to the observed record of river flows. The observed record includes low-flow periods and wet periods of various frequencies, intensities, and durations.

Past practice assumes that the past observed record will be representative of the future. Global warming may cause very different conditions than were observed in the past. This concern has focused attention on trend analysis to determine if hydrologic conditions are changing. The U.S. Army Corps of Engineers, for example, has adopted guidance to consider climate change in project planning (USACE, 2018) and for the detection on nonstationarities in hydrologic records (USACE, 2017b). The procedures can detect abrupt changes or more gradual trends, but are not able to detect 'the potential presence of Long-Term Persistence (LTP) in the discharge time series that are related to oscillations in climate regime over a wide range of temporal scales' (USACE, 2017b). The presence of a trend is often associated with what period of record is used, since there may be episodic wet and dry periods in the observed record (Matalas & Olsen, 2001). Both abrupt changes and gradual trends could be associated with natural DCV patterns. Understanding of the underlying climate processes and DCV patterns may better distinguish trends and attribute the trend to decadal climate shifts or another factor. Kundzewicz *et al.* (2019) reviewed the links between climate variability indices and floods across the globe and found relationships in different regions. Salas *et al.* (2018) provide a review of techniques to address nonstationary extreme events and the challenges in using them.

For flood risk reduction projects, future benefits are estimated from the likelihood of future flooding and the damages that would occur under different flood magnitudes and with alternative flood risk reduction measures. The current engineering practice to determine the likelihood of future flooding is to fit a frequency distribution to the observed record of flood flows. In the United States, the Water Resources Council in the 1970s developed uniform guidelines for U.S. Federal agencies to follow and adopted the log-Pearson Type 3 distribution for flood frequency estimation. The Guidelines for Determining Flood Flow Frequency have been recently revised and published as Bulletin 17C (England *et al.*, 2019). Bulletin 17C says that the problem of multidecadal episodic wet and dry periods is a lingering problem for flood frequency analysis.

> 'In certain locations, flood records may indicate apparent nonrandomness and exhibit strong multi-decadal trends or wet and dry cycles that are not explained by land use change, water management, or climate change. Such records are particularly challenging and this is one of the most vexing problems in flood frequency analysis. The Work Group did not evaluate methods to account for nonrandomness and/or multidecadal trends in flood frequency. Additional work in this area is warranted, as it is a seriously unresolved problem. If multidecadal trends of this sort are identified through appropriate statistical tests and data analysis, it is recommended that the underlying physical mechanisms be investigated to gain hydrological understanding (Lins & Cohn, 2011). How to adjust such a record for flood frequency is an unresolved problem.' (England *et al.*, 2019)

The problem of multidecadal wet and dry periods and their effect on flood frequency analysis will be illustrated with the example of the Red River of the North.

Case study 2: impact of DCV on flood frequency analysis: Red River of the North

The Red River of the North has flooded several times since 1997. Large floods occurred in 1997, 2009, and 2011. The Fargo, North Dakota and Moorhead, Minnesota metropolitan area have suffered flood damage in these events. The Corps of Engineers conducted a feasibility study to investigate flood issues in the Fargo–Moorhead metropolitan area and to identify flood risk management measures that could be implemented (USACE-MVP, 2011a). One of the requirements of a feasibility study is to estimate the future benefits of proposed flood risk reduction alternatives and compare them with the cost of the alternative to estimate annual net benefits and the benefit–cost ratio. To estimate future flood damage, reduction benefits require an estimate of the probability and magnitude of future flooding.

Figure 7 shows the annual average daily flow and the annual maximum daily flow for the Red River at Fargo. The record shows that the recent decades have been wetter than the earlier record. The increase in streamflow raised questions about the stationarity of the record and what methods could be applied in developing a flood frequency distribution. The study team organized an expert elicitation panel to evaluate the apparent non-

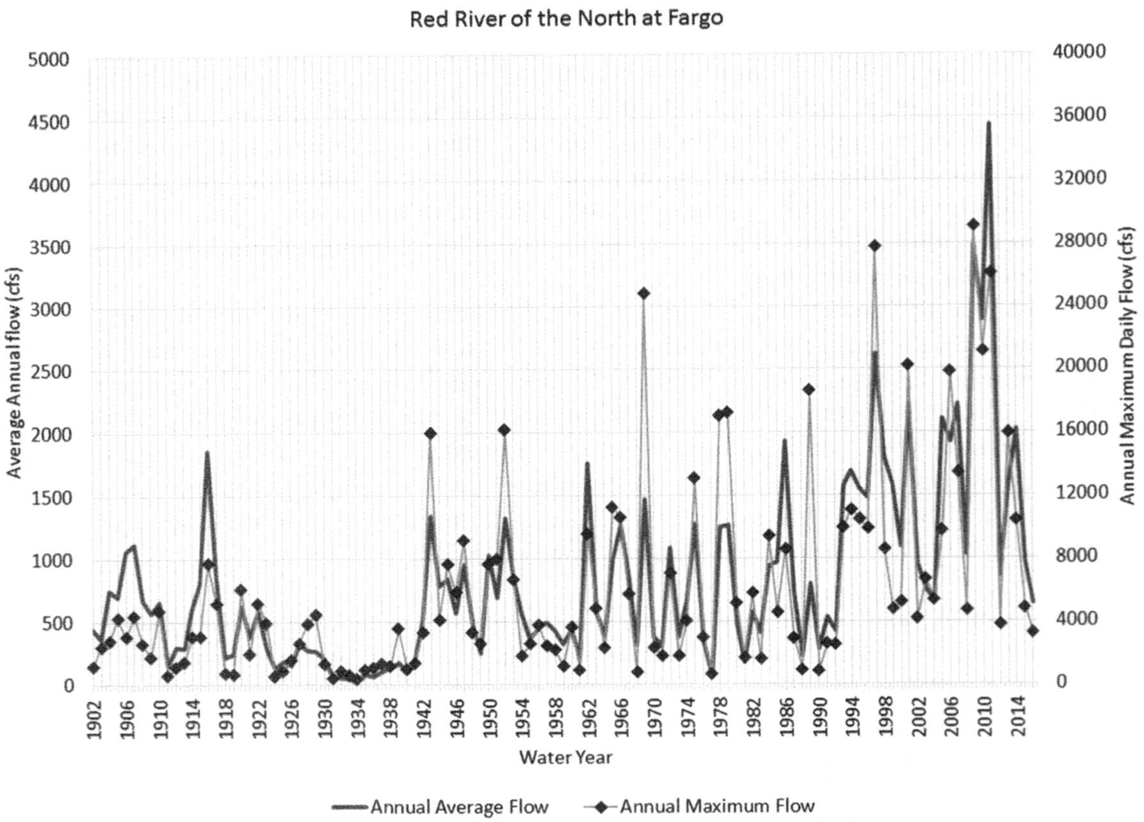

Fig. 7 | Annual average flow (left axis) and annual maximum daily flow (right axis) for the Red River of the North at Fargo, North Dakota.

stationarity. The panel concluded that there was insufficient evidence that the more recent wet period was due to anthropogenic global warming. The historical record indicated that major floods occurred in 1882 and 1897 and there were major floods and wet periods prior to the 19th century. The panel concluded that the flow period could be divided into an earlier dry period and the current wet period. A Pettit test was conducted on the unregulated peak flows and the year 1941 was the break point with the highest statistical significance. A change point test conducted by Villarini *et al.* (2009) on regulated flows showed a significant step change at 1942 (USACE-MVP, 2011b).

One flood frequency distribution was estimated for the dry period and one for the wet period, and then combined to form a composite distribution. For the period 1902–1941, the regulated 1% flood was 9,500 cubic feet per second (cfs), while for 1942–2009, the regulated 1% flood was 34,700 cfs. Using the full period of record plus the historic floods in 1882 and 1897, the regulated 1% flood was 33,000 cfs (Mueller & Foley, 2014). Each period had less variance than the variance of the full record (USACE-MVP, 2011b). It is uncertain how long the current wet period will last until a possible shift to a drier period. USACE used the probability of occurrence of being in a dry period as 35% and of being in a wet period as 65%. This estimate was based on the dry period occurring about 35% of the time during the period of record. These probabilities were used to form the combined distribution (Mueller & Foley, 2014).

The following questions are addressed by this case study: Does decadal climate information provide any insights into the flow record of the Red River of the North? Are different dry and wet periods associated with different climate patterns? Is there any evidence about how long the dry and wet periods last before shifting to a different regime?

Decadal climate patterns are associated with annual and seasonal conditions, rather than with individual weather events. However, Figure 7 shows that large floods generally occur in wet years and major floods are not likely in dry years. An exception is the large flood that occurred in 1969, a year with only medium annual flow. The correlation between the annual maximum flow and the annual average flow is 0.86.

Figure 8 shows the annual average streamflow, the annual average precipitation, and the annual maximum temperature as deviations from the long-term average value. Indices for the PDO and the TAG are also shown. For the full period of record, streamflow, precipitation, and temperature have almost no correlation with the PDO and TAG indices. If we focus on the dry period before 1941, the streamflow record shows the low flows. The PDO pattern is generally positive throughout this period. The TAG is negative in the 1920s, then shifts to generally positive in the late 1920s and 1930s. In this case using this particular statistical analysis, the recent wet period in the 1990s and 2000s does not seem to be associated with either the PDO or the TAG index.

CONCLUSION

The ENSO, the PDO, the TAG variability, the AMO, and the WPWP are patterns of natural climate variability. The phases of the PDO, TAG, and WPWP tend to persist over decadal time periods. During the past couple of decades, empirical data analysis and climate model simulations have improved the understanding of DCV. There are also international efforts to produce decadal climate hindcasts and forecasts.

There is still a great deal of uncertainty involving decadal climate information associated with both the future ocean climate pattern and the impact on local hydrology. The use of DCV information in reservoir management depends critically on the risk if the forecasts are wrong. In reservoirs with flood storage space, there is a danger of increased flood damages if a flood occurs when drought conditions are forecast. Another consideration is that reservoir water management plans are developed to balance the objectives of multiple stakeholders. The beneficiaries of a forecast may not be the same stakeholders who bear the risk of a wrong forecast. For example, a forecast of drought conditions may imply that reservoir managers should store more water and reduce flood storage. The increased storage would benefit water supply, navigation, hydropower, and irrigation stakeholders. However, if a flood occurs, the damages may increase to downstream residents who may be flooded. The need to balance the risk of wrong forecasts with the forecast benefits applies to other water resources decisions beside reservoir management including energy production, agriculture, and inland navigation.

Seasonal forecasts using DCV information could be used to inform drought triggers. If DCV and ENSO indices indicate that the climate has a higher probability of dry conditions, drought plans could be triggered earlier. In drought plans, there is concern about balancing conservation uses of the limited water and not about filling flood storage space. However, some stakeholders may be adversely affected by early triggering of drought plans.

Understanding of DCV processes could also improve hydrologic frequency analysis when there are trends in the observed record. Different DCV patterns may be associated with wet or dry periods in the hydrologic record. For example, conditions may be in a wet period due to natural DCV which show up as an increasing trend in the observed record. This trend may likely be reversed when there is a future phase shift. In the case study described in this paper, the trend could not be attributed to the PDO and the TAG, which implies the trend is due to some other factor.

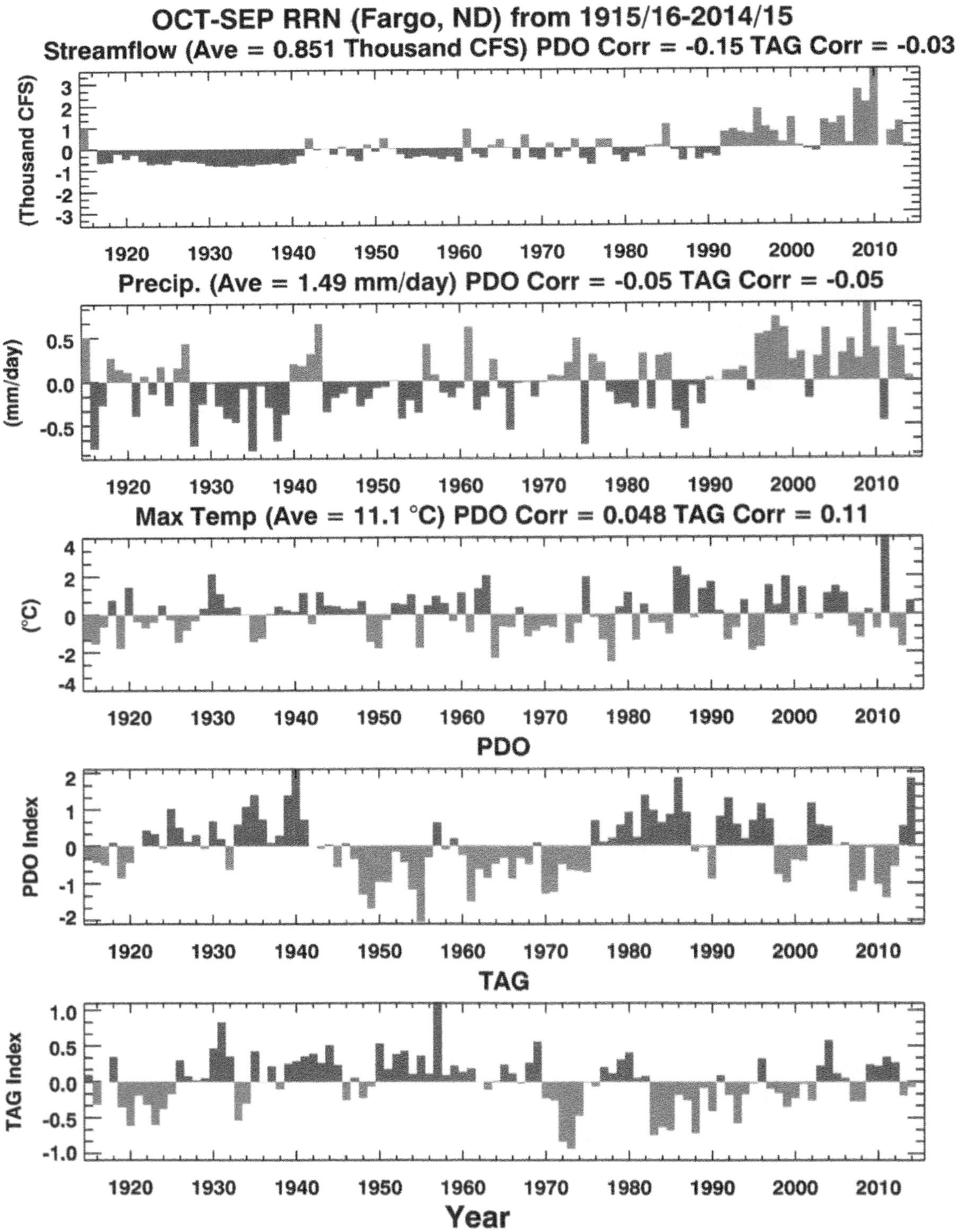

Fig. 8 | Annual average streamflow, annual average precipitation, and temperature for the Red River of the North (RRN) basin and the PDO and TAG indices.

There has been much progress in the understanding of DCV in the last few decades. Future research may provide a better understanding of the predictability of DCV patterns and the uncertainty of these predictions. Future research should also be conducted on how DCV forecasts can be used in water resources management and other fields. Bayesian analysis and other analytical techniques that consider the probabilistic nature of the forecasts and their uncertainty may be appropriate. Water managers and climate scientists should develop and continue collaboration to evaluate how this information could better inform the operation and design of water infrastructure.

DATA AVAILABILITY STATEMENT

All relevant data are available from an online repository or repositories https://waterdata.usgs.gov/nwis/inventory/?site_no=05054000&agency_cd=USGS& and www.crces.org.

REFERENCES

Dracup, J. A. & Kahya, E. (1994). The relationships between U.S. streamflow and La Niña events. *Water Resources Research 30*(7), 2133–2141.

Enfield, D. B., Mestas-Nuñez, A. M. & Trimble, P. J. (2001). The Atlantic Multidecadal Oscillation and its relation to rainfall and river flows in the continental U.S. *Geophysical Research Letters 28*, 2077–2080.

England Jr., J. F., Cohn, T. A., Faber, B. A., Stedinger, J. R., Thomas Jr., W. O., Veilleux, A. G., Kiang, J. E. & Mason, R. R. (2019). *Guidelines for Determining Flood Flow Frequency – Bulletin 17C*. Available at: https://pubs.usgs.gov/tm/04/b05/tm4b5.pdf (Accessed 15 May 2021).

Eyring, V., Bony, S., Meehl, G. A., Senior, C. A., Stevens, B., Stouffer, R. J. & Taylor, K. E. (2016). Overview of the Coupled Model Intercomparison Project Phase 6 (CMIP6) experimental design and organization. *Geoscientific Model Development 9*, 1937–1958.

Fernandez, M., Huang, P., McCarl, B. & Mehta, V. M. (2016). Value of decadal climate variability information for agriculture in the Missouri River Basin. *Climatic Change 139*, 517–533. doi:10.1007/s10584-016-1807-x.

FIRO Steering Committee (FIRO) (2017). *Preliminary Viability Assessment of Lake Mendocino Forecast Informed Reservoir Operations*. Available at: http://cw3e.ucsd.edu/firo-preliminary-viability-assessment-for-lake-mendocino/ (Accessed 15 May 2021).

Folland, C. K., Palmer, T. N. & Parker, D. E. (1986). Sahel rainfall and worldwide sea temperatures. *Nature 320*, 602–606.

Folland, C. K., Colman, A. W., Rowell, D. P. & Davey, M. K. (2001). Predictability of northeast Brazil rainfall and real-time forecast skill, 1987–98. *Journal of Climate 14*, 1937–1958.

Glantz, M. H. (2001). *Currents of Change. Impacts of El Niño and La Niña on Climate and Society*. Cambridge University Press, New York, p. 252.

Glantz, M. H., Katz, R. W. & Nichols, N. (1991). *Teleconnections Linking Worldwide Climate Anomalies*. Cambridge University Press, Cambridge, UK, pp. 309–343.

Hamlet, A. F. & Lettenmaier, D. P. (1999). Colombia river streamflow forecasting based on ENSO and PDO climate signals. *ASCE Journal of Water Resources Planning and Management 125*(6), 333–341.

Hamlet, A. F., Huppert, D. & Lettenmaier, D. P. (2002). Economic value of long-lead streamflow forecasts for Columbia River Hydropower. *ASCE Journal of Water Resources Planning and Management 128*(2), 91–101.

Hill, H. S. J. & Mjelde, J. W. (2002). Challenges and opportunities provided by seasonal climate forecasts: a literature review. *Journal of Agricultural and Applied Economics 34*, 603–632.

Johnson, N. T., Martinez, C. J., Kiker, G. A. & Leitman, S. (2013). Pacific and Atlantic sea surface temperature influences on streamflow in the Apalachicola-Chattahoochee-Flint river basin. *Journal of Hydrology 489*, 160–179.

Kirtman, B., Power, S. B., Adedoyin, J. A., Boer, G. J., Bojariu, R., Camilloni, I., Doblas-Reyes, F. J., Fiore, A. M., Kimoto, M., Meehl, G. A., Prather, M., Sarr, A., Schär, C., Sutton, R., van Oldenborgh, G. J., Vecchi, G. & Wang, H. J. (2013). Near-term climate change: projections and predictability. In: *Climate Change 2013: The Physical Science Basis. Contribution of Working Group I to the Fifth Assessment Report of the Intergovernmental Panel on Climate Change*. Stocker, T. F., Qin, D., Plattner, G.-K., Tignor, M., Allen, S. K., Boschung, J., Nauels, A., Xia, Y., Bex, V. & Midgley, P. M. (eds.). Cambridge University Press, Cambridge, UK and New York, NY, USA.

Kundzewicz, Z. W., Szwed, M. & Pinskwar, I. (2019). Climate variability and floods – a global review. *Water 11*(7), 1–24.

Lins, H. F. & Cohn, T. A. (2011). Stationarity: wanted dead or alive? *JAWRA Journal of the American Water Resources Association 19 47*(3), 475–480.

Mantua, N. J., Hare, S. R., Zhang, Y., Wallace, J. M. & Francis, R. C. (1997). A Pacific interdecadal climate oscillation with impacts on salmon production. *Bulletin of the American Meteorological Society 78*, 1069–1079.

Matalas, N. C. & Rolf Olsen, J. (2001). Analysis of trends and persistence in hydrologic records. In *Risk-Based Decision Making in Water Resources IX Proceedings of the Ninth Conference*. American Society of Civil Engineers (ASCE), Reston, VA, pp. 61–76.

McCabe, G. J., Palecki, M. A. & Betancourt, J. L. (2004). Pacific and Atlantic Ocean influences on multidecadal drought frequency in the United States. *Proceedings of the National Academy of Sciences 101*, 4136–4141.

McCabe, G. J., Betancourt, J. L., Gray, S. T., Palecki, M. A. & Hidalgo, H. G. (2008). Associations of multi-decadal sea-surface temperature variability with US drought. *Quaternary International 188*, 31–40. doi:10.1016/j.quaint.2007.07.001.

Meehl, G. A., Goddard, L., Murphy, J., Stouffer, R. J., Boer, G., Danabasoglu, G., Dixon, K., Giorgetta, M. A., Greene, A., Hawkins, E., Hegerl, G., Karoly, D., Keenlyside, N., Kimoto, M., Kirtman, B., Navarra, A., Pulwarty, R., Smith, D., Stammer, D. & Stockdale, T. (2009). Decadal prediction: can it be skillful? *Bulletin of the American Meteorological Society 90*, 1467. doi:10.1175/2009BAMS2778.1.

Meehl, G. A., Goddard, L., Boer, G., Burgman, R., Branstator, G., Cassou, C., Corti, S., Danabasoglu, G., Doblas-Reyes, F., Hawkins, E., Karspeck, A., Kimoto, M., Kumar, A., Matei, D., Mignot, J., Msadek, R., Navarra, A., Pohlmann, H., Rienecker, M., Rosati, T., Schneider, E., Smith, D., Sutton, R., Teng, H., Jan van Oldenborgh, G., Vecchi, G. & Yeager, S. (2014). Decadal climate prediction: an update from the trenches. *Bulletin of the American Meteorological Society 95*(2), 243–267.

Mehta, V. M. (1998). Variability of the tropical ocean surface temperatures at decadal-multidecadal timescales, part I: the Atlantic Ocean. *Journal of Climate 11*, 2351–2375.

Mehta, V. M. (2017). *Decadal Climate Variability: Societal Impacts*. CRC Press (Taylor & Francis), Boca Raton, p. 325.

Mehta, V. M. (2020). *Natural Decadal Variability Phenomena, Mechanisms, and Predictability*. CRC Press (Taylor & Francis), Boca Raton, p. 355.

Mehta, V. M., Meehl, G., Goddard, L., Knight, J., Kumar, A., Latif, M., Lee, T., Rosati, A. & Stammer, D. (2011). Decadal climate predictability and prediction: where are we? *Bulletin of the American Meteorological Society 92*, 637–640.

Mehta, V. M., Wang, H. & Mendoza, K. (2013). Decadal predictability of tropical basin-average and global-average sea-surface temperatures in CMIP5 experiments with the HadCM3, GFDL-CM2.1, NCAR-CCSM4, and MIROC5 global earth system models. *Geophysical Research Letters 40*. doi:10.1002/grl.50236.

Mehta, V. M., Wang, H. & Mendoza, K. (2019). Predictability of phases and magnitudes of natural decadal climate variability phenomena in CMIP5 experiments with the UKMO-HadCM3, GFDL-CM2.1, NCAR-CCSM4, and MIROC5 global earth system models. *Climate Dynamics 52*, 3255–3275.

Meza, F. J., Hansen, J. W. & Osgood, D. (2008). Economic value of seasonal climate forecasts for agriculture: review of ex-ante assessments and recommendations for future research. *Journal of Applied Meteorology and Climatology 47*(5), 1269–1286.

Mjelde, J. W. & Hill, H. S. (1999). The effect of the use of improved climate forecasts on variable costs, input usage, and production. *Agricultural Systems 60*, 213–225.

Mo, K. C., Schemm, J. K. E. & Yoo, S. H. (2009). Influence of ENSO and the Atlantic Multidecadal Oscillation on drought over the United States. *Journal of Climate 22*, 5962–5982.

Mueller, C. & Foley, P. M. (2014). Discussion of 'Uncertainty of the assumptions required for estimating the regulatory flood: Red River of the North' by Paul E. Todhunter. *Journal of Hydrologic Engineering 19*(1), 275–276.

Olsen, J. R., Zepp, L. & Dager, C. (2005). Climate impacts on inland navigation. In: *World Water Congress 2005, Impacts of Global Climate Change, World Water and Environmental Resources Congress 2005*, May 2005, Anchorage, Alaska.

Philander, S. G. (1990). *El Niño, La Niña and the Southern Oscillation*. Academic Press, San Diego, p. 289.

Rajagopalan, B., Kushnir, Y. & Tourre, Y. M. (1998). Observed decadal mid-latitude and tropical Atlantic climate variability. *Geophysical Research Letters 25*, 367–370.

Rasmusson, E. M. (1991). Observational aspects of ENSO cycle teleconnections. In *Teleconnections Linking Worldwide Climate Anomalies*. Glantz, M. H., Katz, R. W. & Nichols, N. (eds.). Cambridge University Press, Cambridge, UK, pp. 309–343.

Ropelewski, C. F. & Halpert, M. S. (1989). Precipitation patterns associated with the high index phase of the Southern Oscillation. *Journal of Climate 2*, 268–284.

Salas, J. D., Obeysekera, J. & Vogel, R. M. (2018). Techniques for assessing water infrastructure for nonstationary extreme events: a review. *Hydrological Sciences Journal 63*(3), 325–352.

Schlesinger, M. E. & Ramankutty, N. (1994). An oscillation in the global climate system of period 65–70 years. *Nature 367*, 723–726.

Schubert, S. D., Suarez, M. J., Pegion, P. J., Koster, R. D. & Bacmeister, J. T. (2004a). On the cause of the 1930s Dust Bowl. *Science 303*, 1855–1859.

Schubert, S. D., Suarez, M. J., Pegion, P. J., Koster, R. D. & Bacmeister, J. T. (2004b). Causes of long-term drought in the US Great Plains. *Journal of Climate 17*, 485–503.

Singh, S., Srivastava, P., Abebe, A. & Mitra, S. (2015). Baseflow response to climate variability induced droughts in the Apalachicola–Chattahoochee–Flint River Basin, U.S.A. *Journal of Hydrology 528*, 550–561.

Stevens, K. A. & Ruscher, P. H. (2014). Large scale climate oscillations and mesoscale surface meteorological variability in the Apalachicola-Chattahoochee-Flint River Basin. *Journal of Hydrology 517*, 700–714.

Taylor, K. E., Stouffer, R. J. & Meehl, G. A. (2012). An overview of CMIP5 and the experiment design. *Bulletin of the American Meteorological Society 93*, 485–498.

Tootle, G. A., Piechota, T. C. & Singh, A. (2005). Coupled oceanic–atmospheric variability and US streamflow. *Water Resources Research 41*, 1–11.

U.S. Army Corps of Engineers Mobile District (USACE-SAM) (2017). *Master Water Control Manual Apalachicola–Chattahoochee–Flint (ACF) River Basin Alabama, Florida, Georgia, February 1958, Revised March 2017, Mobile, Alabama.*

U.S. Army Corps of Engineers, St. Paul District (USACE-MVP) (2011a). *Fargo-Moorhead Metropolitan Area Flood Risk Management, Final Feasibility Report and Environmental Impact Statement*. Available at: http://www.fmdiversion.com/environmental-impact-statement/ (Accessed 15 May 2021).

U.S. Army Corps of Engineers, St. Paul District (USACE-MVP) (2011b). *Fargo-Moorhead Metropolitan Area Flood Risk Management, Appendix A Hydrology*. Available at: https://fmdiversion.gov/library/supplemental-draft-feasibility-report-and-environmental-impact-statement/ (Accessed 15 May 2021).

U.S. Army Corps of Engineers (USACE) (1977). *Shasta Dam and Lake: Report on Reservoir Regulations for Flood Control*. U.S. Army Corps of Engineers, Sacramento, CA.

U.S. Army Corps of Engineers (USACE) (2005). *Oroville Dam and Lake, Feather River, California: Water Control Manual*. U.S. Army Corps of Engineers, Sacramento, CA.

U.S. Army Corps of Engineers (USACE) (2017a). *Management of Water Control Systems, Engineer Manual (EM) 1110-2-3600.*

U.S. Army Corps of Engineers (USACE) (2017b). *Guidance for Detection of Nonstationarities in Annual Maximum Discharges. Technical Letter No. 1100-2-3.*

U.S. Army Corps of Engineers (USACE) (2018). *Guidance for Incorporating Climate Change Impacts to Inland Hydrology in Civil Works Studies, Designs, and Projects, Engineering and Construction Bulletin No. 2018-14.*

U.S. Water Resources Council (USWRC P&G) (1983). *Economic and Environmental Principles for Water and Related Land Resources Implementation Studies.*

Villarini, G., Serinaldi, F., Smith, J. A. & Krajewski, W. F. (2009). On the stationarity of annual flood peaks in the continental United States during the 20th century. *Water Resources Research 45*, WO8417.

Voisin, N., Hamlet, A. F., Graham, L. P., Pierce, D. W., Barnett, T. P. & Lettenmaier, D. P. (2006). The role of climate forecasts in Western U.S. power planning. *Journal of Applied Meteorology and Climatology 45*, 653–673.

First received 18 November 2021; accepted in revised form 18 November 2021. Available online 1 December 2021

IWA Publishing's authorised EU representative for General Product Safety Regulations is Diane D'Arras, 15 rue Duret, 75116 Paris, France, e-mail: safety@iwap.co.uk.

Printed and bound by CPI Group (UK) Ltd, Croydon, CR0 4YY
12/05/2026
02108880-0020